THE LEAN SIX SIGMA BLACK BELT HANDBOOK

Tools and Methods for Process Acceleration

THE LEAN SIX SIGMA BLACK BELT HANDBOOK

Tools and Methods for Process Acceleration

Frank Voehl • H. James Harrington
Chuck Mignosa • Rich Charron

CRC Press is an imprint of the
Taylor & Francis Group, an informa business
A PRODUCTIVITY PRESS BOOK

CRC Press
Taylor & Francis Group
6000 Broken Sound Parkway NW, Suite 300
Boca Raton, FL 33487-2742

Printed on acid-free paper
Version Date: 20130426

International Standard Book Number-13: 978-1-4665-5468-9 (Hardback)

Library of Congress Cataloging-in-Publication Data

Voehl, Frank, 1946-
The lean six sigma black belt handbook : tools and methods for process acceleration / Frank Voehl, H. James Harrington, Chuck Mignosa, Rich Charron.
pages cm
Includes bibliographical references and index.
ISBN 978-1-4665-5468-9 (hardcover : alk. paper)
1. Total quality management. 2. Six sigma (Quality control standard) 3. Production management. 4. Industrial management. I. Title.

HD62.15.V64 2013
658.4'013--dc23 2013014246

Visit the Taylor & Francis Web site at
http://www.taylorandfrancis.com

and the CRC Press Web site at
http://www.crcpress.com

This book is dedicated to the thousands of Lean Six Sigma Belts that Jim Harrington and I have had the privilege of mentoring and certifying over the past 10 years. And to the Four Horsemen of the Quality Movement—Deming, Juran, Crosby, and Feigenbaum. Thank you for teaching us how to teach, and for encouraging us to find a better way.

Frank Voehl

I dedicate this book to the person who created me and gave me her love—my mother, Carrie, and to the person who helped mold my life and made me what I am—my wife, Marguerite.

H. James Harrington

This book is dedicated to two people who have had the most profound impacts on my life—

my dad and my daughter. To my late dad, Edmund J. Charron, who in my formative years provided a steady stream of learning exercises through quotations, principles, and actions. Unbeknownst to me at the time, these gems of his essence would intermittently emerge throughout my life and guide me just when I needed them the most. To my precious daughter Hali, who as a child taught me that love is the most powerful force in the universe. Like my father before me, may I provide to you through word and deed values to live by.

Richard Charron

Contents

SECTION 2 The Lean Journey into Process Improvement

SECTION 3 SSBB Overview

SECTION 4 LSSBB Advanced Nonstatistical Tools

Acknowledgments

I thank Candy Rogers, who invested endless hours to edit and transform my rough draft into a finished product. I couldn't have done it without her help.

I would be remiss not to acknowledge the contributions made by my many clients to the broadening of my experience base and to the many new concepts that they introduced to me. Every engagement I worked on has been a learning experience that could not be duplicated in any university.

I also recognize my many good friends who have freely shared their ideas with me, helping me to grow and look at things in a new light. I would particularly like to point out the many contributions I received from Chuck Mignosa, Frank Voehl, and Armand Feigenbaum.

H. James Harrington

About the Authors

FRANK VOEHL

Present Responsibilities

Frank Voehl serves as CEO and president of Strategy Associates, Inc., and Chancellor for the Harrington Institute. He is also chairman of the board for a number of businesses and is a Grand Master Black Belt instructor and technology advisor at the University of Central Florida in Orlando. He is recognized as one of the world leaders in applying quality measurement and Lean Six Sigma methodologies to business processes.

Previous Experience

Voehl has extensive knowledge of NRC, FDA, GMP, and NASA quality system requirements. He is an expert in ISO 9000, QS 9000/14000/18000, and integrated Lean Six Sigma quality system standards and processes. He has degrees from St. John's University and advanced studies at New York University, as well as an honorary doctor of divinity degree. Since 1986, he has been responsible for overseeing the implementation of quality management systems with organizations in such diverse industries as telecommunications and utilities, federal, state, and local government agencies, public administration and safety, pharmaceuticals, insurance/banking, manufacturing, and institutes of higher learning. In 2002, he joined the Harrington Group as the chief operating officer and executive vice president. Voehl has held executive management positions with Florida Power and Light and FPL Group, where he was the founding general manager and COO of QualTec Quality Services for seven years. He has written and published/co-published over 35 books and hundreds of technical papers on business management, quality improvement, change management, knowledge management, logistics, and team building, and has received numerous awards for community leadership, service to third world countries, and student mentoring.

Credentials

The Bahamas National Quality Award was developed in 1991 by Voehl to recognize the many contributions of companies in the Caribbean region, and he is an honorary member of its Board of Judges. In 1980, the city of Yonkers, New York, declared March 7 Frank Voehl Day, honoring him for his many contributions on behalf of thousands of youth in the city where he lived, performed volunteer work, and served as athletic director and coach of the Yonkers-Pelton Basketball Association. In 1985 he was named Father of the Year in Broward County, Florida. He also serves as president of the Miami Archdiocesan Council of the St. Vincent de Paul Society, whose mission is to serve the poor and needy throughout South Florida and the world.

Frank's contributions to quality improvement around the world have brought him many honors and awards, including ASQ's Distinguished Service Medal, the Caribbean Center for Excellence Founders Award, the Community Quality Distinguished Service Award, the Czech Republic Outstanding Service Award on behalf of its business community leaders, FPL's Pioneer Lead Facilitator Award, the Florida SFMA Partners in Productivity Award, and many others. He was appointed the honorary advisor to the Bahamas Quality Control Association, and he was elected to the Eastern Europe Quality Hall of Fame. He was also named honorary director of the Association Venezuela de Control de Calidad by Banco Consolidado.

DR. H. JAMES HARRINGTON

In the book *Tech Trending*, Dr. Harrington was referred to as "the quintessential tech trender." The *New York Times* noted his "knack for synthesis and an open mind about packaging his knowledge and experience in new ways—characteristics that may matter more as prerequisites for new-economy success than technical wizardry." The author Tom Peters stated, "I fervently hope that Harrington's readers will not only benefit from the thoroughness of his effort but will also 'smell' the fundamental nature of the challenge for change that he mounts." William Clinton, past president of the United States, appointed Dr. Harrington to serve as an Ambassador of Good Will. It has been said about him, "He writes the books that other consultants use."

Harrington Institute was featured on a half-hour TV program, *Heartbeat of America*, which focuses on outstanding small businesses that make America strong. The host, William Shatner, stated

"You [Dr. Harrington] manage an entrepreneurial company that moves America forward. You are obviously successful."

Present Responsibilities

Harrington serves as the chief executive officer for the Harrington Institute and Harrington Associates. He is also chairman of the board for a number of businesses.

Dr. Harrington is recognized as one of the world leaders in applying performance improvement methodologies to business processes. He has an excellent record of coming into an organization, working as its CEO or COO, resulting in a major improvement in its financial and quality performance.

Previous Experience

In February 2002 Dr. Harrington retired as the COO of Systemcorp A.L.G., the leading supplier of knowledge management and project management software solutions, when Systemcorp was purchased by IBM. Prior to this, he served as a principal and one of the leaders in the Process Innovation Group at Ernst & Young; he retired from Ernst & Young when it was purchased by Cap Gemini. Dr. Harrington joined Ernst & Young when Ernst & Young purchased Harrington, Hurd & Rieker, a consulting firm that Dr. Harrington started. Before that Dr. Harrington was with IBM for over 40 years as a senior engineer and project manager.

Dr. Harrington is past chairman and past president of the prestigious International Academy for Quality and of the American Society for Quality Control. He is also an active member of the Global Knowledge Economics Council.

Credentials

Harrington was elected to the honorary level of the International Academy for Quality, which is the highest level of recognition in the quality profession. Harrington is a government-registered quality engineer, a Certified Quality and Reliability Engineer by the American Society for Quality Control, and a Permanent Certified Professional Manager by the Institute of Certified Professional Managers. He is a certified Master Six Sigma Black Belt and received the title of Six Sigma Grand Master. He has an MBA and PhD in engineering management and a BS in electrical engineering.

Dr. Harrington's contributions to performance improvement around the world have brought him many honors. He was appointed the honorary advisor to the China Quality Control Association, and was elected to the Singapore Productivity Hall of Fame in 1990. He has been named lifetime honorary president of the Asia-Pacific Quality Control Organization and honorary director of the Association Chilean de Control de Calidad. In 2006 Dr. Harrington accepted the honorary chairman position of Quality Technology Park of Iran.

Harrington has been elected a fellow of the British Quality Control Organization and the American Society for Quality Control. In 2008 he was elected to be an honorary fellow of the Iran Quality Association and Azerbaijan Quality Association. He was also elected an honorary member of the quality societies in Taiwan, Argentina, Brazil, Colombia, and Singapore. He is also listed in the "Who's Who Worldwide" and "Men of Distinction Worldwide." He has presented hundreds of papers on performance improvement and organizational management structure at the local, state, national, and international levels.

Recognition

- The Harrington/Ishikawa Medal, presented yearly by the Asian Pacific Quality Organization, was named after H. James Harrington to recognize his many contributions to the region.
- The Harrington/Neron Medal was named after H. James Harrington in 1997 for his many contributions to the quality movement in Canada.
- Harrington Best TQM Thesis Award was established in 2004 and named after H. James Harrington by the European Universities Network and e-TQM College.
- Harrington Chair in Performance Excellence was established in 2005 at the Sudan University.
- Harrington Excellence Medal was established in 2007 to recognize an individual who uses the quality tools in a superior manner.
- H. James Harrington Scholarship was established in 2011 by the ASQ Inspection Division.

Harrington has received many awards, among them the Benjamin L. Lubelsky Award, the John Delbert Award, the Administrative Applications Division Silver Anniversary Award, and the Inspection Division Gold Medal Award. In 1996, he received the ASQC's Lancaster Award in recognition of his international activities. In 2001 he received the Magnolia Award in recognition for the many contributions he has made in improving quality in China. In 2002 Harrington was selected by the European Literati Club to receive a lifetime achievement award at the Literati Award for Excellence ceremony in London. The award was given to honor his excellent literature contributions to the advancement of quality and organizational performance. Also, in 2002 Harrington was awarded the International Academy of Quality President's Award in recognition for outstanding global leadership in quality and competitiveness, and contributions to IAQ as Nominations Committee chair, vice president, and chairman. In 2003 Harrington received the Edwards Medal from the American Society for Quality (ASQ). The Edwards Medal is presented to the individual who has demonstrated the most outstanding leadership in the application of modern quality control methods, especially through the organization and administration of such work. In 2004 he received the Distinguished Service Award, which is ASQ's highest award for service granted by the society. In 2008 Dr. Harrington was awarded the Sheikh Khalifa Excellence Award (UAE) in recognition of his superior performance as an original Quality and Excellence Guru who helped shape modern quality thinking. In 2009 Harrington was selected as the Professional of the Year. Also in 2009 he received the Hamdan Bin Mohammed e-University Medal. In 2010 the Asian Pacific Quality Association (APQO) awarded Harrington the APQO President's Award for his "exemplary leadership." The Australian Organization of Quality NSW's Board recognized Harrington as "the Global Leader in Performance Improvement Initiatives" in 2010. In 2011 he was honored to receive the Shanghai Magnolia Special Contributions Award from the Shanghai Association for Quality in recognition of his 25 years of contributing to the advancement of quality in China. This was the first time that this award was given out. In 2012 Harrington received the ASQ Ishikawa Medal for his many contributions in promoting the understanding of process improvement and employee involvement on the human aspects of quality at the local, national, and international levels. Also in 2012 he was awarded the Jack Grayson Award. This award recognizes individuals who have demonstrated outstanding leadership in the application of quality philosophy, methods and tools in education, health care, public service, and not-for-profit organizations. Harrington also received the A.C. Rosander Award in 2012. This is ASQ Service Quality Division's highest honor. It is given in recognition of outstanding long-term service and leadership resulting in substantial progress toward the fulfillment of the division's programs and goals. Additionally, in 2012 Harrington was honored by the Asia Pacific Quality Organization by being awarded the Armand V. Feigenbaum Lifetime Achievement Medal. This award is given annually to an individual whose relentless pursuit of performance improvement over a minimum of 25 years has distinguished himself or herself for the candidate's work in promoting the use of quality methodologies and principles within and outside of the organization he or she is part of.

Contact Information

Dr. Harrington is a prolific author, publishing hundreds of technical reports and magazine articles. For eight years he published a monthly column in *Quality Digest Magazine*. He has authored 37 books and 10 software packages.

You may contact Dr. Harrington at:

Address: 16080 Camino del Cerro, Los Gatos, California 95032
Phone: (408) 358-2476
E-mail: hjh@harrington-institute.com

CHARLES MIGNOSA

Charles "Chuck" Mignosa has over 30 years of diversified experience in high technology, biomedical devices, telecommunications, and food processing industries and 25 years of experience in IBM, holding patents in solid lubricants. He was a second-level manager in charge of implementing quality systems in five manufacturing areas. He is a certified course developer and has developed courses including Total Quality Management, Continuous Flow Manufacturing, Customer-Driven Quality, Statistical Design and Analysis of Experiments, Team Building, Six Sigma, Conflict Resolution, and Communication Skills.

After leaving IBM he worked as an independent consultant doing all of the TQM training for Spectrian Telecommunications and facilitating its conversion from a DOD to public sector company and attaining its ISO registration.

Mignosa has consulted for and done training with such companies as Siemens Automotive, General Mills, Gatorade, Zea Corporation, Connors Peripherals, HP, IBM, ADAC Labs, Cholestech, Heinz USA, and many more. He has held positions as director of quality for P-com, a telecommunication company, and Cholestech Corporation, a medical device company.

Mignosa is currently president of Business Systems Architects (BSA), a Silicon Valley consulting, training, and documentation company specializing in the design and implementation of business and quality management systems and, with Upward Performance, of which he is also president, implementing Six Sigma and Lean Manufacturing programs.

In addition to a BS in chemistry, Mr. Mignosa has graduate degrees in statistics, systems research, and management training with IBM and is a senior member of ASQ.

RICHARD M. CHARRON

Rich Charron is the founder and president of the Lean Manufacturing Group, Inc., a South Florida company that provides a number of "hands-on" employee learning and Lean implementation programs focused on waste elimination, productivity improvement, and profitability enhancement. He is a Certified Master Black Belt in Lean Six Sigma Excellence from the Harrington Institute. He has trained and coached over 100 teams in Lean Manufacturing, Lean Six Sigma, and Kaizen events, generating savings of over $25 million. In conjunction with Strategy Associates he completed a three-part DVD series on Lean concepts for the University of Central Florida. His expertise is in process performance excellence, Lean Six Sigma, Lean Manufacturing, design for manufacturability, problem solving, product and process failure analysis, products development, and performance testing.

Mr. Charron holds BS and MS degrees in plastics engineering from the University of Massachusetts. His MS thesis, "Product Liability in the Plastics Industry," is a survey of our legal system and the impacts of unsafe products and legal uncertainties. He is the author of over a dozen technical publications on product quality, products performance testing, and products failure analysis.

Section 1

Overview of Lean Six Sigma

ABOUT THIS BOOK

What are the prerequisite knowledge and skills required to using this book? What kind of tools and methodologies are included in this Lean Six Sigma Black Belt (LSSBB) handbook? How is the book organized? Why did we choose to organize the book as we did? The following are our answers to these questions.

The LSSBB methodology has been designed to provide a growth path for the people using it. The following outlines this growth path along with the prerequisites for each step in the path, starting at the beginning level:

- Six Sigma White Belt—No prerequisite
- Six Sigma Yellow Belt—No prerequisite
- Six Sigma Green Belt—Yellow Belt prerequisite
- Lean Green Belt—No prerequisite
- Six Sigma Black Belt—Six Sigma Green Belt or equivalent
- Lean Six Sigma Black Belt—Six Sigma Green Belt and Lean Green Belt or equivalent
- Master Black Belt—Lean Six Sigma Black Belt or equivalent

This book is written on the assumption that the reader has already gained experience in using all the tools and methods required to be certified as a Six Sigma Green Belt (SSGB), or at the least, he or she has been trained in these tools and methodologies. As a result, the tools and methods taught in Six Sigma Green Belt classes, as outlined in our *Six Sigma Green Belt Handbook* (Harrington, Voehl, and Gupta; Paton Press, Chico, CA), will not be repeated in this book. (A list of the tools and methods that the reader should already have mastered before using this handbook can be found in Appendix C.) A list of the additional tools and methods presented in this book and which are required to be classified as a LSSBB includes the following:

Nonstatistical tools:

- 5S
- Benchmarking
- Bureaucracy elimination
- Conflict resolution
- Critical to quality
- Cycle time analysis and reduction
- Fast action solution technique
- Foundation of Six Sigma
- Just-in-time
- Matrix diagrams/decision matrix
- Measurement in Six Sigma
- Organizational change management
- Pareto diagrams
- Project management
- Quality function deployment
- Reliability management systems
- Root cause analysis
- Scatter diagrams
- Selection matrix/decision matrix
- SIPOC (Suppliers, Inputs, Processes, Output, and Customers)
- SWOT (Strengths, Weaknesses, Opportunities, and Threats)
- Takt time
- Theory of constraints
- Tree diagram
- Value stream mapping

Statistical tools:

- ANOVA—one-way
- ANOVA—two-way
- Box plots
- Confidence intervals
- Data transformation
- Design of experiments
- Measurement system analysis
- Method of least squares
- Multivari charts
- Nonparametric statistical tests
- Population and samples
- Regression analysis
- Rolled throughput yield
- Taguchi methods
- Validation

In developing the content for this handbook, we relied heavily upon the core content that we developed for our *Six Sigma Yellow Belt Handbook* (Harrington and Voehl; Paton Press, Chico, CA) and *Six Sigma Green Belt Handbook* (Harrington, Voehl, and Gupta; Paton Press, Chico, CA). Both of these have a core introduction and brief history of the Six Sigma movement. Our primary goal in preparing this book was to develop a comprehensive framework for organizing large amounts of useful knowledge about Lean, quality management, and continuous improvement process applications so that only the tools that the reader will use to solve 90% of the problems are the focus of the books.

The book begins with an overview and brief history of Six Sigma, followed by some tips for using Lean Six Sigma (LSS), along with a look at the LSS quality system, its supporting organization, and the different roles involved. In Section 2 we discuss in detail the theories and concepts required to support a Lean system. It is pointed out that Lean is more a way of managing than a set of tools. Lean focuses on the elimination of all waste. In Section 3 the book presents the new additional skills and technologies that an individual is required to master in order to be certified at the LSSBB level. Next, in Sections 4 and 5, the advanced nonstatistical and statistical tools that are new to the LSSBB body of knowledge are presented in detail.

An LSSBB is an experienced individual who has the capabilities to solve complex problems as well as the capabilities of training Six Sigma Green and Yellow Belts. They also are capable of managing a number of improvement projects simultaneously while giving guidance to Six Sigma Green Belts who are working on improvement projects. The combination of the skills and techniques presented in this handbook are designed to prepare an individual to be certified as an LSSBB. Those individuals who are classified as LSSBBs are required to have mastered both the Lean and Six Sigma methodologies and techniques, making them an extremely valuable resource for any organization.

The format used is that of a handbook; it contains descriptions and diagrams of the tools that a team has practiced or will consider. *The Lean Six Sigma Black Belt Handbook* is an easy read and is designed to be used by readers at every level of the organization hierarchy and any industry. The handbook discussions will cover the following five areas:

- Alignment of individuals and organizational performance
- Implementation of results-oriented process improvement
- Institution of the human capital management function
- Need for sustained management attention for addressing key organization responsibilities in an integrated manner
- Facilitation of the transformation process within the organization

You may have experienced problem solving as a one-step process—just solve it! But there is more to it than that. For starters, you have to know what the problem or waste area really is, what's causing the problem, and then look at new creative ways to make things better. These are covered in the following sections by providing a step-by-step roadmap you can use in your organization. The DMADV roadmap that is presented can be used to find the right solutions and avoid some of the pitfalls you have experienced in the past. The roadmap is set up to be followed one step or "tollgate" at a time, but you may not need to follow each step in every situation. For example, if your need is gathering, you can proceed directly to tollgate 5, or maybe if the root causes are already obvious, you can breeze through that step of analyzing the causes. In other words, each one of you may have a different story to tell based on the tollgate sequences that you decide to use. The process starts where the priority is the strongest.

This book is divided into five sections:

- Section 1: Overview of Lean Six Sigma
- Section 2: The Lean Journey into Process Improvement
- Section 3: SSBB Overview
- Section 4: LSSBB Advanced Nonstatistical Tools
- Section 5: LSSBB Advanced Statistical Tools

It is our hope that this book will serve as the basic reference for your training to become an LSSBB and as an ongoing reference book as you use the LSS methodology. But it is important to realize that the assignment as a LSSBB is a stepping stone, not a career. LSSBBs are needed to solve major problems that have a significant impact on the organization's performance and to help install a new culture where waste is not tolerated. In all well-managed organizations these major problems should have been addressed, solved, and the cultural change implemented within 1 to 3 years. As a result, after a short period of time, there will be no need for full-time LSSBBs, but there will always be a need for the knowledge and skills that the LSSBBs bring to the table. As such, the tools and methodology used by the LSSBBs should become part of the organization's assets. The skills and experience that the LSSBBs acquire provide a logical base for promoting them to a higher level and more responsible positions once the major problems have been addressed. These are exactly the conditions that we have seen in organizations like IBM, Motorola, and GE. Individuals who have tried to make being an LSSBB a career have often received additional training to become a Lean Six Sigma Master Black Belt (LSSMBB) and moved on to another company that had major problems and wanted to install an LSS system, or to become a consultant.

1

Introduction to Lean Six Sigma Methodology

Lean Six Sigma increases the focus on Lean approaches with less emphasis on the statistical rigor included in the Six Sigma methodology alone.

—H. J. Harrington

Lean Six Sigma is a synergistic process that creates a value stream map of the process identifying value add and non-value add costs, and captures the Voice of the customer to define the customer Critical To Quality issues. Projects within the process are then prioritized based on the delay time they inject. This prioritization process inevitably pinpoints activities with high defect rates (Six Sigma tools) or long setups, downtime (Lean tools), with the result often yielding savings of $250,000 and a payback ratio between 4–1 and 20–1.

—Michael George
*Lean Six Sigma expert and one of the founding fathers**

* Source: An interview with Michael George (Chairman and CEO of George Group) by Frank Voehl, February 2008. During the course of the interview, George went on to state that while classical projects involving Six Sigma are most closely associated with defects and quality, Lean itself is linked to speed, while eliminating all forms of waste and inefficiency. George has written extensively that Lean provides tools to reduce lead time of any process and eliminate no-value-added cost, and he explained that Six Sigma did not contain any tools to control lead time (e.g., pull systems), or tools specific to the reduction of lead time (e.g., setup reduction). Since companies "must become more responsive to changing customer needs, faster lead times are essential in all endeavors." Lean is an important complement to Six Sigma, George insisted, and fits well within the Six Sigma DMAIC and even the DMADV processes. Finally, George points to the Lean Kaizen approach, which he extols as a great methodology that can be used to accelerate the rate of improvement and even breakthrough. In our writings in other books, the authors advocate both the need to improve quality so you can achieve maximum speed and the need to do the things at the same time that allow maximum speed in order to reach the highest Sigma levels of quality. In other words, George concluded, you need both Lean (speed) and Six Sigma (quality) principles and tools to drive improvements and gain and maintain competitive advantage.

IN A NUTSHELL

As with most new concepts, each has evolved from a previous concept. Total Quality Control is an expansion of the quality assurance concept. Total Quality Management (TQM) is an expansion of the Total Quality Control concept. Process redesign is an enlargement and expanded use of some of the key Total Quality Management concepts, applying them to the service and support processes. The Six Sigma methodology just set a new performance standard using tools that were part of the Total Quality Management concepts. The metaphor of "standing on the shoulders of giants" is often used to illustrate the notion of building upon the systems and programs that are already in place, as opposed to "reinventing the wheel." While it is considered by most experts essential for a company to create familiarity with both Lean and Six Sigma disciplines by training employees, it is even more important to incorporate an integrated Lean Six Sigma philosophy into the company in order to change the culture.

The Lean concepts were based upon Henry Ford's production process that was refined in the early 1900s and which became part of Total Quality Management and the Six Sigma methodologies. The Lean Six Sigma (LSS) methodology increases the focus on Lean approaches, with less emphasis on the statistical analysis requirements normally included in the Six Sigma methodology. The Six Sigma approach was directed at reducing variation, but most of the real contributions were made related to the elimination of no-value-added activities (elimination of wastes). The LSS methodology focuses on waste reduction with less emphasis on reducing variation. It recognizes nine forms of waste, which are an expansion of Taiichi Ohno's original seven wastes. The nine forms of waste are:

1. Overproduction
2. Overprocessing
3. Motion
4. Transportation
5. Inventory
6. Waiting
7. Underutilized employees
8. Defects
9. Behavior

Taiichi Ohno was the father of the Toyota Production System and the creator of the original "seven deadly wastes." Ohno's training and insight led him to the conclusion that Toyota's productivity should not be any lower in any way than that in the Detroit and European automobile manufacturers' shops. As a worker and supervisor, he set out to eliminate the waste and inefficiencies in the part of the production process that he could control the results for, and these efforts ultimately led to the core beliefs of the Toyota Production System (TPS). Over the past 30 years, several elements of the TPS system have become adapted and adopted in the Western world, like muda (the elimination of waste), jidoka (the injection of quality), and Kanban (the pull system of just-in-time inventory stock control). For more information, see *Guide to Management Ideas and Gurus*, by Tim Hindle (Economist Books; Profile Books, London, UK). This guide has the lowdown on more than 50 of the world's most influential management thinkers, past and present, and over 100 of the most influential business-management ideas in one volume.

Often just giving new terms or new names to already established approaches rekindles interest in that approach. This is evident in the recent enthusiasm of using Japanese terms for American established terms. For example, using the word *muda* (the Japanese word for waste) or *Kaizen* (the Japanese term for continuous improvement) can revive the executive team's interest because it is not viewed as the same old thing.

INTRODUCTION

To avoid any confusion, let's start out by offering our basic operational definitions of Lean, Six Sigma, and Lean Six Sigma.

- **Lean:** The Lean methodology is an operational philosophy with a focus on identifying and eliminating all waste in an organization. Lean principles include zero inventory, batch to flow, cutting batch size, line balancing, zero wait time, pull instead of push production control systems, work area layout, time and motion studies, and cutting cycle time. The concepts are applied to production, support, and service applications. Lean focuses on eliminating waste from processes and increasing process speed by focusing on what customers actually consider quality, and working backwards from that.

- **Six Sigma:** The Six Sigma methodology is a business-management strategy designed to improve the quality of process outputs by minimizing variation and causes of defects in processes. It is a subset of the TQM methodology with a heavy focus on statistical applications used to reduce costs and improve quality. It sets up a special infrastructure within the organization that is specifically trained in statistical methods and problem solution approaches that serve as the experts in these approaches. The two approaches that these experts use in their problem analysis and solution activities are Define, Measure, Analyze, Improve, and Control (DMAIC) and Define, Measure, Analyze, Design, and Verify (DMADV).

  Six Sigma aims to eliminate process variation and make process improvements based on the customer definition of quality, and by measuring process performance and process change effects.
- **Lean Six Sigma (LSS):** The LSS methodology is an organization-wide operational philosophy that combines two of today's most popular performance improvement methodologies: Lean methods and the Six Sigma approach. The objective of these approaches is to eliminate nine kinds of wastes (classified as defects, overproduction, transportation, waiting, inventory, motion, overprocessing, underutilized employees, and behavior waste) and provide goods and services at a rate of 3.4 defects per million opportunities (DPMO).

Note: Six Sigma, Lean, and LSS are all methodologies that contain a number of tools, techniques, and concepts that are designed to improve organizational performance.

Throughout this book, in keeping with common practice, the Six Sigma methodology, the Lean methodology, and the LSS methodology will also be referred to as just Six Sigma, Lean, and/or LSS.

It is important to note that Six Sigma, Lean, and LSS methodologies are all organization-wide operational philosophies/strategies with accountability and strategic focus. The real value of LSS starts to show when it is integrated with the organization's strategic plan, helping to implement that plan with a focus on the end-use customers. In order to achieve the true benefits of LSS, projects will cross organizational boundaries and be focused on business processes. Sustained strategic results can be achieved when this is done. When applied to a business process, the benefits obtained move the organization toward world-class performance in that business process.

At the heart of all successful LSS programs is an effective infrastructure that translates the strategic goals and activity areas of the organization into specific short-term action plans that maximize value and provide proper governance and management, along with the monitoring of results.

The Notion of Standing upon the Shoulders of Giants

> A dwarf standing on the shoulders of a giant may see farther than a giant himself.
>
> —**Didacus Stella**

This saying reminds us that for growth to occur, you grow from what has already been learned and/or created. If you ignore what has already been learned, at best you will get what has already been learned/created. On the other hand, if you learn from and start from what has already been done, then you can go farther. In our LSS leadership training that means we want clients to learn as much as possible from others' experiences. We encourage them to study leadership through reading lots of books, audio books, book summaries, and through personal discussions with other leaders that they know. For our business ventures, that means we gather mentors (giants) that have already been successful in the LSS project venture that we are getting into, and to learn from what they have done—learn from what is working or has worked for them, and learn from what did not work for them. This allows us to skip ahead in our own learning process and gives us a better chance for success.

LSS Cultural Building Blocks

All too often these methodologies are treated as a group of tools to be applied rather than a new cultural behavioral pattern that starts with the executive team. Too much focus has been applied to teaching the tools related to these methodologies in a two-level focus on embedding the Lean and Six Sigma cultures into the organization. As a result, these initiatives are often treated as projects where people are trained, problems are solved, and the immediate problem is put to bed while the organization goes back to business as usual. When this occurs, savings are short-lived and the problems that were put to bed soon wake up and have to be addressed again and again.

The person who makes the LSS methodology a success within an organization is not an LSS Black Belt (LSSBB) or even an LSS Master Black Belt

(LSSMBB). The people who make LSS a success are the executives who will not be satisfied with any waste within the organization and who won't allow anything but exceptionally good products and services to be provided within or as output from the organization he/she is responsible for. It is the executive who doesn't have excellence as a goal, but as a standard of today's performance for himself/herself and everyone within his/her organization, who is the ideal role model for the organization.

These are hard requirements to meet, but it's what is required to be successful in today's highly competitive environment. LSS Green Belts (LSSGBs) and LSSBB can bring about significant changes and improvement in organizational performance, but these gains often last for only a short period of time unless there is a significant change in the fundamental culture within the organization. Business excellence in an organization encompasses the areas of strategic focus or intent, customer loyalty/advocacy, employee delight, and seamless process integration. All the business excellence models like Malcolm Baldrige, EFQM, etc., have these areas incorporated in their models in different ways. This handbook clearly shows how these tools, along with the analysis/reduction of constraints, inventive problem solving, and the importance of human behavior modifications like applied behavior analysis, among others, can be used effectively to build upon the shoulders of the TQM and quality management programs of the past.

Connecting the Tools with Engineering Goals

Each of the LSS tools has been often connected with well-known engineering goals, including cause-effect analysis, variability reduction, bottleneck reduction, waste reduction, and the theory of inventive problem solving, which has its roots in TRIZ. In fact, many of the tools have had their beginnings in the TQM programs of 20 years ago, which was adopted by a majority of the Fortune 1000 companies, a statistic that was validated by our 2002 survey, which showed that over 80% of the Six Sigma approaches were carryovers from the TQM methodology.

WHAT CAME FIRST—SIX SIGMA OR LEAN?

It may surprise you to learn that Lean came before Six Sigma. Since the very earliest production systems management has fought to eliminate waste. Near perfection, on the other hand, was not a requirement for most consumers. Functionality for most consumers was the *meet requirements* quality standard that provided the maximum value (cost versus quality) to the consumer. The father of the Lean Production Systems, as we know them today, was Henry Ford Sr. There are very few of our products and methodologies that are truly breakthrough approaches. Almost all of our approaches are evolutionary rather than revolutionary. The Lean approach was born in the early 1990s with F.B. Gilbreth's time and motion studies, where he believed that there was one best way of doing everything. This, in conjunction with Henry Ford Sr.'s Lean Production System, started the Lean focus. Toyota made use of this as a foundation and improved upon it to develop Toyota's manufacturing system. Today's approaches to Lean are based upon Toyota's very successful manufacturing system.

In 1974 Motorola sold its TV production and design facilities to Matsushita, a Japanese manufacturer. Motorola had been one of the early developers of the TV concepts and one of the leading manufacturers under the brand name of Quasar Electronics, Inc. It promoted its TV as a TV with the work-center drawer for easy repair. Matsushita restructured the manufacturing process, applying Total Quality Control to it. As a result, internal and external defect rates and cost were decreased significantly. (See Figure 1.1.)

This major turnaround in the Quasar brand name reflected poorly on Motorola's reputation. In addition, Motorola's other operations were losing market share at a very rapid rate. In 1981, William J. Weisz, Motorola's COO, directed that all processes within the company should show a 10-fold improvement within 5 years. To do this, Motorola embraced the TQM concept. In 1986 Weisz required all measurements to improve by

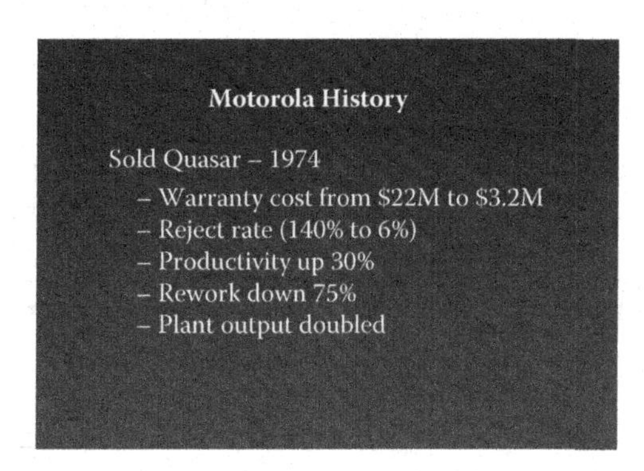

FIGURE 1.1
Results of TQC on Quasar's performance.

a factor of 10, this time in just 3 years. This called for a radical change in the way Motorola's processes functioned. To bring about such a drastic change, Motorola implemented what it called the Six Sigma Program. The program set an objective for all processes to statistically perform at an error rate no greater than 3.4 errors per defect per million opportunities. Six Sigma Quality became popular in the United States immediately following Motorola winning the 1988 Malcolm Baldrige National Quality Award. The information package that Motorola distributed to explain its achievements stated: "To accomplish its quality and total customer satisfaction goals, Motorola concentrated on several key operational initiatives. At the top of the list is Six Sigma Quality, a statistical measure of variance from a desired result. In concrete terms, Six Sigma translates into a target of no more than 3.4 defects per million opportunities. At the manufacturing end, this requires robust designs that accommodate reasonable variation in component parts while providing consistently uniform final products. Motorola employees record the defects found in every function of the business and statistical technologies are made part of each and every employee's job."

Also during the early 1990s Motorola's Six Sigma methodology was gaining momentum, as GE was promoting it as one of its major improvement drivers. Although Six Sigma was originally designed as an approach to reduce variability, quality professionals and consultants added to the basic statistical approaches a number of additional techniques that focus on process improvement. (See Appendix C for a list of Six Sigma Green Belt tools.)

Motorola's initial focus was on reducing variation in a single measurement. (See Figure 1.2.) Although the concept of focusing on reducing

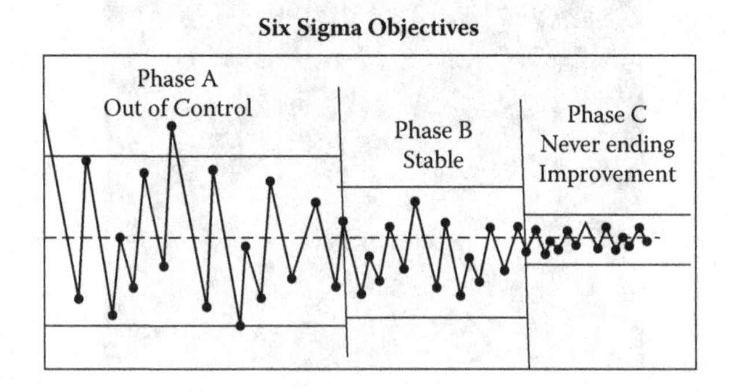

FIGURE 1.2
Results of focusing on variation reduction.

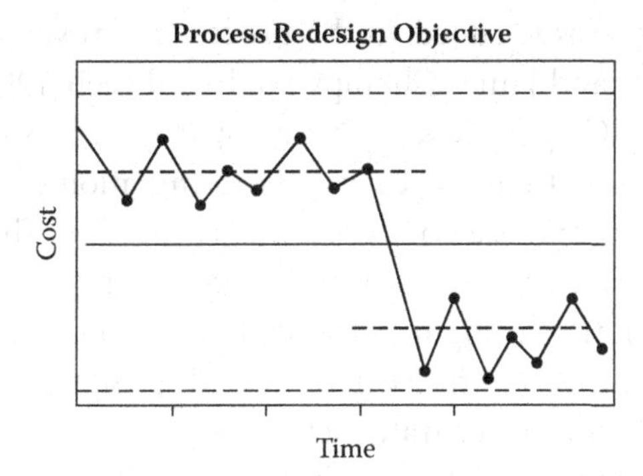

FIGURE 1.3
Results from redesigning a process to set new performance levels.

variation was a sound one, Motorola continued to lose a major portion of its market share. By the time GE embraced Six Sigma, it realized that the major gains from a performance improvement initiative would be reached by focusing on streamlining the processes by reducing cost and cycle time. As a result, the Six Sigma program was expanded to focus on setting new levels of performance. (See Figure 1.3.)

Along with this new emphasis came a new set of measurements focusing on cost reduction, decreased cycle time, inventory turns, etc. As a result, the basic Six Sigma approach of Define, Measure, Analyze, Improve, and Control (DMAIC) was modified with the addition of an approach of Define, Measure, Analyze, Design, and Verify (DMADV). With Six Sigma roots based upon variation reduction and the manufacturing environment, many of the Lean tools became part of the Six Sigma body of knowledge.

With the economy in the United States changing from a production to a service economy, the focus on performance improvement transferred from the manufacturing process to the service and support areas in the early 1980s. This resulted in a great deal of focus on reviewing the support and service processes and redesigning them to make them more efficient, effective, and adaptable. Process redesign methodologies roots stem from the poor-quality cost studies that IBM conducted in the indirect (support) areas during the 1970s. This evolved into the business process improvement methodologies that they developed during the early part of the 1980s. These approaches were further defined and developed by Ernst & Young and published in the 1991 book entitled *Business Process Improvement—The Breakthrough Strategy for Total Quality Productivity and Competitiveness*, published by McGraw-Hill, New York.

Additional depth was added to the process improvement focus when Michael Hammer and James Champy published their 1993 book entitled *Reengineering the Corporation*, published by Harper Business, New York. The basis of these methodologies was the elimination of no-value-added (NVA) activities from these critical business processes. These approaches, which were developed to redesign or re-engineer processes, became key building blocks in the Six Sigma methodology during the 1990s.

The end result is that over the years Lean and Six Sigma have been viewed and utilized as distinctly separate methodologies to analyze and improve processes. Rather than employing them separately, however, many process gurus now advocate a merger of the two for more dramatic process improvement. While we agree that this merger or marriage is valid, project leaders of process improvement efforts that forcibly combine the two methodologies without understanding what they are trying to improve will achieve limited success. In the final analysis, process professionals must first understand their level of process maturity to choose the appropriate blend of Lean and Six Sigma methods and tools, and employ some type of hierarchy or Belt System to help make the marriage last.

TECHNICAL COMPETENCY LEVELS

Due to the heavy focus on statistical applications and extensive amount of time and training that was required to prepare individuals to use these tools, a unique organizational structure was established that allowed different titles to be used in support of the successful deployment at GE. It created an innovative recognition system called Black Belt Program to support its Six Sigma Quality Program. Individuals progressed through various expertise levels as follows:

- **Blue Belts:** Individuals who are trained in basic problem solving and team tools, thereby establishing a common improvement approach throughout the organization. All employees should be at a minimum at the Blue Belt level.
- **Yellow Belts:** Individuals who have been trained to perform as members of Six Sigma Teams. They are used to collect data, participate in problem solving, and assist in the implementation of the individual improvement activities.

(**Note:** The Yellow Belt level was added on later to account for those who would become members of the Six Sigma Teams and assist with the process mapping, data gathering, brainstorming, communications, and implementation of solutions.)

- **Green Belts:** Individuals who have completed Six Sigma training, are capable of serving on Six Sigma project teams, and managing simple Six Sigma projects.
- **Black Belts:** Individuals who have had advanced training with specific emphasis on statistical applications and problem-solving approaches. These individuals are highly competent to serve as on-site consultants and trainers for application of Six Sigma methodologies.
- **Master Black Belts:** Individuals who have had extensive experience in applying Six Sigma and who have mastered the Six Sigma methodology. In addition, these individuals should be capable of teaching the Six Sigma methodology to all levels of personnel and to deal with executive management in coaching them on culture change within the organization.

These definitions have changed over time. It's now accepted that Green Belt training is different and less complex than Black Belt training. The same levels of expertise are used to distinguish background levels for LSS methodologies. Although the basic theory behind Six Sigma involved everyone in the organization, in many organizations that was not the case. In these organizations the Six Sigma initiative was made up of a few highly trained Green Belts or Black Belts whose total objective was to solve problems that would result in large savings to the organization. In many cases the Black Belts were expected to save the organization a minimum of $1 million a year or they would be reassigned. In these cases big improvement opportunities were acted upon and solved within 2 to 3 years and there was no need for the Black Belts' service anymore. As these major problems were solved, many executives began to realize that everyone needed to focus on the elimination of waste, not just in the production areas, but also in the support areas. This has resulted in LSSGB and LSSBB becoming facilitators of waste reduction as well as problem solvers. It also focuses the organization on waste reduction versus reducing variation. (Note: LSS does not ignore measurement where it is required, but does not rely upon it absolutely as Six Sigma does.)

Blue Belts keep the Six Sigma culture alive in the organization year after year.

LSS BELT LEVELS

LSS builds upon the technical competency structure that was developed for Six Sigma. The following are the competency titles and requirements related to the LSS methodology.

Lean Six Sigma Master Black Belt (LSSMBB)

The standard practice is one Lean Six Sigma Master Black Belt (LSSMBB) for every 15 to 20 Lean Six Sigma Black Belts (LSSBBs) or one for the total organization, if the organization is less than 200 employees. The LSSMBB is a highly skilled project manager, who should be Project Management Institute certified. LSSMBBs are the heart of the organization's LSS process. They must be more skilled and experienced than regular LSSBBs. They should be experienced teachers and mentors who have mastered the LSS tools.

The LSSMBB is responsible for:

- Certifying LSSBB and Lean Six Sigma Green Belts (LSSGBs)
- Training LSSBBs and LSSGBs
- Developing new approaches
- Communicating best practices
- Taking action on projects that the LSSBB is having problems in defining the root causes and implementing the change
- Conducting long-term LSS projects
- Identifying LSS opportunities
- Reviewing and approving LSSBB and LSSGB project justifications and project plans
- Working with the executive team to establish new behavioral patterns that reflect a Lean culture throughout the organization

Typically, an LSSMBB will interface with 15 to 20 LSSBBs to provide mentoring and development service in support of their problem-solving knowledge. When most organizations start an LSS process, they don't have people who are experienced enough to take on the role of an LSSMBB even when they have completed the LSSBB and LSSMBB training.

Training alone does not provide the required experience that is needed to function as an LSSMBB. As a result, organizations normally hire a consultant to serve as the LSSMBB for the first 6 to 12 months, and then they select one of the organization's LSSBBs to undergo the additional LSSMBB training and experience for the subsequent projects. No LSSMBB can be successful unless he or she is able to influence the executive team to embrace Lean concepts resulting in a cultural change throughout the organization.

Lean Six Sigma Black Belt (LSSBB)

Lean Six Sigma Black Belts are the workhorses of the Lean Six Sigma System.

One LSSBB for every 100 employees is the standard practice. (Example: A small organization with only 100 employees needs only one LSSBB or two part-time LSSBBs.)

LSSBBs are highly skilled individuals who are effective problem solvers and who have a very good understanding of the most frequently used statistical tools that are required to support the LSS system. Their responsibilities are to lead Lean Six Sigma Teams (LSSTs) and to define and develop the right people to coordinate and lead the LSS projects. Candidates for LSSBB should be experienced professionals who are already highly respected throughout the organization. They should have experience as a change agent and be very creative. LSSBBs should generate a minimum of US$1 million in savings per year as a result of their direct activities. LSSBBs are not coaches. They are specialists who solve problems and support the LSSGBs and LSSYBs. They are used as LSST managers/leaders of complex, simple, and important projects. The position of LSSBB is a full-time job; he/she is assigned to train, lead, and support the LSST. They serve as internal consultants and instructors. They normally will work with two to four LSSTs at a time. The average LSSBB will complete a minimum of eight projects per year, which are led by the LSSBB himself/herself or by the LSSGBs that they are supporting. The LSSBB assignment usually lasts for 2 years.

A typical LSSBB spends his/her time as follows:

- 35% running projects that he/she is assigned to lead
- 20% helping LSSGBs who are assigned to lead projects
- 20% teaching either formally or informally
- 15% doing analytical work
- 10% defining additional projects

The LSSBB must be skilled in the following six areas:

- Project management
- Leadership
- Analytical thinking
- Adult learning
- Organizational change management
- Statistical analysis

Most of LSSBB training focuses on analytical skills, so selecting the LSSBB often is based solely upon the candidate's analytical interests. This is all wrong. Other traits to look for in selecting an LSSBB are:

- Trusted leader
- Self-starter
- Good listener
- Excellent communicator
- Politically savvy
- Has a detailed knowledge of the business
- Highly respected
- Understands processes
- Customer focused
- Passionate
- Excellent planner
- Holds to schedules
- Motivating
- Gets projects done on schedule and at cost
- Understands the organization's strategy

- Excellent negotiation skills
- Embraces change

LSSBBs should be specialists, not coaches. It's important to build a cadre of highly skilled LSSBBs. However, they shouldn't be placed in charge of the managing and improvement process. LSSBBs are sometimes responsible for managing individual projects, but not directing the overall improvement process; that should be the job of management.

For organizations that do not have an LSSMBB, the LSSBB is responsible for working with the executive team to bring about a cultural change in the organization where waste is not accepted and excellence is the standard for everyday operations.

The American Society for Quality (ASQ) recommends a 4-week class to train LSSBBs. Typical subjects that are covered are:

1. Define and measure phase tools
2. Introduction to Minitab
3. Introduction to iGrafx
4. Lean overview
5. Probability concepts
6. Basic statistics
7. Documenting the process
8. Measurement systems evaluation (gauge R&R)
9. Basic statistics and introduction to process capability
10. Advanced process capability concepts
11. Process simulation
12. Graphical analysis
13. Project management
14. Program and training expectations
15. Analysis phase tools
16. Failure modes and effects analysis
17. Central limit theorem
18. Confidence intervals
19. Introduction to hypothesis testing
20. T-tests
21. Hypothesis testing with discrete data
22. Power and sample size
23. Correlation and regression
24. Logistic regression

25. Testing for equal variances
26. Analysis of variance (ANOVA)
27. Nonparametric statistics
28. Analyze phase deliverables
29. Design of experiments
30. Full factorial designs
31. Fractional factorial experiments
32. Simulating designed experiments
33. Creating future state maps
34. Center points in two-level designs
35. Response surface designs (supplement)
36. Analyzing standard deviation
37. Statistical process control
38. Husky bracket exercise (Lean/flow and work-in-process (WIP))
39. Design for Six Sigma
40. Creating acceptance sampling plans
41. Standard work
42. Statistical tolerancing
43. Mistake proofing
44. Developing control plans

Lean Six Sigma Green Belt (LSSGB)

One Lean Six Sigma Green Belt (LSSGB) for every 20 employees and 5 LSSGBs per every LSSBB is the standard practice. (Example: A small organization with 100 employees needs 1 LSSBB and 5 LSSGBs.)

Being an LSSGB is a part-time job. An LSSGB is assigned to manage a project or work as a member of an LSST by the LSS champion and his/her manager. Sometimes an LSSGB is the manager of the area that is most involved in the problem. However, it is very difficult for managers to lead or even serve on an LSST unless they are relieved of their management duties. They will need to spend as much as 50% of their time working on the LSS project. In most cases, it is preferable that the LSSGB is a highly skilled professional who has a detailed understanding of the area that is involved in the problem. LSSGBs work as members of LSSTs that are led by LSSBBs or other LSSGBs. They also will form LSSTs when projects are assigned to them. When that happens, the LSSGB's primary responsibility is to manage (coordinate) the project LSST's activities during the entire product cycle.

The LSSBB will support the LSSGB by providing just-in-time training to the project team when the LSSGB feels it is necessary. Normal LSST members will receive some basic problem-solving and team orientation training before they are assigned to a LSST. When LSSGBs are leading a LSST, they should have at least 50% of their workload assigned to another individual.

A typical annual cycle for an LSSGB would be as follows:

- Six months as an LSST team member on two different LSSTs spending 25% of his/her time on the LSST projects
- Three months as an LSST team leader spending 50% of his/her time on the LSST project
- Three months not working on any Lean Six Sigma projects

LSSGBs are also expected to identify other LSS opportunities and bring them to management's attention.

ASQ conducts a 2-week course on LSS for Green Belts. Typical subjects that are covered during this class include:

1. Process mapping
2. Introduction to Minitab
3. Probability and basic statistics
4. Rolled throughput yield
5. Process capability
6. Failure mode and effects analysis
7. Basic tools
8. Confidence intervals
9. Measurement system analysis (gauge R&R)
10. Hypothesis testing
11. Project management
12. Correlation and regression
13. Analysis of variance
14. Randomized blocks
15. Design of experiments
16. Full factorial experiments
17. Acceptance sampling plans
18. Statistical process control
19. Control planning and application
20. Mistake proofing

Lean Six Sigma Yellow Belt (LSSYB)

One Lean Six Sigma Yellow Belt (LSSYB) for every five employees and four LSSYBs for every LSSGB is the standard practice. (Example: A small organization with 100 employees needs only 1 LSSBB, 5 LSSGBs, and 20 LSSYBs.)

LSSYBs will have a practical understanding of many of the basic problem-solving tools and the DMAIC methodology. Team members are usually classified as LSSYBs when they have completed the 2 or 3 days of LSSYB training and passed an LSSYB exam. They will work part-time on the project and still remain responsible for their normal work assignments. However, they should have some of their workload re-assigned to give them time to work on the LSST. They usually serve as the expert and coordinator on the project for the area they are assigned to.

Lean Six Sigma Blue Belt

Blue Belts keep the Six Sigma culture alive in the organization year after year.

All employees should be trained as LSS Blue Belts as a standard practice. LSS Blue Belts are the normal workforce and may never be assigned to a LSST. However, they need to be part of the LSS culture and know how to apply LSS concepts to their day-to-day activities. They will receive 2 to 3 days of training covering the following subjects:

- How teams function
- What the Six Sigma processes are about
- How Six Sigma applies to them
- How to define who their customers are
- The seven basic problem-solving tools
- How to flowchart their process
- Area activity analysis
- How to participate in the suggestion program
- How to participate in "quick and easy Kaizen"

It is very important to note that once the major problems and opportunities have been addressed by the LSSTs, the organization's LSS culture is sustained through the LSS Blue Belts using area activity analysis, suggestion programs, and quick and easy Kaizen.

LSS Blue Belt activities drive continuous improvement throughout the organization. Their efforts should result in a 5 to 15% improvement in all the organization's measurements.

FIVE PHASES OF AN LSS IMPROVEMENT PROJECT

There are five phases that an LSS improvement project must go through:

- Phase I: The selling phase
- Phase II: The planning and training phase
- Phase III: The rollout phase
- Phase IV: The measurement of results phase
- Phase V: Sustaining the concept and holding the gain phase

Phase V is a very difficult, but the most important, phase, and the one that has been the least successfully completed. With methodologies like TQM, Six Sigma, and LSS to have a lasting effect within any organization, it needs to change the habit patterns for the employees and management. It is easy for an individual to do the *in thing*. During the rollout phase the changes are very visible. Big successes are celebrated. Everyone is getting credit for helping to implement the project. But with time (i.e., 1 to 3 years) the excitement wears off. Management stops following up to see that the changes are still working. Disruption in the status quo forces everyone's attention on other things, and the gains that were made through the LSS initiative begin to disappear. For example, we once visited a client to check on how the LSS initiative was still working 2 years after implementation. We were shocked to see that much of the effort that was put in to making the transformation was wasted. The in-boxes in the support area that were completely empty at the end of the day 2 years ago are now piled high with paper. The production area that was spotlessly clean is now in disarray and dirty. And this was just a superficial view that any observer would notice. Without continuous focus on the elimination of waste it is awfully easy for management and the employees to slip back into their old bad habits.

This handbook focuses on the tools and techniques that make up the tools of the trade for LSSGBs and LSSBBs. But the most difficult part of the job for these highly trained technical people is mastering the organizational change management methodology that is required to transform the culture within the organization to one where waste of any kind is not tolerated and excellence is accepted as *meets requirements.* Success for an LSSBB or an LSSMBB is not measured by the dollars saved—that is only a temporary measurement. The real measurement of the success of the LSS program is a permanent change in the culture within the organization.

SUMMARY

> Lean Six Sigma is a lifetime commitment. If you are not ready to make this commitment, don't waste the organization's money by trying to implement Lean Six Sigma.
>
> **—H. James Harrington**

The process of incorporating LSS as a key component of an organization's change infrastructure requires more than just training, for it takes a strong infrastructure built to support all projects and change initiatives from the ground up. LSS programs, when all is said and done, are a path to business excellence.

As we have discussed in this chapter covering the marriage of Lean and Six Sigma, both approaches have a few shortcomings that were especially evident a few years ago, before so many previous programs merged into the LSS program as we know it today. As a result, LSS evolved, getting stronger than some of the "giants" that had come before it.

> Lean Six Sigma is like a river of change, with the bridge being a change in attitude rather than a change in the way something is done.
>
> **—Frank Voehl**

2

Process Improvement and Lean Six Sigma

IN A NUTSHELL

This chapter describes the need for an LSS quality focus on business processes. It introduces process improvement (PI), describing the characteristics of business processes and explaining the advantages of using a cross-functional focus. When you complete this chapter, you should be able to:

- Explain what is meant by LSS quality focus on the organization's processes
- Explain the need for PI
- Describe the objectives of PI
- Explain the advantages of PI
- Explain what a business process is
- Explain the difference between traditional views of management and the process view
- Explain how the owner of a business process is selected
- Describe the authority and responsibilities of a business process owner
- Describe the composition and function of the Process Management Committee (PMC) and the Process Improvement Team (PIT), also known as the Six Sigma Team (SST)

INTRODUCTION

It is no secret that the 21st-century business environment is changing rapidly. The present environment is more competitive than it was in the past, with competition coming from both domestic and international global sources. In addition, there is increased emphasis on the quality of the products and services that businesses provide to their customers. Because of these changes, many businesses are discovering that the old ways of doing business no longer work so well. Many businesses have found it necessary to re-evaluate their corporate goals, procedures, and support structure. In one such corporate re-evaluation, the discovery that the support structure was not keeping pace with the changes led to the development of a management approach known as *process improvement* (PI). PI emphasizes both quality and excellence. It looks at the function of business management on the basis of processes rather than organizational structure.

AN LSS QUALITY FOCUS ON THE BUSINESS PROCESS

In 1978 many large manufacturers had relatively stable product lines that were offered through two or maybe three marketing channels, and usually shipped through one distribution channel. By 1994, however, manufacturers had almost tripled their number of products, offering them through seven or more marketing channels and shipping them through four to five distribution channels. Such rapid growth is not unusual in many organizations.

Due in large part to such growth, many corporations found that their sales representatives were spending far more time on administrative details and paperwork. In fact, the productive sales time of the corporation's marketing staff had dropped from an average of 40% in 1978–1980 to only 5% in 1994! Why? What could have caused this change? Compared to similar periods in the past, the corporation's growth rate was unprecedented. Moreover, systems and procedures that were adequate in 1978 were clearly out-of-date by 1994. More products meant more procedures, more complexity, and more internal competition for limited resources—human and other. In turn, this meant an even greater workload for an already strained support structure.

In addition, the company's various operating units were struggling to meet their own objectives, with little or no attention to the relationship of their work product to that of any other operating unit in the business. Driven partly by the resulting problems and partly by a re-emphasis on quality, corporate management decided to act. It provided the leadership and set the tone for making the needed changes by introducing the concept of process management in the corporate-wide instruction. Figure 2.1 shows the first page of a sample corporate instruction with an LSS focus.

SUBJECT: Lean Six Sigma Quality Focus on the Business Process

OBJECTIVE:

The objective of a Lean Six Sigma Quality Focus on the process is to improve the operational effectiveness, efficiency, and adaptability of corporate business processes.

BACKGROUND:

A key requirement for the achievement of our corporate goals is the ability of our business processes to meet the changing needs of the business. The traditional approach of managing through key business indicators has proven valuable, as has the focus on functional productivity. However, the rate of change created by growth and diversity now calls for increased attention to the effectiveness, efficiency, and adaptability of the business processes. This is to be accomplished through quality management.

CONCEPT:

Quality Management Lean Six Sigma consists of methodical approach to remove defects from an organization's processes and to improve its efficiency. It includes an understanding of supplier and customer requirements, process definition, defect measurement, root cause removal, and adapting the process to assure relevancy to business needs. It has been successfully applied in our corporation to improve both product and non-product processes.

Our Lean business operations can be characterized as a set of interrelated processes (e.g. Billing, Distribution, Accounting, etc.). Most business processes consist of subprocesses (e.g. Purchase Billing within Billing, Ship-to-Plan Distribution within Distribution, Inventory, Accounting within Accounting, etc.).

Each corporate operating unit will apply quality management to its key functional and cross-functional processes. Line management will define and own these processes. They will have responsibility for and authority over the process results. An executive is named as the single process owner and must operate at a level high enough in the organization to:

- Identify the impact of new business direction on the process.
- Influence change in policy/procedures affecting the process.
- Commit a plan and implement change for process improvement.
- Monitor business process effectiveness and efficiency.

FIGURE 2.1
First page of a sample corporate instruction with a LSS quality focus.

SOME BASIC DEFINITIONS

A corporate instruction like that in Figure 2.1 is important because it demonstrates top management commitment to LSS and provides basic guidelines for implementation. This philosophy is based primarily on the following statement in the corporate instruction: "Our business operations can be characterized as a set of interrelated processes."

LSS seeks to focus management attention on the fundamental processes that, when taken together, actually drive the organization. Figure 2.1 also introduces several important terms that are key to the LSS process improvement approach. Among these are customer, requirements, and quality.

Definition: A *customer* is any user of the business process's output. Although we usually think of customers as those to whom the corporation sells a product or service, most customers of business processes actually work within the same corporation and do not actually buy the corporation's products or services.

Definition: *Requirements* are the statement of customer needs and expectations that a product or service must satisfy.

Definition: *Quality* means conformance to customer requirements. In other words, since requirements represent the customer's needs and expectations, a quality focus on the business process asks that every business process meet the needs of its customers.

Definition: A *business process* is the organization of people, equipment, energy, procedures, and material into the work activities needed to produce a specified end result (work product). It is a sequence of repeatable activities that have measurable inputs, value-added activities, and measurable outputs. Some major business processes are the same in almost all businesses and similar organizations. Some typical business processes are listed in Table 2.1. Of course, some business processes are unique to a particular business and may be industry related. For example, steelmaking obviously has some processes, such as the foundry or slag disposal, not found in other industries.

OBJECTIVES OF PROCESS IMPROVEMENT

Process improvement is a disciplined management approach. It applies prevention methodologies to implement and improve business processes in order to achieve the process management objectives of effectiveness,

TABLE 2.1

Typical Business Processes

Accounts Receivable	Production Control
Backlog Management	Field Parts
Billing	Inventory Control
Service Reporting	Commissions
Customer Master Records	Personnel
Distribution	Payroll
Procurement	Order Processing
Finance/Accounting	Field Asset Management
Research and Development	Material Management
Public Relations	Traffic
Customer Relations	Supply Chain Mgt.

efficiency, and adaptability. These are three more key LSS terms introduced in Figure 2.1.

- An *effective* process produces output that conforms to customer requirements (the basic definition of quality). The lack of process effectiveness is measured by the degree to which the process output does not conform to customer requirements (that is, by the output's degree of defect). This is LSS's first aim—to develop processes that are effective.
- An *efficient* process produces the required output at the lowest possible (minimum) cost. That is, the process avoids waste or loss of resources in producing the required output. Process efficiency is measured by the ratio of required output to the cost of producing that output. This cost is expressed in units of applied resources (dollars, hours, energy, etc.). This is LSS's second aim—to increase process efficiency without loss of effectiveness.
- An *adaptable* process is designed to maintain effectiveness and efficiency as customer requirements and the environment change. The process is deemed adaptable when there is agreement among the suppliers, owners, and customers that the process will meet requirements throughout the strategic (3- to 5-year outlook) period. This is LSS's third aim—to develop processes that can adapt to change without loss of effectiveness and efficiency.

CROSS-FUNCTIONAL FOCUS

Essential to the success of the process management approach is the concept of *cross-functional focus*, one of the few truly unique concepts of the LSS approach. Cross-functional focus is the effort to define the flow of work products according to their sequence of activities, independent of functional or operating-unit boundaries. In other words, LSS recognizes that a department or similar operating unit is normally responsible for only a part of a business process. A cross-functional focus permits you to view and manage a process as a single entity, or as the sum of its parts.

CRITICAL SUCCESS FACTORS

While not altogether unique, the concept of *critical success factors* also plays an important role in the LSS approach. Critical success factors are those few key areas of activity in which favorable results are absolutely necessary for the process to achieve its objectives. In other words, critical success factors are the things that *must* go right or the entire process will fail. For example, payroll as a process has numerous objectives and performs many activities, but certainly its primary objective is to deliver paychecks on time. If payroll fails to deliver checks on time, it has failed in its major purpose, even if all its other objectives are met. Therefore, on-time check delivery is the most critical success factor for payroll. Payroll has other critical success factors, but these are invariably less important. In implementing LSS, it is important to identify and prioritize the critical success factors of the process and then to use the LSS methodology to analyze and improve them in sequence.

NATURE OF LSS PROCESS IMPROVEMENT

In addition to the key concepts just introduced, LSS uses many accepted standards of management and professional excellence. Among these standards are:

- Management leadership
- Professionalism
- Attention to detail
- Excellence
- Basics of quality improvement
- Measurement and control
- Customer satisfaction

What is new in process management is the way LSS applies these standards to the core concepts of ownership and cross-functional focus. In the following chapters the key elements of managing a process are introduced and topics already introduced are elaborated. However, it should be noted that identifying and defining a business process almost always precedes the selection of the process owner. That is, you first identify the process, and then you identify the right person to own it. This seemingly out-of-sequence approach is taken in order to firmly establish and reinforce the key concepts of LSS before getting into how-to-do-it details.

Advantages of LSS Process Improvement

Management styles and philosophies abound within the framework of LSS. There are some similarities to such approaches as value stream management, participative management, excellence teams, and management by objectives. That is not unusual, since all of these approaches are based on the management standards mentioned earlier. However, LSS has advantages that make it both useful and relatively easy to implement:

- LSS, through its cross-functional focus, identifies true business processes. It bypasses functional and operating-unit boundaries that may have little to do with the actual process or that may obscure the view of the overall process.
- LSS, through its emphasis on critical success factors, gives early and consistent attention to the essential activities of the process. It provides a means of prioritizing so that less important activities are identified, but action on them is delayed until the more important ones have been addressed.
- LSS improves process flow by requiring the total involvement of suppliers, owners, and customers, and by assuring that they all have the same understanding of process requirements.

- LSS is more efficient than other methods in that it can be implemented process by process. It does not require simultaneous examination and analysis of all of a corporation's processes. With management leadership and guidance, workers implement LSS within their process while continuing their regular duties.

DETERMINING PROCESS OWNERSHIP

As previously described, the nature of business processes is how they often cross organizational boundaries. It is important to describe process ownership, show how the process owner is selected, and describe the responsibilities and authority of the process owner. LSS stresses a cross-functional focus instead of the traditional hierarchy-of-management approach. LSS requires a process owner with authority to cross functional or operating-unit boundaries in order to ensure the overall success of the process.

Because the role of the process owner is at the heart of LSS, it is extremely important that top management understand and accept the principles presented in this chapter. Without the active support of top management, the process owner cannot function properly, and thus LSS will fail.

Selection of a process owner follows the identification and definition of the process, which is covered later. The purpose in discussing process ownership first is to give a broad portrayal of the environment over which the owner must have authority and responsibility. This, in turn, will provide an understanding of process improvement that is far more comprehensive than any simple definition of the term.

The Nature of Business Processes

By definition, LSS emphasizes the management of organizational processes, not business operating units (departments, divisions, etc.). Although a set of activities that comprise a process may exist entirely within one operating unit, a process more commonly crosses operating-unit boundaries. In LSS, a process is always defined independently of operating units.

Management's Traditional Focus

The traditional management of a business process has been portrayed in the form of an organization (pyramid) chart. This "chain of command" management hierarchy has worked very well for many corporations over the years. So why does LSS seem to question this traditional approach? Actually, the traditional approach is not being questioned so much as it is being redirected. Process ownership is intended not as a replacement of the traditional organizational structure, but as a complement to it, a complement with a specific goal. The goal is to address the fact that the traditional structure often fosters a somewhat narrow approach to managing the whole business.

Typically, a manager's objectives are developed around those of some operating unit of the business in which he or she is working at the moment. This emphasis on operating-unit objectives may create competition rather than cooperation within the business as a whole as managers vie for the resources to accomplish those objectives. With rare exceptions, these managers are responsible for only a small portion of the larger processes that, collectively, define a particular business in its entirety. Examples of these major business processes are shown in Table 2.1. In addition, the recognition and rewards system is almost always tied directly to the degree of success achieved in the pursuit of those operational objectives. So there is no particular motivation for managers to tie their personal goals to the larger—and more important—process goals. LSS seeks to address these common deficiencies through the concept of process ownership.

Cross-Functional Focus

In one large manufacturing corporation an internal review of its special/custom features (S/CF) process disclosed some serious control weaknesses and identified the lack of ownership as the major contributing cause. (S/CFs are devices that can be either plant or field installed to modify other equipment for some specialized task.) The S/CF process was a multimillion-dollar business in its own right and touched almost every other department in the corporation—order entry, manufacturing, inventory control, distribution, billing, service, and so on. Yet the review found that the process had no central owner—no one person who was responsible for the process from initial order to final disposition, no one who provided overall direction and a cohesive strategy for this cross-functional process.

Because business processes such as S/CF cross operating-unit lines, the management of these processes cannot be left entirely to the managers of the various departments within the process. With no one in overall charge of the S/CF process, the disruption caused by lost orders, incorrect

FIGURE 2.2
Owner's cross-functional perspective of the S/CF process.

shipping, and wrong billings is difficult to eliminate. Each manager can pass the buck, saying someone else is responsible. No one manager has the authority to solve process-wide problems.

If no one person is in charge of the S/CF process flow shown in Figure 2.2, the success of the process depends on the four managers of the four departments, for whom S/CF may be only a minor task in their departments. However, if one person is in charge, he or she can view the process with a cross-functional focus, emphasizing the process rather than the departments. The figure shows an owner's cross-functional perspective of the S/CF process.

PROCESS OWNERSHIP

As shown above, process management requires an understanding of the fundamental importance of cross-functional focus. Clearly, there is a need to have someone in charge of the entire process—a *process owner.* While this chapter deals primarily with ownership of the major processes and sub-processes of a business, the concept of ownership applies at every level of the business. For example, the purchasing process can be broken down into various levels of components, such as competitive bidding and the verification of goods and services received. The components of a process are examined further in Section 2.

Because business processes often cross functional or operating-unit boundaries, it is generally true that no one manager is responsible for the entire process. LSS addresses this point through another unique concept—*process ownership.*

Definition

A process owner is a manager within the process who has responsibility and authority for the overall process result.

The Process Owner

The process owner is the manager within the process who has responsibility and authority for the overall process results. The process owner assumes these duties in addition to his or her ongoing functional or operating-unit duties, and is supported by a PMC, described below. The process owner is responsible for the entire process, but does not replace managers of departments containing one or more process components. Each component continues to exist within its own function or operating unit and continues to be managed by the managers of that function or operating unit.

Normally, the term *process owner* is used in connection with major business processes, either corporate-wide (for example, the general manager of a plant owns the process of manufacturing a particular product). On a practical level, however, an executive is often somewhat removed from the day-to-day operation of the process. Consequently, it has been found useful to implement a two-tier approach, in which the executive (vice president of finance) is referred to as the *focus* owner, and a subordinate, but still high-level, manager (for example, the controller) is referred to as the *functional* owner. In general, the functional owner has all the authority and responsibility of the focus owner except for the ultimate responsibility for the success of the process.

The Process Management Committee

Headed by the *focus* owner, the PMC is a group of managers who collectively share the responsibilities related to the committee's mission. The committee has the following characteristics:

- All process activities are represented.
- The senior/top manager of the parent function for each activity is a member.
- The members are peer level, to the degree possible.

In practice, an owner should never be a lower-level manager than other members of the committee. When differences of opinion arise, a lower-level owner would have difficulty exercising his or her authority.

The mission of the PMC is to:

- Steer and direct the process toward quality objectives
- Support and commit the assignment of resources
- Ensure that requirements and measurements are established
- Resolve conflicts over objectives, priorities, and resources

The PMC mission can be expanded to accommodate unique process conditions. Issues related to the PMC mission are raised by the *functional* owner when, as often happens, he or she has been unable to resolve the issue with the Process Quality Team, discussed below. It is important for the credibility and effectiveness of the functional owner that issues raised to the focus owner and PMC be addressed and resolved promptly.

The Process Quality Team

Headed by the *functional* owner, the Process Improvement Team (PIT) is made up of lower-level managers. Preferably, they are peers of the functional owner and work for members of the PMC. The members of the PIT are the designated implementers of process management actions, which are to:

- Establish the basics of process quality management
- Conduct ongoing activities to ensure process effectiveness, efficiency, and adaptability

The functional owner and the PIT may be unable to resolve some issue, such as resource allocation, expense reductions, or process change. In that case, the issue should be brought to the focus owner for resolution through the PMC.

SELECTION, RESPONSIBILITIES, AND AUTHORITY OF THE PROCESS OWNER

The rest of this chapter covers process owner selection, responsibilities, and authority. It concentrates on the role of the *focus* owner, but adds perspective on the PMC and the *functional* owner as well. But first, it is important to remember that all three elements—selection, responsibilities, and authority—are needed. A process owner, with detailed and specific responsibilities (which will be described shortly), cannot function without commensurate authority. Just as important, all of the concerned parties within the process must be made aware of both the owner's role and their own. And once ownership is established, in all its elements, the owner must get ongoing and visible support from both executive and peer managers.

Selection of the Process Owner

As mentioned earlier, process ownership is not intended to override the existing management structure. The process owner is defined, in part, as a manager within the process. The process owner continues in his or her current position, and simply takes on additional duties as head of the overall process. The other managers within the process become members of the PMC. Top management normally appoints the process owner. The question of which manager to choose as the process owner can often be answered by applying two words: *most* and *best*. That is, which manager in the process has the most resources invested, does the most work, feels the most pain when things go wrong, gets the most benefit/gain/credit when things go right? Which manager has the best chance of affecting or influencing positive change—perhaps because he or she is in the best position to do so?

For the most part, answers to these questions should clearly identify the best candidate for the ownership role. At the very least, the answers should narrow the field, in which case, the concept of *critical success factors* comes into play once again. That is, of the remaining candidates, which one currently manages the activities most critical to the success of the overall process? In addition, the process owner should be acknowledged as having the personal and professional qualities necessary for the job and a track record to accompany them. He or she should be comfortable dealing at high levels, be able to create and execute a plan, be a skilled negotiator successful at gaining consensus, and be a team player.

Responsibilities of the Process Owner

In general, the process owner is responsible for making sure that the process is effective, efficient, and adaptable, and as discussed in later chapters, for getting it certified. More specifically, Figure 2.3 lists the 12 responsibilities of a process owner, above and beyond that person's normal job responsibilities.

It is also important for the process owner to create an environment in which defect prevention is the most highly regarded "quality" attribute. In many enterprises it is all too common to find that the most visible rewards go to the "troubleshooters" or "problem fixers." Those individuals who consistently strive for defect-free results often go unrecognized and unrewarded.

1. Determine and document process requirements and secure customer concurrence.
2. Define the sub-process, including information used by the process.
3. Designate line management ownership over each sub-process.
4. Identify implementers and assure application of quality management principles.
5. Ensure documentation of task-level procedures.
6. Identify critical success factors and key dependencies in order to meet the needs of the business during the tactical and strategic timeframe.
7. Establish measurements and set targets to monitor and continuously improve process effectiveness and efficiency.
8. Rate the process/sub-process against defined quality standards and control criteria.
9. Report process status and results.
10. Identify and implement changes to the process to meet the needs of the business.
11. Ensure that information integrity exists throughout the process, including integrity of measurements at all levels.
12. Resolve or escalate cross-functional issues.

FIGURE 2.3
Ownership responsibilities.

Authority of the Process Owner

Given the key elements of ownership selection, it is obvious that the process owner already has a great deal of inherent authority within the process. That is, if a manager meets all or most of the selection criteria, he or she is probably already in a fairly authoritative position in the process. Nevertheless, the issue of authority cannot be taken for granted. The process owner must have the right combination of peer and top management support to get the job done. Top management provides the greatest contribution by validating the owner's authority through various forms of direct and indirect concurrence and support, including the following six areas:

1. Approving resource allocation
2. Financing the process
3. Providing recognition for the owner
4. Running interference
5. Cutting the red tape
6. Backing the focus owner or functional owner when issues have to be escalated

It is not enough for the process owner to have the necessary authority, however. All concerned parties within the process must clearly understand the authority being vested in the process owner. The process owner can first be told of his or her responsibilities and authority through a performance plan or job description. The next step is a formal announcement or, at the very least, an announcement to all managers throughout the process. If necessary, or if considered helpful, the PMC can create a written agreement defining and clarifying the various relationships. However it is done, there must be clear communication to all concerned parties regarding who has been selected, what authority is being vested in that manager, and what that manager's responsibilities are. And top management must continue to support the process owner by doing whatever necessary to show that the owner's authority is real. Only with this sort of strong, continuing support from senior management can LSS work.

PROCESS DEFINITION AND THE PROCESS MODEL

Once the *focus* owners of major business processes are identified, the existing managers within the process would begin to define and model the lower-level processes and sub-processes. This is accompanied by meeting in logical work groups from within the major process. For example, within the major process of finance, logical groups may include accounting managers for the lower-level accounting process and payroll managers for the lower-level payroll process.

A process model is a detailed representation of the process as it currently exists. When preparing the model, it is important to avoid the temptation to describe the process as it should be, or as one would like it to be. The model may be written, graphic, or mathematical. The model can take several forms, but the preferable, recommended form is a process flowchart, described in Section 2. Whatever form is used, the model must contain supporting textual documentation, such as a manual or a set of job descriptions.

The definition of a business process must clearly identify the following:

1. The boundaries of the process and all of its components
2. All of the suppliers and customers of the process and the inputs and outputs for each

3. The requirements *for* the process suppliers and *from* the process customers
4. The measurements and controls used to insure conformance to requirements

The model and supporting documentation are used as the base tools for analyzing how the process can be simplified, improved, or changed.

Definition of Process Mission and Scope

The first thing to be done, once ownership has been established, is to define the mission, or purpose, of the process, and to identify the process scope. But defining the mission and identifying the scope are not as easy as they sound. The mission of a process is its purpose—what the process is in existence to do. The mission clearly identifies exactly what the process does, from beginning to end, especially for someone not working within the process; the mission also describes how the process helps to attain the corporate goals.

The mission statement does not have to be lengthy or elaborate. It should be concise and to the point. Figure 2.4 shows a mission statement for a procurement process.

The scope of a business process is defined by identifying where the process begins and ends. These are the boundaries of the process. By identifying the boundaries, you identify the scope, just as boundaries tell you how much land a piece of property covers. There may be disagreements over the exact scope of a process. Such a disagreement might occur, for example, over whether a particular activity is the last activity of one process or the first activity of the next process. In settling disagreements about the scope of a process, it helps to identify the process whose mission is most clearly related to the activity in question. It may also require the help of some of the customers and suppliers of the process. Most importantly, you should concentrate on how the activity in question affects, or is affected by, the critical success factors of the process. As you remember, critical success factors are those few, key areas of activity that must succeed in order for the entire process to achieve its goals.

MISSION OF THE PROCUREMENT PROCESS

The procurement process covers how "off the shelf" parts, equipment, or supplies are purchased from external suppliers.

FIGURE 2.4
Mission statement: Procurement process.

To define the scope of a process, then, all parties must agree on the answers to four questions:

1. Where does the process start?
2. What does the process include?
3. What does the process not include?
4. Where does the process end?

The question of what a process does not include should be addressed even if there appears to be complete agreement on what it does include. Often, when discussing what the process includes, people make assumptions. (For example, if the process includes *X*, it must include *Y* and cannot include *Z*.) Only by agreeing on what the process does not include can you be sure that you have avoided making any such assumptions.

Even in well-established processes, there often is disagreement on answers to these four questions. Moreover, the questions deal only with a broad overview of the scope. The task of a process definition gets more complex later on as you detail all of the activities and tasks that go on within the process. That is another reason for beginning the definition of mission and scope by focusing on critical success factors. The mission and scope should both be written by the owner and the team. Neither has to be elaborate.

Figure 2.5 shows the scope of the accounts receivable process, as written by an accounts receivable group. Figure 2.6 shows the mission

SCOPE OF THE ACCOUNTS RECEIVABLE PROCESS

The accounts receivable process begins when an invoice is issued and the accounts receivable system is updated.

The accounts receivable process includes the following:

- Collection or other settlement of customer accounts.
- Accurate maintenance of customer accounts.
- Assessment of the credit worthiness of customers.
- Initiation and/or processing of credit notes.
- Preparation of regular reports for staff action.
- Notification to management of out-of-line situations.
- Monitoring resolution of management directives.

The accounts receivable process ends when the invoice is cleared and the accounts receivable system is appropriately updated.

FIGURE 2.5
Scope statement: Accounts receivable process.

MISSION AND SCOPE OF THE PROCUREMENT PROCESS

- The procurement process covers how "off the shelf" parts, equipment, or supplies are purchased from external suppliers.
- The process begins when a requestor submits a purchase requisition.
- The process then includes part requisition, requisition processing, vendor build, Information Systems support, and receiving/shipping.
- The process does not include inspection for concealed damage, shortages, or shipment errors, nor does it include payment of the vendor invoice.
- The process ends when the requested part is delivered to the requestor.

FIGURE 2.6
Mission and scope statement: Procurement process.

and scope of the procurement process, as written by a procurement group. These figures give a clear picture of the broad components of each process.

SUMMARY

This chapter has given you a general introduction to LSS and its connection to process improvement. Process improvement has been the focus of quality initiatives since the 1970s. Initially the focus was on manufacturing processes, but as the service industry became a major driver of the nation's revenue stream and the overhead costs in any production organization exceeded the direct labor costs, emphasis shifted to business process improvement. In the 1980s organizations like IBM, Hewlett-Packard, Boeing, and AT&T developed programs to make step function improvements in their business processes, resulting in decreasing costs of the business processes as much as 90%, while decreasing cycle time from months to days or even hours. Methodologies like benchmarking, business process improvement, process redesign, and process re-engineering became the tools of the trade for the process improvement engineer during the 1990s. This focus on process improvement was driven by the realization that most of the major processes within an organization run across functions, and that optimizing the efficiency and effectiveness of an individual function's part of these processes often resulted in sub-optimizing the process as a whole.

The designers of the LSS methodology recognized early in the cycle the importance of focusing upon process improvement and, as a result, incorporated the best parts of the methodologies that were so effectively used in the 1980s and 1990s. Throughout this chapter we pointed out how LSS has built upon the business process improvement techniques used in the 1980s and 1990s, modifying them to meet the high-tech, rapid changing environments facing organizations in the 21st century.

> In most organizations there is a bigger opportunity for improvement in streamlining their processes than there is in solving their problems.
>
> **—H. James Harrington**

EXERCISE

1. What does the term *business process* mean?

   ____

   ____

2. If an *effective* process is one that conforms to customer requirements, what is an *efficient* process?

   ____

   ____

3. What is an *adaptable* process?

   ____

   ____

4. What is cross-functional focus?

____

____

5. If your corporation has official goals, what are they? If not, what do you think they might be?

____

____

____

____

6. List five or six of the business processes that exist in your corporation.

____

____

____

____

7. List what you believe are the three or four critical success factors of your corporation.

____

____

____

____

8. If your corporation has a formal quality policy statement, list the three most important elements of that policy (from your point of view).

   ____

   ____

   ____

9. What is the difference between functional owner and focus owner?

   ____

   ____

10. What is the role of the Process Quality Team (PQT)?

    ____

    ____

11. What bearing do critical success factors have on the selection of a process owner?

    ____

    ____

12. What is the relationship between process ownership and cross-functional focus?

    ____

    ____

    ____

    ____

13. What is the mission of the Process Management Committee (PMC)?

____

____

14. What are the cross-functional aspects of the process that your job is part of?

____

____

____

15. Who is the owner of the process that your job is part of? What is his or her title?

____

____

16. Why is it important for top management to accept and support the LSS concepts of process ownership and cross-functional focus?

____

____

17. What is the most common form of process modeling used in your corporation?

____

____

____

18. What is most often used as supporting material for that process model?

____

____

19. What is the difference between process mission and process scope?

____

____

____

20. What are the four questions necessary to answer to fully identify the process scope?

____

____

____

21. Using the scope statement in Figure 2.5 as a guide, write a mission statement for the accounts receivable process.

____

____

____

____

____

____

____

____

22. Select a top priority business process activity in which you are directly involved. Write a mission statement for and define the scope of that activity.

____

____

____

____

____

____

____

____

____

____

____

____

Section 2

The Lean Journey into Process Improvement

INTRODUCTION

Every accomplishment starts with a decision to try.

—Gail Devers
Three-time Olympic champion, track and field

Understanding and implementing Lean and LSS is often referred to as a journey. The word *journey* has been aptly chosen since it implies that this transformation is neither quick nor easy. It can also describe changes that occur in individuals, departments, functional areas, divisions, and the entire organization during the LSS learning experience. First, for the individual, it can be described as an evolution in your way of learning and thinking. We learn to better understand our processes. We learn about our fellow employees. Ultimately, we learn about ourselves and how we approach our job. For departments or functional areas, the journey has more to do with getting a collective group of individuals to arrive at a common understanding of the improvement process and working together to achieve it. Finally, for organizations, the journey has to do with promoting a culture where individuals, departments, divisions, and suppliers can come together with a common set of beliefs and behaviors that are focused on improving performance for customers.

This journey starts with the individual—you. It can take on an exciting and almost mystical meaning. What will we encounter on the way? What new things will we learn and experiences will we have? What unknowns will we encounter and how will we face those? All of these are present during the change management that occurs when we commit to adopting a LSS philosophy. Let the journey begin!

This section includes three chapters that will become the fundamental building blocks for what you will experience and learn. In Chapter 3, "Waste Identification," you will be introduced to a new meaning of what waste is and how to identify it. In Chapter 4, "Lean Concepts, Tools, and Methods" you will be introduced to new ways of thinking about your processes and the activities required to complete those value-added processes. In Chapter 5, "Three Faces of Change—Kaizen, Kaikaku, and Kakushin," you will be introduced to the process of change plus improve. In this chapter we tie it all together by defining three specific improvement approaches to show you how to focus on the identified wastes with newly learned ways of thinking and apply the concepts and tools to achieve new levels of personal and organization performance. Think of each of these chapters as companions in your daily work life. The true LSS practitioner carries these companions wherever he/she goes. With each new experience you become more proficient at recognizing waste. With every use of the concepts or tools, you become more familiar with their use. With every change in your processes, you become more open to change and more comfortable and confident in your ability to produce positive change with LSS. You begin to apply the sequence of see the waste, understand the LSS concepts, and apply the tools, so effortlessly; it becomes second nature.

Together these chapters are about learning, changing, and improving. They describe concepts, tools, and activities that will allow us to change how we think and how we act. Once completed, our hope is that your point of view will be irreversibly changed. Oliver Wendell Holmes once said, "Man's mind, once stretched by a new idea, never returns to its original dimension." He was telling us that every experience we have expands our understanding of our surroundings. Once our mind entertains a new idea, it cannot remain the same. This is the essence of becoming an LSS organization—a learning organization that (1) observes processes differently, (2) identifies weaknesses in these processes, (3) conducts better measurements to quantify where we are, and (4) constantly improves, adapts, and changes processes to better meet customer needs. These three chapters define a learning and improvement cycle that can be adopted by employees at all levels of the organization.

OVERVIEW

In Section 2 we undertake a significant challenge—to blend Lean and Six Sigma into an integrated synergistic approach to process improvement. This approach can be readily understood and applied regardless of company size or industry type. One of the primary challenges for organizations is how do we integrate Lean with Six Sigma. How do the concepts and tools of each process improvement philosophy mesh? In a rudimentary definition, Lean is identifying and eliminating waste, while Six Sigma is identifying and eliminating variation. Lean uses predominantly Kaizen as an instrument of improvement. Six Sigma uses the DMAIC and DMADV methodologies.

The concepts and tools in Section 2 are presented in the sequential order of the learning and change management process. Most of these tools are first defined, then you are instructed on how to use them, and next you are provided with examples of how they have been used before. The contents of Section 2 are:

- Chapter 3, which describes the nature of waste, is dependent on our reference point. One person may consider something wasteful, while the next may consider the very same thing essential to complete the job. So which is correct? How do we identify what is truly waste? To understand this, we need a new understanding of how to classify waste and what activities produce waste in our organization.

  In Chapter 3 we present several topics:

  - What is waste?
  - What is variation?
  - Value-added activities
  - No-value-added activities
  - The power of observation
  - Nine wastes

1. Overproduction
2. Excess inventories
3. Defects
4. Extra processing
5. Waiting
6. Motion
7. Transportation
8. Underutilized employees
9. Behavior

- Chapter 4 describes a fundamental understanding of Lean concepts and tools. This new understanding is critical to successfully apply Kaizen improvement and change management philosophies (covered in Chapter 5). When I was growing up, my dad attempted to teach me some concepts to live by, using familiar quotations. One of his favorites was, "Knowing where you are going is all you need to get there." The Lean concepts presented in the chapter are "where you are going." Learn them well!

  In Chapter 4 the topics include:
  - Lean concepts
  - Waste identification
  - Waste elimination
  - Value-add
  - No-value-add
  - Business-value-add
  - Standard work
  - Value stream
  - Value stream management
  - Continuous flow
  - Point of use storage
  - Quality @ source
  - Pull systems
  - Just-in-time
  - Takt time
  - Kaizen
  - Lean tools
  - 5S Workplace Organization and Standardization
  - Mistake proofing
  - Cellular manufacturing

- Overall equipment effectiveness
- Kanban
- Value stream mapping
- Single-minute exchange of dies
- Visual controls
- Total Productive Maintenance

- Chapter 5 describes the three fundamental approaches of Kaizen, providing a comprehensive look at applying this concept on a daily basis across your organization. Through *Kaizen and You* we learn to identify waste and apply the Lean concepts in our individual and immediate sphere of influence. With *Kaizen Teams* we focus on larger areas with group improvement activities. With *Kaizen and Process Troubleshooting* we learn how to apply the concepts and tools and improve a process while it is operating.

  In Chapter 5 we present:
  - Kaizen and You method
  - Kaizen and Project Teams
  - Kaizen and Process Troubleshooting
  - Kaikaku—Transformation of Mind
  - Kakushin—Innovation

INTEGRATING LEAN WITH SIX SIGMA

Integrating Lean and Six Sigma concepts is not only beneficial, but it can often be a critical first-step requirement of process improvement projects. For example, one of the first steps of many Six Sigma projects is process assessment to understand if the process is "stable and in control" (a requirement to complete much of the statistical analysis described in later sections of this handbook). Typically processes that are unstable or out of control have many "special causes" associated with process variation. In these instances, the use of Lean concepts and tools is invaluable in standardizing a process prior to embarking on statistical projects. This is a fundamental reason why our Standardize-Do-Check-Act (SDCA) cycle is a precursor to the Plan-Do-Check-Act (PDCA) cycle that is prominent in process improvement.

In Table S2.1 we show how to integrate Lean concepts and tools into typical phases of Six Sigma projects using the DMAIC and DMADV

TABLE S2.1

Basic Lean Concepts and Tools for Lean Six Sigma Project Success

Lean Concepts	**D**	**M**	**A**	**I**	**C**	**D**	**M**	**A**	**D**	**V**
Waste	X	X	X			X	X	X		
Value-added	X	X	X			X	X	X		
No-value-added	X	X	X			X	X	X		
Business-value-added	X	X	X			X	X	X		
Waste identification	X	X	X			X	X	X		
Waste elimination				X	X				X	X
Standard work	X	X	X	X	X	X	X	X	X	X
Value stream management	X	X	X	X	X	X	X	X	X	X
Continuous flow	X	X	X	X	X	X	X	X	X	X
Pull systems	X	X		X	X	X	X		X	X
Point of use storage (POUS)	X			X	X	X			X	X
Quality @ source		X	X	X	X		X	X	X	X
Takt time	X	X	X	X	X	X	X	X	X	X
Just-in-time (JIT)	X	X	X	X	X	X	X	X	X	X
Kaizen	X	X	X	X	X	X	X	X	X	X
Materials, machines, man, methods, and measurements (5M's)	X	X	X	X	X	X	X	X	X	X
Lean Tools										
5S	X	X	X	X	X	X	X	X	X	X
Overall equipment effectiveness (OEE)		X	X	X	X		X	X	X	X
Mistake (error) proofing			X	X	X			X	X	X
Cellular manufacturing	X	X	X	X	X	X	X	X	X	X
Kanban	X	X	X	X	X	X	X	X	X	X
Value stream mapping	X	X	X	X		X		X	X	
Visual controls				X	X				X	X
Single-minute exchange of dies (SMED) or quick changeover	X	X	X	X	X	X	X	X	X	X
Total Productive Maintenance	X	X	X	X	X	X	X	X	X	X

methodologies. There are no hard-and-fast rules. The table gives a basic reference of where these may fit into your LSS projects, that is, what phase in the two Six Sigma methodologies that each Lean concept or tool is most apt to be used. This does not mean that the tool *should* be used each time a project goes through the indicated phase, but rather what *could* be used in that phase. In Appendix B we give a complete LSS body of knowledge that identifies which concepts and tools are used in typical Green Belt, Black Belt, and Master Black Belt programs.

In Table S2.1 many of these Lean concepts are used in the early phases of a project—Define, Measure, and Analyze. They are a natural fit in these phases for improving your understanding of the current state of your process. Many of the Lean tools are predominantly used in the Improve and Control phases where you are taking action to improve a process. Taken together, these Lean concepts and tools are an invaluable component of defining your improvement plan and the future state improved process.

As you proceed through Section 2, refer back to this table for a clearer understanding of applying Lean concepts and tools. As you move further through later sections of the handbook and ultimately embark on LSS projects, this table will serve as a reference check for integrating important Lean concepts and tools into DMAIC or DMADV projects.

3

Waste Identification

There is nothing so useless as doing efficiently that which should not be done at all.

Peter F. Drucker

IN A NUTSHELL

Waste is generally composed of unnecessary activities that can be described either qualitatively or quantitatively. In its most basic form, LSS encompasses both these descriptors. Waste identification is also called learning to see muda, which is a traditional Japanese term for an activity that is wasteful and doesn't add value or is unproductive, value-none, trivial, or unuseful.* It is also one of three key concepts in the Toyota Production System (TPS). The other two are mura, which means irregularity, unevenness, or variability, and muri, which refers to overburden or strenuous work.† These three terms describe the waste that infiltrates organizations and allows us to begin "learning to see" waste.

Waste identification and reduction is an effective way to increase profitability. Toyota merely picked up these three words beginning with the prefix *mu-*, which in Japan are widely recognized as a reference to a product improvement program or campaign.

In this chapter, we present fundamentals on how to recognize muda, mura, and muri in the workplace. The chapter discusses the importance

* Muda, 無駄; translation to English on Sanseido: "exceed." *Japanese-English Dictionary*.

† *Lean Lexicon*, 4th ed., Lean Enterprise Institute, Cambridge, MA, March 2008.

of looking at your organization in new ways, developing an inquisitive approach that encourages questioning of current beliefs and practices, and taking a look at everything you do from the customer viewpoint. One approach is to start at the end; that is, walk backward through your organization assessing process steps in reverse and asking questions. To this end, we include a series of checklists to get the LSS practitioner thinking in new ways.

OVERVIEW

Until recently, when we talk about learning to see waste, we are considering organizational wastes, not behavioral wastes. However, behavioral wastes can severely hinder Lean initiatives. This chapter describes what waste is and how to identify waste. It also discusses the mindsets that are the root causes of waste. These mindsets or belief systems are put into context here and described with terms like *just-in-case logic*. Almost all organizational wastes or process wastes are related to an employee in the organization that holds a traditional belief system.

Learning to see variation (mura) or waste (muda) requires a shift in how we view our organization. How do we view our processes? How do we measure our processes? What questions we ask about process performance, people performance, and equipment performance all indicate how we look at variation and waste. Throughout this chapter we present checklists that help you to begin to question everything you do in an effort to learn to see variation and waste in a new light.

WHAT IS VARIATION?

The way in which numbers differ is called variation.* Virtually everything that is measured is subject to variation. Our equipment is subject to variation. Our employees are subject to variation. The instrumentation that

* H. James Harrington, Glen D. Hoffherr, Robert P. Reid Jr., *Statistical Analysis Simplified*, McGraw Hill, New York, 1998.

we often use to measure a process is subject to variation. How we measure our measures can be a key source of variation. There is variability inherent in many methods completed in our value-adding processes. For example, one of the most critical and prevalent tools used in every organization today is the computer and its software packages. How we gather information, analyze information, and report information can be subject to tremendous variation. Based upon the simple observations, one can see why the study and understanding of variation is a critical component to an LSS organization.

Variation can also be described as "a measure of the changes in the output from the process over a period of time."* As you collect data over time, you can measure and view the variation of process input variables, process methods, or process output variables. Understanding, controlling, and limiting process variation is a primary goal of any LSS practitioner. As we begin our journey toward being an LSS organization, we must become proficient at measuring variation, analyzing root causes of this variation, and taking corrective actions to eliminate variation from all of our processes.

The entire study of variation is an endeavor to quantify and chart process behavior. At the beginning of our value-added processes, we can quantify and chart our process input variables. These include the 5M's: materials, machines, manpower, methods, and measurements. The objective here is to minimize variation in our supply chain inputs to our value-added processes. For example, by measuring variation in material specifications, we are able to better control our value-adding process steps, thereby assuring a predictable outcome for product performance.

How Do We Chart Variation?

One of the most common process output variables in an LSS environment is process lead time. In many instances, customers are sensitive to the amount of time it takes us to add value for them. How we organize our materials, deploy our human resources, set up our equipment, and sequence our value-added steps has a tremendous influence on our process output lead time. The average lead time chart in Figure 3.1 shows

* H. James Harrington, Praveen Gupta, Frank Voehl, *The Six-Sigma Greenbelt Handbook*, Paton Press, New York, 2009.

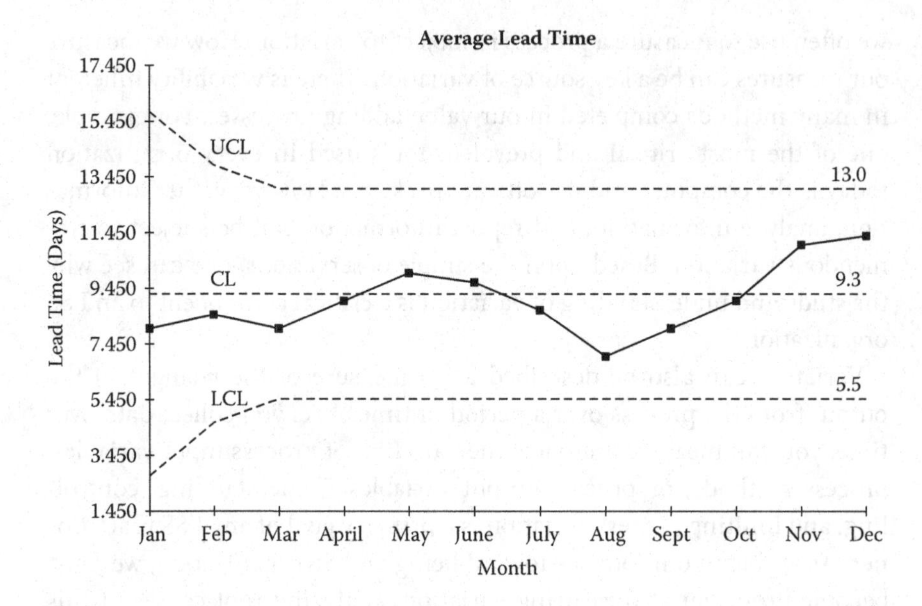

FIGURE 3.1
Average lead time chart.

how the lead time varies over time, specifically, how month-to-month lead time varies during the year. It also shows the upper control limit (UCL) and lower control limit (LCL) for the data set.

Why Is Understanding and Controlling Variation So Important?

Simple charting like this can help us to understand, control, and improve lead time for our customers. The importance of understanding, charting, and controlling process variation cannot be overstated. Understanding variation and decreasing variation is the fundamental underlying foundation of all LSS organizations. First, it allows us to understand, control, and improve our entire supply chain, which includes many activities that are conducted outside of our physical facilities. Second, it allows us to uncover valuable insight concerning the interactions between materials and our processing equipment. Perhaps, more importantly, it provides a fundamental foundation for assessing our performance output behavior that is critical for customer satisfaction. Virtually all of our outputs are key performance indicators and subject to variation. As a consequence, our ability to understand and chart variation is paramount for improved performance from a customer viewpoint. The remainder of this chapter and several chapters that follow in this handbook are dedicated to understanding waste identification and process variation and applying LSS tools for process improvement.

WHAT IS WASTE?

Describing waste is not as easy as one might think. Waste appears throughout organizations and is often mixed with nonwaste. There are times and conditions within our organizations where deciding what is waste versus nonwaste can be somewhat of a moving target. For example, in today's organization, e-mail is virtually impossible to live without as a communication tool. In and of itself, it has great capacity to assist with many processes. Yet it can also be a significant source of extra processing waste. The telephone can produce a similar waste, but are all telephone calls wasteful? There are times when some organizations consider inventory an asset, that is, right up until the customer no longer wants to purchase the inventory. Clearly, one or more definitions describing just what is waste (all no-value-added activities) and what is not waste (value-added activities) are needed.

Defining the Value-Added Work Components

LSS organizations are constantly searching for more effective ways to deliver value for the customer. How do we define value and distinguish it from activities that produce no value?

To better understand this term, we have provided you with a practical definition of value-added (VA), no-value-added (NVA), and no-value-added but necessary. (See Figure 3.2.) Value-added is an activity that transforms or shapes raw material or information to meet customer requirements.

Organizations that strive to eliminate NVA work while increasing their VA work are the ones that will be the most successful.* There are a number

* As a management technique, companies seek to provide additional value-added in their products as a way of distinguishing them from competitors; value-added in this sense is a means of avoiding commoditization and maintaining profit margins.

- **Value-added** is an activity that transforms or shapes raw material or information to meet customer requirements.
- **No-value-added** is an activity that takes time, resources, or space, but does not add to the value of the product or service itself from the customer perspective.
- **No-value-added but necessary** is an activity that does not add value to the product or service but is required (e.g., accounting, health and safety, governmental regulations, etc.). In the business process management methodology this is called business value-added.

FIGURE 3.2
Definitions of VA, NVA, and NVA but necessary.

of ways to accomplish this. One of the most effective ways is to first evaluate the practices used, so you can recognize any NVA work and then take steps to reduce it and be more efficient in your work. The basic characteristics include VA components such as customer VA and operational VA, as well as NVA components such as idle time, rework, and bureaucracy. Detailed analysis of these factors is a fundamental part of waste identification and the foundation of LSS initiatives.

In the case of a manufacturing operation, VA means all those activities that turn raw materials into value (the product) for your customer. In the case of a service organization, VVA means all those activities that are required to deliver the intended service. In essence, the service is your product. Your VA product or service is what you end up with, or what the customer wanted. Conversely, NVA is anything that the customer in not willing to pay for. NVA entities can be employee activities, materials, information exchanges, and equipment. The difficulty comes in separating NVA from VA activities and still providing what the customer wanted. The remainder of this section is about identifying anything that is NVA in your organization.*

* Many of the TPS/Lean techniques work in a similar way. By planning to reduce manpower, or reduce changeover times, or reduce campaign lengths, or reduce lot sizes, the question of waste comes immediately into focus upon those elements that prevent the plan being implemented. Often it is in the operations area rather than the process area that muda can be eliminated and remove the blockage to the plan. Tools of many types and methodologies can then be employed on these wastes to reduce or eliminate them.

HOW DOES WASTE CREEP INTO A PROCESS?

Waste can creep into any process over time and usually does.* In the 1980s there was a popular story about a tire manufacturing plant in the Midwest. The story goes that they were conducting a continuous improvement effort when one of the observers posed a simple question to an operator. "Why are we wrapping these tires in protective white plastic?" The operator was not sure; his reply was that they had done it for the 3 years that he had been with the company. So they went to the shift supervisor and posed the same question. His response was that the machine was there since he joined the company 7 years earlier. He said that Charlie in maintenance may know: "He's been here for 25 years." So they headed off to maintenance to find Charlie. When they posed the question to him, he replied that it is to protect the whitewall tires in shipping. They hadn't made whitewall tires in the plant for years; however, they continued to wrap the newer all-black tires as if they had whitewalls. This is one example where a product change without a process change allows waste to creep in to your process.

THE POWER OF OBSERVATION

From the Renaissance period in the 1500s to the emergence of many of the pure and applied sciences in the 1700s to 1800s, there were limited technical tools compared to today. There were no computers, no Internet with instant information, no instant communication, no telephone, and no mass transportation. The sharing of knowledge was slow and difficult. In relative isolation, science was advanced by disciplined individuals committed to observation and experimentation.

---

* Shigeo Shingo divides process-related activity into process and operation. He distinguishes process, the course of material that is transformed into product, from operation, which are the actions performed on the material by workers and machines. This distinction is not generally recognized because most people would view the operations performed on the raw materials of a product by workers and machines as the process by which those raw materials are transformed into the final product. He makes this distinction because value is added to the product by the process but not by most of the operations. He states that whereas many see process and operations in parallel, he sees them at right angles (orthogonal); this throws most of operations into the waste category. See value stream mapping for a further explanation.

It was during these times that the power of observation was the dominant tool for improvement. As mechanical scientific instruments were developed, these highly trained and skilled observers applied these tools, coupled with keen observation capabilities, to make astounding discoveries. However, in almost every organization that we go into, the power of observation is almost nonexistent. Employees at all levels wander through the organization focused on their individual worlds and completely ignoring the blatantly obvious signs of waste that engulf their organization.

Some have called these *organizational cataracts.* These cataracts can grow and hinder our vision and render the power of observation an obsolete tool. Managers fail to see the waste of rework associated with the poor scheduling or haste with which they initially produced a product. Employees focused exclusively on a daily production deadline completely miss a multitude of opportunities to improve their environment in favor of producing the daily production quantity.

> The range of what we think and do is limited by what we fail to notice. And, because we fail to notice that we fail to notice, there is little we can do to change until we notice how our failing to notice shapes our thought and deeds.
>
> **—R.D. Laing**

This approach (observation and experimentation), which has been used by science for hundreds of years, is the key to advancing knowledge and improving our understanding of our surroundings. We must be able to accurately observe our surroundings, document what we see, investigate and analyze our observations to find out what is causing what we see, and ultimately take effective action to improve our environment.

> Science is "the desire to know causes."
>
> **—William Hazlitt (1778–1830), English essayist**

This emergence of the power of observation is a key ingredient in the formation of a learning environment. The remainder of this chapter is about igniting the power of observation in our employees. More importantly, it's about learning to see waste and variation with new eyes, eyes that know what to look for.

SEEING WITH NEW EYES

Traditionally, Lean has classified waste into eight major categories. These categories were developed based upon visual symptoms in the organization. We have added a ninth waste, behavior waste, which revolves around individual and collective belief systems and how they influence daily behavior. The remainder of this chapter discusses each category in detail.

What types of waste are present? What are typical causes of each waste? How can waste be identified? Checklists are included to assist you with learning to see waste and variation. However, you are encouraged to expand these checklists by looking at each process step in your organization and developing your own questions. The nine waste categories are:

1. Overproduction
2. Excess inventory
3. Defects
4. Extra processing
5. Waiting
6. Motion
7. Transportation
8. Underutilized people
9. Employee behavior

Waste 1: Overproduction

Overproduction means making more of a product than is needed by the next process or the end customer.* It can also be described as making the product earlier in time than is needed or making a product at a faster rate than is needed. Overproduction has been labeled by some as the worst waste because typically it creates many of the other wastes. For example, overproduction leads to excess inventory, which in turn leads to the wastes of motion and transportation. In addition, excess inventory requires more

* Overproduction happens each time you engage more resources than needed to deliver to your customer. For instance, large batch production, because of long changeover time, exceeds the strict quantity ordered by the customer. For productivity improvement, operators are required to produce more than the customer needs. Extra parts will be stored and not sold. Overproduction is a critical muda because it hides or generates all others, especially inventory. Overproduction increases the amount of space needed for storing raw material as well as finished goods. It also requires a preservation system.

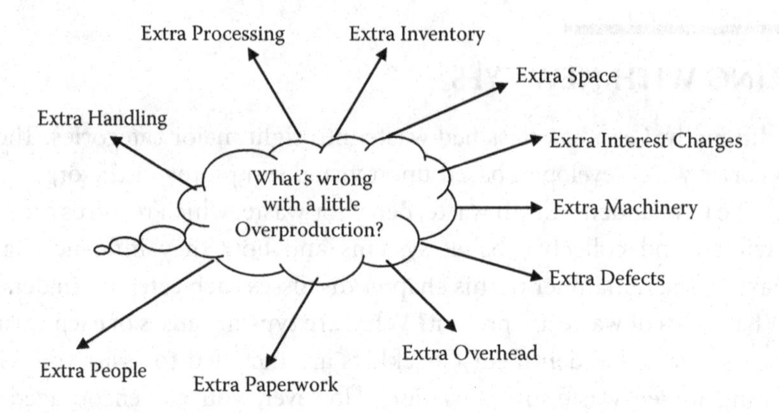

FIGURE 3.3
Waste of overproduction.

people, equipment, and facility space, all of which reduce company productivity and profitability. This is shown in Figure 3.3.

What Causes Overproduction?

Overproduction can be traced to many management and employee behaviors. Some of the most common causes are:

- Just-in-case logic
- Unleveled scheduling
- Unbalanced workloads
- Misuse of automation
- Long process setup times

The Just-in-Case Logic Trap

Just-in-case logic is exactly what it sounds like. You make more product *just-in-case*—you fill in the blank. For example, just-in-case the machine breaks down, just-in-case our suppliers don't send enough raw materials, just-in-case our customer orders more than we can make or deliver on time, etc. There are many reasons for using just-in-case logic, and they are all bad!

Just-in-case logic is one of the most common non-Lean employee behaviors present in companies today. It is responsible for productivity losses in any type of organization by robbing employee time when working on NVA overproduction. It is commonly found in other waste categories, such as the waste of motion, transportation, inventory, waiting, and defects. Most importantly, it reveals an inherent weakness in your current process capability and reliability. Managers that practice just-in-case logic invariably have poorly understood processes and poor process control. Instead of fixing the process, they prefer to mask the system with just-in-case overproduction. They have fallen into the just-in-case logic trap. Don't do it!

Unleveled Scheduling and Unbalanced Workloads

Unleveled scheduling and unbalanced workloads can both lead to overproduction. When these conditions occur and employees continue to produce, even when there is no customer demand, overproduction occurs. In forecast-driven environments unleveled scheduling frequently occurs. In areas where the workload is not balanced properly between two or more process steps, one step will have excess capacity while the next may have excess demand.

Misuse of Automation

Another common mistake is that owners, senior managers, and/or department managers want to see expensive equipment running, not sitting idle. This misuse of automation can cause severe overproduction. Not very often does customer demand exactly meet machine capacity. One of the most difficult challenges for LSS practitioners is to change the misconception that the machine must always be running. In environments where there is a combination of automated and manual production, the misuse of automation at one process step creates the unleveled scheduling and unbalanced workloads at downstream process steps. In this case we have one cause of overproduction (misuse of automation) forcing overproduction at another process step.

Long Process Setup Times

The length of time required to set up equipment has long been a primary justification for overproducing and carrying excess inventory. The traditional thought is that if your setup times are long, then you must build larger batches than are required. One traditional approach is to define an

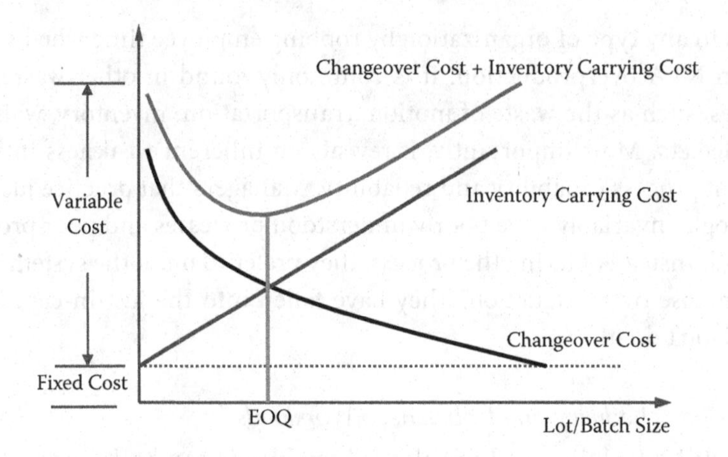

FIGURE 3.4
Batch size selection based upon changeover cost and inventory carrying costs.

economic order quantity (EOQ) where the changeover cost + the inventory carrying cost are the lowest, and then you build a batch this size. This concept is shown in Figure 3.4. Do these assumptions on batch size selection make sense if changeover time can be significantly reduced? The answer is no. As you reduce changeover time, you reduce both changeover cost and the inventory carrying cost, and the EOQ moves toward the left on the chart. In this case, instead of using EOQ, target your processes to build just what the customer wants. The most cost effective EOQ is always what the customer wanted.

This is a classic example of bad measures driving bad behavior. The primary assumption that you have to live with long changeovers and high inventory levels and inventory carrying costs, rather than try to eliminate them, was made based upon these two measures. In an LSS environment we focus on the process (long changeover time), identify the waste, and eliminate the waste by simplifying the setup process.

How to Identify Overproduction

The learning to see overproduction checklist in Figure 3.5 presents several questions designed to help you identify overproduction.

Waste of Overproduction Checklist			
Process:	**Date:**		
Description	**Yes**	**No**	**Apparent Cause**
Do we make more product than is required by the next process step?			
Do we make more product than is required by the customer?			
Do we make product faster than is required and store it for later use?			
Do we keep machinery running even when there is no demand?			
Do we create "busy work" for employees when demand falls?			
Are we producing more reports than needed?			
Are we making more copies than needed?			
Are we printing, faxing, and e-mailing more than what is needed?			
Are we entering repetitive information on multiple work documents or forms?			
Are we ordering more tests or services than what is required by the customer or patient?			

FIGURE 3.5
Learning to see overproduction checklist.

Waste 2: Excess Inventories

Excess inventory is "any supply in excess of a one-piece flow through your manufacturing process."* One-piece flow is often referred to as a *make one–move one environment*. Excess inventory could also refer to any finished goods inventory. Most organizations today run a mixed model of both "build to order" and "build to stock" products. Although some amount of raw materials and finished goods is required, many organizations use inventory to cover up poor process performance. They keep raising the level of inventory until they cover process problems.

Like all of the nine wastes, living with excess inventory creates the "more syndrome." For example, in an excess inventory environment, your company requires more people, more equipment, and more facility space. All the while you're making more products (that you may or may not have customers for), more defects, more write-offs, etc. The "more syndrome" robs your company of productivity and profitability. In an LSS environment we reduce the sea of inventory and use the Lean or Six Sigma tools to identify the root causes of why the inventory was needed and then eliminate the root causes once and for all.

What Causes Excess Inventory?

- Poor market forecast
- Product complexity
- Unleveled scheduling
- Unbalanced workloads
- Unreliable or poor-quality shipments by suppliers
- Misunderstood communications
- Reward system

Poor Market Forecast

Many organizations decide what they will build based upon a market forecast. Basically they take a sales and marketing forecast and convert it to a manufacturing forecast and then in turn set up a build schedule. Unfortunately, the only thing we can say about a forecast with a high

* Inventory, be it in the form of raw materials, work-in-process (WIP), or finished goods, represents a capital outlay that has not yet produced an income either by the producer or for the consumer. Any of these three items not being actively processed to add value is waste.

degree of certainty is that it will be wrong. When this occurs, organizations are typically left with large amounts of inventory, much of which may be unsaleable.

Product Complexity

In a rush to get to market, many products are moved from the product development to full production before sufficient design for manufacturability has been completed. When product complexity is high, there are several issues that lead to excess inventory. These include raw materials performance issues, engineering changes that lead to supplier changes, production issues, and in-service performance, to name a few. In a competitive product cost environment, product complexity and high quality are often at odds with each other and are another source of excess inventory.

Unleveled Scheduling and Unbalanced Workloads

Similarly with overproduction, unleveled scheduling and unbalanced workloads can both lead to excess inventory. These conditions typically occur in forecast-driven environments. In areas where the workload is not balanced properly between two or more process steps, one step will have excess capacity while the next may have excess demand. In the end, you wind up with excess inventory.

Unreliable or Poor-Quality Shipments by Suppliers

LSS organizations can only be sustained with an LSS supply chain. Inferior materials can, and often do, produce myriad troubles during your VA activities. Unreliable suppliers that deliver materials of poor quality or insufficient quantities only serve to help your competitors. To achieve LSS performance, focus on developing relationships with LSS suppliers.

Misunderstood Communications

Poor communication invariably leads to excess inventory. In the age of information overload, it is staggering how much bad information our employees are using and how much good information is being unused or misused. There are basically three fundamental areas in all organizations. (See Figure 3.6.) These are product development or service delivery, operations management, and information management. Most companies are

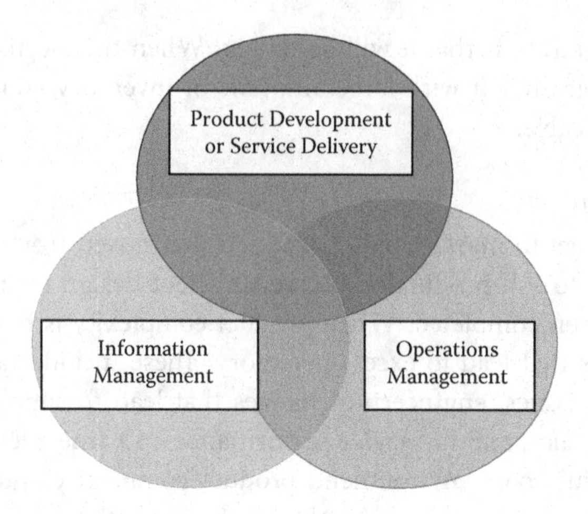

FIGURE 3.6
The organizational universe.

good performers in one or two of these categories, but rarely all three. Depending on the nature and structure of the senior management team, more emphasis usually goes to one area. For example, companies with a perceived technology advantage tend to pay more attention to product or service development at the expense of the other two areas. When communication breakdowns occur, inventory increases, quality decreases, and profitability is hurt. These are signs that you are in a poor communication environment:

- Poorly understood customer requirements
- Product or service is frequently delivered late
- Poor customer satisfaction
- Incomplete or inaccurate documentation
- Poor work instructions
- Inadequate information management system
- Barriers between departments
- Conflicting measurements system

Rewards System

There are several factors of company-wide rewards systems that can contribute to excess inventory. These factors can originate from senior management or from most departments. Since we know that measures drive behavior, poorly defined measures tied to rewards often result in excess inventory and many other wastes.

One example could be if an operations group has a measure of "on-time delivery" without regard for inventory levels. Another may be how the sales group gets compensation. Still another may arise from inadequate knowledge of the true cost of carrying inventory. Regardless of the reasons, if a large level of inventory exists in your facility, review the rewards program for an inadequate measurements system.

How to Identify Excess Inventory

The learning to see excess inventory checklist presents several questions designed to expose inventory waste. (See Figure 3.7.)

Waste of Inventory Checklist			
Process:	**Date:**		
Description	**Yes**	**No**	**Apparent Cause**
What does the customer want?			
How much do they want and when?			
What are your purchasing signals—when, how much, how often?			
How do you structure your organization to meet these needs?			
How responsive is your inventory control and purchasing process to fluctuations in customer demand?			
Can you adequately describe the range of your customer demand for products or services?			

FIGURE 3.7
Waste of inventory checklist.

Waste 3: Defects

Definition

A defect can be described as anything that the customer did not want. Defects include product or service attributes that require manual inspection and repair or rework at any point in the value stream. Defects can be detected and identified before your product or service reaches the customer or post-consumer in the form of warranty returns.*

What Causes Defects?

Defects can result from myriad causes. These causes can be classified into a few basic areas listed below. Each is followed by a brief description.

- Customer needs not understood
- Poor purchasing practices or quality materials
- Inadequate education/training/work instructions
- Poor product design
- Weak process control
- Deficient planned maintenance

Customer Needs Not Understood

Establishing comprehensive customer requirements is essential to defect-free products. More often than not we think we know what the customer wants or we make many assumptions about how he or she will use our product or service or what's important to him or her in terms of product or service performance. The more we can know in this area, the better we can develop our processes to respond to customer requirements.

Poor Purchasing Practices or Quality Materials

In the global marketplace, controlling the supply chain is an ever-increasing challenge. Purchasing departments typically have their own stand-alone measures based upon dead materials costs. Material costs

* Whenever defects occur, extra costs are incurred reworking the part, rescheduling production, etc. Defects cause scrap, repair, rework, backflow, and warranty/replacements; consume resources for inspection, correction, and replacement; cause opportunity loss (capacity and resources used to fix problems); cost 5% of sales for Six Sigma and 40% of sales for One Sigma processes; and reduce variability, lock gains, implement controls, and error proofing.

are typically very visible in financial statements and a common target for cost reductions. This never-ending pressure for cost reduction frequently pushes product quality below levels expected by customers.

What's *not* present on most financial statements is the cost of quality, which includes repair and rework. Oh, it's present on the bottom line; however, there is no individual expense line item that can be targeted. Many defects can be traced to inferior quality materials. Repair and rework costs for these defective materials increase dramatically the further into the value stream your product gets before the defect is discovered. Numbers for how much this costs vary greatly across industries. Some of the components of this cost may include:

- Cost of communication with supplier
- Cost of storage until a disposition can be made
- Cost of employee time for physical moves or quarantine
- Cost of employee doing this NVA activity instead of a VA activity
- Cost of repair if required
- Cost of returns to suppliers
- Cost of re-engineering
- Cost of re-inspection
- Cost of productivity losses on new products due to staff re-assignment to complete rework

Inadequate Education/Training/Work Instructions

Here is an important rule of thumb: At any given point in time, you should have cross-training capacity at 150% of full production at each process step. To accomplish this, there needs to be a well-defined and executed cross-training program and effective work instructions to carry out the program. Often employees are asked to produce a quality product without adequate education, training, or visual work instructions to complete the task. One of the most effective means of defect reduction is the preparation of visual work instructions.

Poor Product Design

Many defects can be traced to poor product design. In examining product design failures, look for cost restrictions, poorly understood in-service product performance requirements, poor materials' selection, little or no product performance testing, and poor supplier performance. Regardless of the root cause of poor product design, the cost for a part design change increases dramatically the further into the value stream the product is before the defect is detected. The relative cost to mitigate a defect detected along the value stream using design engineering as a baseline of $1 is:

- $1 product design engineering
- $2 product manufacturing engineering
- $4 production
- $5 to $10 if the product reaches the customer

In some industries the cost could be significantly more. Many pharmaceuticals, for example, have limited shelf lives. If defect detection occurs at the customer, there may be insufficient time for return and repair or rework, requiring a complete write-off of the shipment. In this case, material/labor/facilities' costs plus profit are lost, not to mention the bad will created with the customer.

Weak Process Control

In all process environments either you control the process or the process controls you. Weak process control can stem from several sources, including deficiencies in materials, machines, manpower, methods, or measurements. It is easy for weak process control to creep into your processes. The three telltale signs you need to work on process control are defects, rework, and high scrap rates.

Deficient Planned Maintenance

Poor equipment maintenance is often a cause for defective products. The justification for not completing planned maintenance can range from not enough time to do Total Productive Maintenance or autonomous maintenance, to can't afford to have production down, to equipment repairs are too expensive, to name a few. In the long run, effective equipment maintenance is always less expensive than equipment breakdowns due to poor maintenance, the cost of scrapping defective parts, or the added cost of rework.

How to Identify Defects

Defects are often only defined as something that an employee can tangibly see in the product. However, a defect is better described as anything that contributes to a product not meeting exactly what the customer wants. The list of questions in Figure 3.8 should help you begin to expose a number of possible defects. This list could be greatly expanded; however, it

Defect Detection Questions

- Materials
  - ☑ Are the proper materials being used?
  - ☑ Are the material specifications adequate?
  - ☑ How many materials are needed?
  - ☑ Are we purchasing excessive supplies of any kind?
  - ☑ When are they needed?
  - ☑ Where are they stored?
  - ☑ How are they handled?
  - ☑ How are they moved to where we create value for the customer?
- Machines
  - ☑ Is current machinery adequate? Optimal?
  - ☑ Where are they located?
  - ☑ Do we have obsolete equipment in the area?
  - ☑ Is there a defined maintenance program and schedule?
  - ☑ Is time allowed for proper equipment maintenance?
  - ☑ Do you have a daily 5S clean and inspect procedure?
- People
  - ☑ Are special personnel needed?
  - ☑ How many?
  - ☑ What specific skills are needed?
  - ☑ Do you have people cross-trained at each position?
- Methods
  - ☑ What methods do we use?
  - ☑ Do you have visual work instructions for every operation?
  - ☑ Are files (or work) awaiting excessive signatures or approvals?
  - ☑ Are files awaiting task completion by others?
  - ☑ Do we have any obsolete files in the area?
  - ☑ Do we have data entry errors?
  - ☑ Do we have standardized pricing, quoting, billing, or coding?
  - ☑ Do we forward partial documentation to the next process?
  - ☑ Do we ever lose files or records?
  - ☑ Do we ever encounter incorrect information on a document?
  - ☑ Are methods easy to understand, learn, and use?
  - ☑ How do we train our staff?
  - ☑ Does poor performance signal a requirement for retraining?
- Measurements
  - ☑ What measures do we use?
  - ☑ Are there clear strategic measures?
  - ☑ Do all tactical process measures "roll up" to strategic measures?
  - ☑ Are all of these measures performance based?
  - ☑ Any dead cost measures?
  - ☑ Do we have key process input (KPI) measures?
  - ☑ Do we have key process output (KPO) measures?
  - ☑ Do we have good in-process measures?

FIGURE 3.8
Defect detection questions.

should give you an idea of how to begin your search for anything that may possibly affect your product or service and be considered a defect.

Waste 4: Extra Processing

Processing waste is described as any effort that adds no value to the product or service from the customers' viewpoint. These are steps that may not be necessary. Many examples of processing waste are present in any product or service delivery. For example, let's consider a product with 15 steps. If a sub-assembly at process step 3 is not assembled correctly, the product moves through the facility and the problem is initially detected at assembly step 13. Unfortunately, steps 5, 7, 9, and 11 may need to be disassembled and the correction made before step 13 can proceed. These repeated steps are rework and take valuable time away from employees who could be working on new products. This extra effort is called *processing waste.*

What Causes Processing Waste?

Processing waste can stem from many sources and is often present regardless of the activity type. Processing waste is predominantly waste that is found in front-office areas, such as order processing, information gathering and dissemination, and all accounting functions. It is also dominant in service industries where service-delivery requirements may be ill-defined or difficult to achieve. Industries like the medical field, janitorial, or food-service industries may have extensive processing waste from several apparent causes. These causes can be classified into a few basic areas listed below. Each is followed by a brief description.

- Product changes without process changes
- Just-in-case logic
- True customer requirements undefined
- Overprocessing to accommodate downtime
- Poor communication
- Redundant approvals
- Extra copies/excessive information

Product Changes without Process Changes

When a product or service is changed, production staff or service personnel need to be properly informed. For example, visual work instructions or service-delivery instructions need to be modified and training conducted for the new process. In many growing companies, products or services are changed frequently, often with little or cursory regard for production or service-delivery personnel. This can be a major source of processing waste for a range of product or service quality issues.

Just-in-Case Logic

Just-in-case logic is exactly what it sounds like. You make more product *just-in-case*—you fill in the blank, for example, just-in-case the machine breaks down, just-in-case your suppliers don't send enough raw materials, just-in-case your customer orders more than you can make or deliver on time, etc. There are many reasons for using just-in-case logic, and they all contribute to decreased company profitability! Just-in-case logic is a primary cause for overprocessing waste.

True Customer Requirements Undefined

When customer requirements are poorly understood or not documented properly and employees are not adequately trained on requirements, extra processing is bound to occur. AN LSS process starts with a clear fundamental understanding of customer requirements. This typically involves a critical to quality (CTQ) assessment and definition of all product or service requirements from the customer standpoint.

Overprocessing to Accommodate Downtime

In traditional organizations, one belief is that people must be busy on production at all times. Consequently, managers order people to produce products even when none are required by a downstream customer. This results in overprocessing and creates overproduction and excess inventory. Alternatively, this time should be used for additional LSS training, cross-training programs, or other continuous improvement activities.

Poor Communication

Poor communication is typically one of the top reasons that organizations lose effectiveness. Communicating information along the entire value stream is critical for a great customer experience. The earlier in any process that the communication breaks down, the worse is the resulting waste. A typical communication cycle includes:

- Identifying critical to quality (CTQ) customer requirements
- Transitioning customer requirements into product or service specifications

- Engineering the product or service
- Creating instructions for producing the product or service
- Product or service delivery to the customer

Redundant Approvals

Although there is a need to have some cost and quality control approvals in any process, it is easy to stifle the process by requiring redundant approvals that can dramatically increase lead time and increase total product cost. After reviewing many approval procedures over the years, this has been identified as a significant example of extra processing.

Extra Copies/Excessive Information

Information sharing can be a significant source waste. How many reports are printed and not read? If they are read, how many items are actions taken on? Then there are charts, graphs, memorandums, e-mail distributions, etc., leading to information overload for employees. One example is what can essentially be described as the e-mail soap opera. The saga begins with one controversial statement or aspect that was sent to too many employees. It quickly evolves into long series of clarifications and reclarification e-mails, with each e-mail raising more questions than it answers. These types of e-mail dialogues rapidly consume significant employee time and energy of everyone involved.

How to Identify Processing Waste

The learning to see processing waste checklist presents some basic questions to uncover process waste.* (See Figure 3.9.)

* Having a direct impact to the bottom line, quality defects resulting in rework or scrap are a tremendous cost to most organizations. Associated costs include quarantining inventory, re-inspecting, rescheduling, and capacity loss. In many organizations the total cost of defects is often a significant percentage of total manufacturing cost. Through employee involvement and continuous process improvement (CPI), there is a huge opportunity to reduce defects at many facilities. In the latest edition of the Lean Manufacturing classic *Lean Thinking*, underutilization of employees has been added as an eighth waste to Ohno's original seven wastes. Organizations employ their staff for their nimble fingers and strong muscles but forget they come to work every day with a free brain. It is only by capitalizing on employees' creativity that organizations can eliminate the other seven wastes and continuously improve their performance. Many changes over recent years have driven organizations to become world-class organizations or Lean enterprises. The first step in achieving that goal is to identify and attack the nine wastes. As many world-class organizations have come to realize, customers will pay for value-added work, but will never knowingly or willingly pay for waste.

Waste of Extra Processing Checklist			
Process:	**Date:**		
Description	**Yes**	**No**	**Apparent Cause**
Is there visible rework being conducted?			
Do we measure the amount of rework?			
Do we collect data on labor and materials associated with rework?			
Are we duplicating reports or information?			
Are we entering repetitive data?			
Do we have many forms with duplicated data?			
Are we doing more work than is required for that process?			

FIGURE 3.9
Waste of extra processing checklist.

Waste 5: Waiting

Waiting waste is often described as time waiting for something to happen or occur. This could be human waiting time, machine waiting, or materials waiting to be processed. When this waste occurs, ultimately it is the customer who is left waiting as lead times expand to accommodate the numerous waiting steps in your processes.

What Causes Waiting Waste?

Waiting time waste may be caused by several sources. Some examples include materials conveyance delays, machinery or equipment breakdowns, operators working too fast or too slow, and an insufficient number of employees, to name a few. Causes of waiting waste include:

- Raw material outages
- Unbalanced scheduling or workloads
- Unplanned downtime for maintenance
- Poor equipment or facility layout
- Long process setup times
- Misuses of automation
- Upstream quality (flow) problems

Raw Material Outages

A prevalent root cause of waiting waste is raw material outages. Poor purchasing practices or purchasing measures can often lead to inadequate raw materials' inventories. Without raw materials you cannot add value for your customers and you bear all the material-related liability. For example, if you make a product that has 25 components and you are out of 2, you cannot build your product. However, you have the inventory carrying costs for the 23 components in stock. In addition, allowing outages to occur almost guarantees some of the other wastes, such as overproduction and extra processing. You cannot build the entire product, so you start building parts, and soon mountains of incomplete sub-assemblies begin to appear around the facility as work-in-process (WIP). Raw materials outages and management can be an LSS project focus topic.

Unleveled Scheduling and Unbalanced Workloads

Similarly with overproduction, unleveled scheduling and unbalanced workloads can both lead to the waste of waiting. These conditions typically occur in forecast-driven environments. In areas where the workload is not balanced properly between two or more process steps, one step will have excess capacity while the next may have excess demand. In the end, you wind up with equipment, materials, and/or manpower waiting.

Unplanned Downtime for Maintenance

When a machine breaks down unexpectedly, there is a significant opening for the waste of waiting. In addition, during the waiting period overproduction or extra processing can follow when management decides the result of finding things for employees to do until the equipment is back on-line. These extra activities are often viewed as steps that would be necessary to complete and not as waste. However, as shown in overproduction and extra processing, these are truly wastes and should be measured and subsequently eliminated.

Poor Equipment or Facility Layout

Equipment placement and facility layout are primary sources of the waste of waiting. The position of equipment within a facility is frequently decided based upon (1) shortest run from electrical service, (2) currently open floor space, and (3) a position near similar equipment, or in an expansion location. None of these criteria are based upon a proper manufacturing sequence or limiting any of the nine wastes. Poor equipment and facility layout can result in significant motion, transportation, and waiting wastes.

Long Process Setup Times

When the time to change equipment over to a different product is long, this can be a contributor to the waste of waiting. Although long process setup times can vary depending on the equipment and complexity of the transition, process setups are necessary for most equipment. Every minute, hour, or day consumed by setup is time permanently lost to waiting and contributes to lower productivity and profitability.

Misuse of Automation

A common mistake that owners or managers often make is that they want to see expensive equipment running, not sitting idle. This misuse of automation can cause the waste of waiting. Not very often does customer demand exactly meet machine capacity. One of the most difficult challenges for LSS practitioners is to change the misconception that the machine must always be running. In environments where there is a combination of automated and manual production, the misuse of automation at one process step creates the unleveled scheduling and unbalanced workloads at downstream process steps.

Upstream Quality (Flow) Problems

Product quality issues can lead to a number of wastes. Two prominent wastes are extra processing and waiting. In the case of many complex products that contain sub-assemblies, as soon as quality issues are uncovered upstream, remaining downstream steps are caught in a waiting game for completed quality sub-assemblies. Every process step should target 100% first-pass quality.

How to Identify Waiting Waste

Waiting waste can be present across the entire value stream. Regardless of the reason for the waste of waiting, the objective of learning to see is to identify where and when in the process waiting waste occurs. The checklist in Figure 3.10 is an effective tool to identify where, when, and how the waiting waste is occurring in a process.

Waste of Waiting Checklist			
Process:	**Date:**		
Description	**Yes**	**No**	**Apparent Cause**
Is work delayed from a previous process?			
Is there misuse of automation?			
Do you have unbalanced workload?			
Do you have unleveled scheduling?			
Are there materials shortages?			
Do you have absenteeism—too few workers?			
How about too many workers?			
Are there frequent unexpected machine downtimes?			
Is your facility layout effective?			
Do you have upstream product quality issues?			
Do you have long process setups?			

FIGURE 3.10
Waste of waiting checklist.

Waste 6: Motion

Waste of motion occurs when there is any movement of people or information that does not add value to the product or service.* The ultimate objective in an LSS organization is to properly connect materials, machines, man/woman power, and methods. When this is achieved, there is a state of continuous flow. Continuous flow is often credited with the highest levels of quality, productivity, and profitability. Wherever there are disconnects between two entities, for example, materials and people, the waste of motion is inevitable.

What Causes Motion Waste?

There are many possible causes for the waste of motion. Some of the major sources are:

- Poor people, materials, and machine effectiveness
- Inconsistent work methods
- Poor information management
- Unfavorable facility or cell layout
- Poor workplace organization and housekeeping

Poor People or Machine Effectiveness

Employee interactions with materials and machinery may result in the waste of motion. This happens when employees have to walk distances to pick up or deliver materials by hand. It can also occur when information must be hand delivered from one process step to another. One example may be delivering a completed order back to accounting to complete the billing cycle. Another example may be delivering a completed order to shipping for scheduled delivery.

Inconsistent Work Methods

Whenever work methods are not documented properly, a number of inconsistent and poor practices slip into any process. The best counter to inconsistent work methods is the creation of standard operating procedures or visual work instructions. These become the foundation of all

* This waste is related to ergonomics and is seen in all instances of bending, stretching, walking, lifting, and reaching. These are also health and safety issues, which in today's litigious society are becoming more of a problem for organizations. Jobs with excessive motion should be analyzed and redesigned for improvement with the involvement of plant personnel.

effective employee training programs. In their absence employee learning occurs through the passing down of "tribal knowledge" known only to "experts" in your organization. Learning that occurs under these conditions is open to interpretation by the employee on "what to do next" or "how to do" specific activities in the process. This frequently results in several employees doing the exact same activity differently. Inconsistent work methods not only result in the waste of motion, but also are frequently the root cause of product- or service-delivery quality issues.

Poor Information Management

The transition of information between employees, departments, and customers often leads to the waste of motion. Information management systems that are not set up to make required information available to employees when and where it's needed often results in employees doing printouts and manual document transfers around the organization. As with many wastes, the waste of motion can also cause several other wastes. For example, when shipping instructions for a specific customer are not completely defined in the information management system, an employee in shipping must stop to track down the proper information, which can require going to order processing or customer service to obtain the information. This initial waste of motion produces the waste of waiting and the waste of extra processing before the order can be properly shipped.

Unfavorable Facility or Cell Layout

If the facility layout is weak, the waste of motion will be present. By facility we mean any department in an organization, wherever value is created for the customer. The layouts of administrative areas, such as order processing, customer service, accounting, and warranty claims departments, are seldom considered as areas where waste can occur, but often are significant sources for the waste of motion. This is due to the frequent manual transportation of documents necessary in these areas, as well as an inordinate amount of information exchange required to produce your product or service.

In a production environment poor facility layout results in excess waste of motion regarding moving raw materials in a position to add value, securing tools and fixtures, or delivering materials to the "next process step."

Poor Workplace Organization and Housekeeping

It never ceases to amaze me how little attention is paid to workplace organization and housekeeping. Managers would rather employees spend hours searching for tools, materials, documentation, etc., than allow 30 minutes/day to maintain an organized work area. This philosophical fixation that every employee activity must be producing product is responsible for many of the wastes observed in organizations today. Every day managers can walk by piles of obsolete materials, in-process rework, and mountains of defective warranty returns, while continuing to allow no time for employees to correct the conditions that produced these results. All process improvement programs begin with workplace organization and housekeeping.

How to Identify Motion Waste

Motion waste can be present across the entire value stream. Regardless of the reason for the waste of motion, the objective of learning to see is to identify where and when in the process motion waste occurs. The checklist in Figure 3.11 is an effective tool to identify where, when, and how the motion waste is occurring in the process.

Waste 7: Transportation

Transportation waste is any activity that requires transporting parts and materials around the facility. Unlike motion waste that typically involves only people, transportation waste is usually reserved for action involving equipment to move materials or parts.* This equipment comes in many forms, such as carts, rolling racks, forklifts, golf carts, and bicycles, to name a few.

What Causes Transportation Waste?

Transportation waste can be caused by a number of factors. The major causes are:

- Poor purchasing practices
- Large batch sizes and storage areas

* Transporting product between processes is a cost incursion that adds no value to the product. Excessive movement and handling cause damage and are an opportunity for quality to deteriorate. Material handlers must be used to transport the materials, resulting in another organizational cost that adds no customer value. Transportation can be difficult to reduce due to the perceived costs of moving equipment and processes closer together. Furthermore, it is often hard to determine which processes should be next to each other. Mapping product flows can make this easier to visualize.

Waste of Motion Checklist			
Process:	**Date:**		
Description	**Yes**	**No**	**Apparent Cause**
Are all materials where needed?			
Do you have the proper material quantities?			
Are materials specifications correct?			
Are tools in good working order?			
Are all tools available?			
Is order documentation complete?			
Is shipping information complete?			
Do you have to search for files on the computer?			
Are you searching for documents in file cabinets or drawers?			
Are you hand-carrying paperwork to another process or department regularly?			
Are you constantly reviewing the same manuals for information?			

FIGURE 3.11
Waste of motion checklist.

- Inadequate facility layout
- Limited understanding of the process flow

Poor Purchasing Practices

The largest contributor to transportation waste is poor purchasing practices. Many organizations measure their purchasing effectiveness on the dead cost/piece for raw materials purchased. This can lead to incredible waste throughout the organization, not just transportation waste but also the waste of overproduction, inventory, extra processing, and defects.

Let's consider a real-world example to demonstrate how non-Lean measures can drive non-Lean behavior within an organization. Let's say your organization has a program in place to drive the cost of raw materials down and John's, the purchasing department manager, bonus is dependent on a 10% reduction in raw materials costs. He has his heart set on that 60-inch plasma TV with surround sound installed before football season, so consequently he sees little else except achieving the dead materials cost reduction.

John begins to think: How can he achieve this predefined material cost reduction? Two actions immediately come to mind; both are non-Lean. First, he can go to current suppliers and try to get price decreases. These decreases usually require that the organization buy in larger volumes, which he does immediately. In fact, at the next manager's meeting John is eager to get a pat on the back from the boss and reports that he has secured a 3% material cost reduction in the first month of the program; the unnecessary raw materials, along with the corresponding inventory and transportation waste, begin to show up in receiving the very next week.

Second, he can search for secondary suppliers that are willing to provide supposedly equal raw materials. At first glance they appear to be equal in every way—specification, function, and quality. He begins to substitute some of these raw materials and again achieves more raw materials cost reductions that are, of course, well received by management. John achieves his bonus and spends Sunday afternoons in bliss with his favorite beverage and gridiron action—an apparent happy ending. Not so fast. In the next few weeks during production, some inconveniences arise because the new materials aren't exactly like the original parts. This leads to some in-process defects that require rework or the waste of extra processing. In addition, weeks later returns begin from customers for poor product service in the field.

Because the sales price is based upon typical labor standards, these extra production costs and warranty return costs don't appear on management's radar and consequently don't exist. Only after months, when management realizes the shrinking profit margins, does another costly search for the reasons begin. This is a classic non-Lean example of how non-Lean traditional measures can drive non-Lean behavior. In this case, what did John learn to see? Certainly not the waste! His behavior was being completely driven by his measurement system.

Large Batch Sizes and Storage Areas

The waste of transportation can also occur when you process large batches of product or set up large storage areas. Both of these decisions require that the materials be moved at some time. These moves invariably require people (materials handlers) and equipment (forklifts, carts, pallet jacks, flatbeds, etc.). This situation is almost always the symptom of a poor purchasing decision that was based upon a non-Lean traditional management belief that organizations save money when they buy large batches of materials.

Inadequate Facility Layout

One of the primary causes of transportation waste is poor facility layout. Where you place equipment, how and where your materials storage areas are set up and regularly accessed, and your organization's purchasing philosophy all affect productivity and profitability. Proper facility layout can reduce lead time by up to 40%, and dramatically reduce the waste of waiting, transportation, and motion.

Limited Understanding of Process Flow

In every process there needs to be a thorough understanding of the materials, machines, man/woman power, or methods required to add value for the customer. A primary component of Lean—and a constant goal for LSS practitioners—is continued process development and deeper understanding of process knowledge. It is important to understand the best sequence of process steps to meet customer demand, such as: How are activities conducted? How fast is product needed? Where do materials get consumed? What are the fluctuations in manpower requirements? Is the correct type of equipment available? Is the equipment in working order? Having well-defined answers for these factors contributes to improved process understanding.

How to Identify Transportation Waste

The learning to see transportation waste checklist presents some specific questions that can help you uncover transportation waste. (See Figure 3.12.)

Waste 8: Underutilized Employees

The waste of underutilized employees often occurs when we fail to recognize and harness people's mental, creative, innovative, and physical

Waste of Transportation Checklist			
Process:	**Date:**		
Description	**Yes**	**No**	**Apparent Cause**
Are materials moved between buildings?			
Do you make large batches?			
Do you buy bulk raw materials?			
Do you have lots of forklifts?			
Do you have many other types of transportation equipment?			
Are materials stored long distances from where they are used?			
Are there multiple temporary storage areas?			

FIGURE 3.12
Waste of transportation checklist.

skills or abilities. This is present to some extent in almost every company, even organizations that have been practicing Lean behaviors for some time. Much of this employee misuse stems from the management concepts previously discussed regarding traditional organization belief systems. Although many Western managers pay lip service to "our employees are our most valuable asset," they are the first to philosophically look at employees as a liability, not an asset. Many are often quick to practice management by head count—this is the practice of stating that we will operate with a specific number of employees regardless of the number required to provide good performance for the customer.

What Causes Underutilized Employees Waste?

There are a number of causes of underutilized employees or people waste. Each of these stems from some aspect of traditional belief systems.

- Old guard thinking, politics, the business culture
- Poor hiring practices
- Low or no investment in training
- Low-pay, high-turnover strategy

Old Guard Thinking, Politics, and Business Culture

Old guard thinking, politics, and general business culture often stifle using employees' creative skills or producing innovative assignments that could result in significant process improvements. Unfortunately, in many organizations an employee's perceived importance to the organization is generally directly proportional to his or her salary or directly linked to his or her title. This is common in the United States and seldom seen in Japanese companies.

Poor Hiring Practices

Most human resource departments are faced with the difficult task of how to attract and retain skilled employees. Poor hiring practices usually stem from the structure of the department and management's mandate for critical components of the hiring process, such as pay level, required skills, or required experience. All of these could hinder getting the best candidate for the position.

There are many factors that can go into poor hiring practices. A few common mistakes include:

- Inadequate job advertisements
- Position definition
- Nepotism
- Not matching skills to position requirements
- Not understanding the technical aspects of job requirements
- Inability to identify the skills necessary to add value to a position

Low or No Investment in Training

Good data on training time, although readily available from many sources, is often difficult to translate to an organization. The American Society for Training and Development (ASTD) puts the dollars per year per employee at about $1,400. For a $40,000/year employee, this equates to about 3.5% of annual salary. It has been cited in the literature that top-performing companies spend approximately 4 to 6% of annual salary on training. It has also been reported that average American companies spend less than 5% on employee training. LSS organizations often approach 10%, with 3 to 4% of annual salary direct spending on new training and 6% employee time committed to improvement activities.

All companies tend to view training differently. One observation is unavoidable—poor-performing companies tend to invest little or nothing in training, while higher-performing organizations invest in training and focus on process improvement.

Low-Pay, High-Turnover Strategy

Another common trait of traditional organizations is the "low-pay, high-turnover rate" philosophy. This is characterized by hiring to a specific hourly or salary level regardless of skills, and living with the performance that pay rate returns. Because the conditions are poor, employees either leave to pursue a better opportunity or are let go by the company for myriad reasons. This is an internally focused philosophy and completely ignores the voice of the customer.

How to Identify Underutilized Employees Waste

The learning to see underutilized people checklist in Figure 3.13 points out some questions you can use to assess your current employee utilization and expose some apparent causes for lack of effective employee use.

Waste 9: Behavior

Behavior waste is any waste that results from human interactions. It is present to some extent in all organizations. It can be minimal in truly LSS organizations; however, it can be pervasive and devastating in traditional organizations. Behavior waste naturally flows from an individual's or a company's inherent beliefs. "The concept of waste has not yet been effectively extended to the self-defeating behaviors of individuals and groups of people in the workplace."*

Behavior waste is a root cause of the other eight common wastes. Many of the previously described wastes alluded to employee beliefs and behaviors

* M. L. Emiliani, "Lean Behaviors," *Management Decision* 36/9 (1998) 615–631.

Waste of Underutilized Employees Checklist			
Process:	**Date:**		
Description	**Yes**	**No**	**Apparent Cause**
Do we know the true experiences and capabilities of our employees?			
How easy is it to move employees to special assignments?			
Is your process so fragile that employees cannot be assigned to special projects?			
Are employees in positions they were trained to do?			
Is there active improvement-idea generation from all employees?			
Are employees allowed to experiment with process improvements?			
Can employees assist in other areas as needed?			
Do managers place obstacles or restrictions on employees?			
Are employees empowered to take action in their area?			
Is there a "can't do" atmosphere?			
In there a "can do" atmosphere?			

FIGURE 3.13
Waste of underutilized people.

as causes for waste generation. The identification and elimination of behavior waste is critical to any successful LSS initiative.

How to Identify Behavior Waste

Behavior waste is classified as either personal (yourself) or people (between two or more employees). Identifying these behavior wastes in your organization is the first step to elimination of this disruptive waste.

Personal Behavior Waste

Personal waste is waste that comes from within oneself. It stems from the way you view yourself, your goals and objectives, or possibly your position in the organization. Oftentimes personnel who prefer a Theory Y organization (empowered-employee environment) and are working in a Theory X organization (command and control environment) feel underappreciated. As a consequence, they become an underutilized employee and can exhibit low morale. The personal waste they generate comes directly from their individual belief system. Gossip, self-imposed barriers, deceptions, and ego are a few of the many examples of personal waste.

Personal waste has been described as the little voice inside your head that provides constant running (negative) commentary. It can control an employee's inability to suspend judgment and projects unresolved internal conflicts of the employee. It does not take much personal waste to bring continuous improvement to a screeching halt. In fact, personal waste will restrict process improvement and Lean deployment at any process step that touches this employee, which is basically your entire value stream.

People Behavior Waste

People waste has to do with relationships between fellow employees. This includes between department managers and senior managers, as well as the manager-employee relationship. Some categories of people waste include turf wars, fiefdoms, or politics. Some specific examples of what people say when they are exhibiting people waste are:

- "Bill's initiative is so stupid!"
- "Forget about what Jane says!"
- "John is impossible to work with!"

One can see how personal waste, when coupled with people waste, can stifle all process improvement.

SUMMARY

Learning to see variation and waste is a critical first step to improving quality, productivity, and profitability. Only after employees begin to learn to see waste and variation with new eyes can they identify previously unnoticed waste in the organization and effectively begin to eliminate the sources of waste and variation.

Processes add either value or waste to the creation of goods or services. The seven wastes originated in Japan, where waste is known as muda. The eighth waste is a concept tool to further categorize muda and was originally developed by Toyota's chief engineer Taiichi Ohno as the core of the Toyota Production System, which also became known as Lean Manufacturing. The ninth waste—behavior waste—is by far the most damaging of all the wastes. The reason is simple: *Everything we think, everything we say, and everything we do* shape the behavior of all employees in our organization and gets them going in a direction of either creating value for the customer or creating waste.

To eliminate variation or waste in a process, it is important to understand exactly what waste is and where it exists, and to clearly view, measure, and limit variation. While activities can significantly differ between factories and the office workplace may seem to be a different world, or in service organizations where the product is actually a service, the typical wastes found in all these environments, and in fact in all business environments, are actually quite similar.

All forms of the nine wastes are highly costly to an organization because waste prohibits the smooth flow of materials and actually degrades quality and productivity. The Toyota Production System mentioned in this chapter is also referred to as just-in-time (JIT) because every item is made just as it is needed. Conversely, overproduction is referred to as just-in-case. This creates excessive lead times, results in high storage costs, and makes it difficult to detect defects. The simple solution to overproduction is turning off the tap; this requires a lot of courage because the problems that overproduction or behavior wastes are hiding will be revealed. The concept is to schedule and produce only what can be immediately sold and shipped and improve machine changeover/setup capability.

For each waste, there is a strategy to reduce or eliminate its effect on an organization, thereby improving overall performance and quality, while at the same time lowering costs. Learning to see is all about learning to use these strategies and tools in a productive manner.

4

Lean Concepts, Tools, and Methods

Management's reliance solely on Lean Six-Sigma Tools for organizational performance improvement is fool's gold! Lean Six-Sigma Tools are not the answer; they are the instrument used to apply Lean Six Sigma Concepts.

Learn the Concepts.... Apply the Tools.... Get the Results.

—Richard Charron

OVERVIEW

Both Lean concepts and Lean tools are an integral part of LSS process improvement projects. Contrary to popular belief, Lean tools alone will not provide the results that are often attributed to Lean implementation programs. Several Lean concepts, when taken together, form a Lean operational philosophy, which is a primary requirement for successful Lean programs. In this chapter we present some basic Lean management concepts and then the most common Lean tools. (See Table 4.1.) When we apply Lean tools in pursuit of these concepts, a Lean organization emerges. If we endeavor to apply Lean tools in the absence of these Lean management concepts, process improvement projects often fall short of desired improvement expectations.

LSS concepts + LSS tools = Performance improvement

TABLE 4.1

Glossary of Lean Concepts and Tools

Lean Concepts	Just the Facts
Waste	Anything in your processes that your customer is unwilling to pay for: extra space, time, materials quality issues, etc.
Value-added	All activities that create value for the external customer.
No-value-added	All activities that create no value for the external customer.
Business-value-added	Activities that do not create value to the external customer, but are required to maintain your business operations.
Waste identification	The primary fundamental Lean concept is the ability to see waste in your organization. This encompasses being able to readily identify the nine wastes and consistently manage to avoid these occurrences. More importantly, from a management standpoint, it means possessing the ability to proactively lead and mentor employees to conduct waste-free activities daily. To do this the Lean manager must master the concept of waste identification by understanding waste creation in all its forms.
Waste elimination	The ability to apply Lean Six Sigma concepts and tools to eliminate identified wastes. This entails continuous learning of how to apply concepts and tools either independently or in groups to achieve process improvement results.
Standard work	Standard work is a systematic way to complete value-added activities. Having standard work activities is a fundamental requirement of Lean Six Sigma organizations.
Value stream	A conceptual path horizontally across your organization that encompasses the entire breadth of your external customer response activities. That is, anything that transpires from the time your organization realizes you have an external customer request until that external customer receives its product or service.
Value stream management	A systematic and standardized management approach utilizing Lean Six Sigma concepts and tools. Value stream management results in an external customer-focused response to managing value-added activities.
Continuous flow	The Holy Grail of manufacturing, often referred to as make one–move one. One-piece flow or continuous flow processing is a concept that means items are processed and moved directly from one processing step to the next, one piece at a time. One-piece flow helps to maximum utilization of resources, shorten lead times, and identify problems and communication between operations. During any process improvement activity, the first thing on your mind should be: "Is what I'm about to do going to increase flow?" Achieving continuous flow typically requires the least amount of resources (materials, labor, facilities, and time) to add value for the customer. Achieving continuous flow has been credited with the highest levels of quality, productivity, and profitability.

TABLE 4.1 (CONTINUED)

Glossary of Lean Concepts and Tools

Lean Concepts	Just the Facts
Pull systems	Systems that only replenish materials consumed by external customer demand. These systems naturally guide purchasing and production activities, directing employees to only produce what the external customer is buying. The Kanban tool is used to achieve this Lean concept.
Point-of-use storage (POUS)	Locating materials at the point of value-adding activities.
Quality @ source	Building quality into value-adding processes as they are completed. This is in contrast to trying to "inspect in quality," which only catches mistakes after they have been made. An effective quality @ source campaign can minimize or eliminate much of the expense associated with traditional quality control or quality assurance programs.
Takt time	Takt time is the demand rate of your external customer for your products. It signifies how fast you have to make products to meet your customer demand. Once you calculate Takt time, you can effectively set your value-added processes to meet this customer demand. In essence, Takt time is used to pace lines in the production environments. Takt time is an essential component of cellular manufacturing.
Just-in-time (JIT)	A concept that espouses materials sourcing and consumption to meet external customer demand. Properly executed, it helps to eliminate several wastes, including excess inventory, waiting, motion, and transportation.
Kaizen	The Japanese term for improvement; continuing improvement involving everyone—managers and workers. In manufacturing Kaizen relates to finding and eliminating waste in machinery, labor, or production methods. Kaizen is a versatile and systematic approach to change + improve all processes. In virtually all Lean Six Sigma organizations the concept of Kaizen is practiced by all employees at all levels of the organization.
Materials, machines, manpower, methods, and measurements (5M's)	The five key process inputs. The pursuit of Lean Six Sigma is an exercise in the effective use of the 5M's to achieve customer requirements and overall organization performance.
Lean accounting	A method of accounting that is aligned horizontally across your organization with the value stream. Traditional costing structures can be a significant obstacle to Lean Six Sigma deployment. A value stream costing methodology simplifies the accounting process to give everyone real information in a

(continued)

TABLE 4.1 (CONTINUED)

Glossary of Lean Concepts and Tools

Lean Concepts	Just the Facts
	basic understandable format. By isolating all fixed costs along with direct labor we can easily apply manufacturing resources as a value per square footage utilized by a particular cell or value stream. This methodology of factoring gives a true picture of cellular consumption to value-added throughput for each value stream company-wide. Now you can easily focus improvement Kaizen events where actual problems exist for faster calculated benefits and sustainability.
Lean supply chain	The process of extending your Lean Six Sigma activities to your supply chain by partnering with suppliers to adopt one or more of the Lean concepts or tools.
Lean metric	Lean metrics allow companies to measure, evaluate, and respond to their performance in a balanced way, without sacrificing the quality to meet quantity objectives, or increasing inventory levels to achieve machine efficiencies. The type of lean metric depends on the organization and can be of the following categories: financial performance, behavioral performance, and core process performance.
Toyota Production System	The Toyota Production System is a technology of comprehensive production management. The basic idea of this system is to maintain a continuous flow of products in factories in order to flexibly adapt to demand changes. The realization of such production flow is called just-in-time production, which means producing only necessary units in a necessary quantity at a necessary time. As a result, the excess inventories and the excess workforce will be naturally diminished, thereby achieving the purposes of increased productivity and cost reduction.
Lean Tools	**Just the Facts**
5S	A methodology for organizing, cleaning, developing, and sustaining a productive work environment. Improved safety, ownership of work space, improved productivity, and improved maintenance are some of the benefits of the 5S program.
Overall equipment effectiveness (OEE)	An effective tool to assess, control, and improve equipment availability, performance, and quality. This is especially important if there is a constraining piece of equipment.

Mistake (error) proofing	Mistake proofing is a structured approach to ensure a quality and error-free manufacturing environment. Error proofing assures that defects will never be passed to the next operation. This tool drives the organization toward the concept of quality @ source.

TABLE 4.1 (CONTINUED)

Glossary of Lean Concepts and Tools

Lean Tools	Just the Facts
Cellular manufacturing	A tool used to produce your product in the least amount of time using the least amount of resources. When applying the cellular manufacturing tool, you group products by value-adding process steps, assess the customer demand rate (Takt time), and then configure the cell using Lean Six Sigma concepts and tools. This is a powerful tool to allow the use of many Lean concepts and tools together to achieve dramatic process improvements. (See cell example at the end of this chapter.)
Kanban	A Kanban is a "signal" for employees to take action. It can be a card with instructions of what product to make in what quantity, a cart that needs to be moved to a new location, or the absence of a cart that indicates that an action needs to be taken to replenish a product. This is a fundamental tool used to establish a "more continuous flow." Kanban is a simple parts-movement system that depends on cards and boxes/containers to take parts from one workstation to another on a production line. The essence of the Kanban concept is that a supplier or the warehouse should only deliver components to the production line as and when they are needed, so that there is no storage in the production area. Kanban can also be effective in Lean supply chain management. In document intensive environments (e.g., medical devices, bio-pharma, healthcare, aviation industries, etc.) with government document management regulations such as Food & Drug Administration (FDA) or other federal agency, Kanban can be an effective tool to improve document flow by establishing which document to review or approve next.
Value stream mapping	A process mapping technique that consists of a current state map describing initial conditions of a process and a future state map that defines an improved process. The current state map typically includes some descriptions of the 5M's that will be targets for modifications in the future state.

Visual controls	Visual controls are tools that tell employees "what to do next," what actions are required. These often eliminate the need for complex standard operating procedures and promote continuous flow by eliminating conditions that would interrupt flow before it happens.

(continued)

TABLE 4.1 (CONTINUED)

Glossary of Lean Concepts and Tools

Lean Tools	Just the Facts

In the early stages of your Lean journey, you should try to understand where you currently stand as an organization. Does the company have a strategic plan with high-level Lean measures that department and process-level measures roll up to? Does the organization have a Lean operational philosophy, starting at the executive management group, which flows down to the senior management group, department managers, and ultimately to associate employees? Does the organization understand the difference between the traditional organization and what it means to operate as a Lean organization?

In today's business world, Lean is for everyone. It is a true operational philosophy that, to produce real sustainable results, must be adopted by employees at all levels of the organization. Also, a Lean operational philosophy can be used in just about any organization. Although Lean was originally defined and developed for the manufacturing community, today Lean or LSS initiatives can be found in almost every organization and every industry. For example, LSS is prominent in service organizations, manufacturing industries, government agencies, health care, and nonprofit organizations. Of course, each of these industries requires that we take a different look at how we apply the tools—not that we will be using tools differently, but how we will be applying those tools differently given the type of industry that we're in and what it is we're trying to achieve in terms of process improvement.

Basically, there are two management philosophies used in most organizations today. The first uses a traditional approach, while the second uses a Lean, LSS, or world-class approach. Understanding your organization's current management philosophy goes a long way to helping you identifying which Lean concepts you may be lacking in and which Lean tools you want to apply within your organization. The remainder of this chapter presents an overview of both Lean concepts and Lean tools.

TRADITIONAL ORGANIZATION OPERATIONAL PHILOSOPHY

The term *traditional organization* is a term we use to describe a set of operational philosophies, policies, and behaviors that drive all daily activities that occur within your organization. We use the term *traditional organization* as a reference point to more clearly define what it means to be a Lean organization. By describing both traditional (non-Lean) and Lean philosophy and beliefs, you will develop some idea of where your organization currently stands. What is your current organizational operational philosophy? What are your collective beliefs as both individuals and the organization as a whole?

Table 4.2 depicts some of the basic operational philosophies and beliefs of both traditional and Lean organizations. We present this in a point-counterpoint format to try and give readers a good description of where it is they currently stand and where it is they're trying to drive their organization to be.

Table 4.2 identifies a number of traditional organizational beliefs and their corresponding Lean organizational beliefs. During the organizational transformation from traditional organization to a Lean organization, one of the primary objectives of the LSS practitioner is to identify these traditional operational beliefs, next use the tools and techniques described in this handbook to remove those beliefs from the organization, and finally replace those beliefs with the Lean beliefs also shown in this table.

For example, let's discuss one of the traditional beliefs and then present the corresponding Lean belief. In this example, we will compare and contrast the concepts of "management by head count" and "employees as needed." Often in traditional organizations, managers use a concept called management by head count. In this environment the traditional manager only uses a very specific number of employees to complete certain components of the process that he/she is working with. What this means is that regardless of how many employees are needed to meet customer demand, the manager fixes the number of employees based upon another measure, usually labor dollars. This is done regardless of its impact on customer satisfaction or performance of the customer. Conversely, the Lean manager adds employees as

TABLE 4.2

Traditional versus Lean Operational Philosophies

Traditional Organization	Lean Organization
Functional focus	Business focus
Management directs	Managers teach
Delegate	Support
Forecast driven	Customer driven
Fear of failure	Share success
Blame people	Improvement opportunities
Heroes and goats	Real teams
Us versus them	Community
Results focus	Process focus
Me (producer)	You (customer)
Dedicated equipment	Flexible equipment
Slow changeover	Quick changeover
Narrow skills	Multiskilled
Managers control	Workers control
Pure production environment	Learning environment
Supplier is enemy	Supplier is ally
Guard information	Share information
Customer as buyer	Customer as resource
Management by head count	Employees as needed
Volume lowers cost	Analyze cost drivers
Internal focus	External focus
Shallow process knowledge	Deep process knowledge
Quality problem detection	Quality problem prevention
Hierarchy	Flat organization
Short-term thinking	Balanced thinking
Worker accountability	Executive accountability
Rewards = money	Rewards = Pride, then money
Competition	Cooperation
Complex	Simple

needed to achieve the level of performance that the organization defined for the customer. In this environment, the Lean manager puts the voice of the customer first, and subsequently builds all of the required internal value-added activities to meet customer performance expectations.

The difference between these two concepts illustrates how easy it is for waste to be introduced into your operational processes when a management by head count philosophy is used. By using less than the number of employees that are required to meet customer performance, some or all of the traditional nine wastes are manifested in your organization. It is these operational wastes that drag down product quality, diminish employee productivity, and ultimately hinder company profitability. The impact that operational philosophy has on performance cannot be overstated. Everything we think, say, or do either creates value for the customer or creates waste.

LEAN OPERATIONAL PHILOSOPHY

A Lean operational philosophy is one that flows from an employee's basic fundamental understanding of Lean concepts and beliefs, and should be spread across your organization until it becomes a living, breathing, daily method of operation for all your employees. The Lean operational philosophy is literally for everyone. In many organizations today trying to implement Lean, it is often viewed as a set of tools for process improvement.

However, Lean is a true operational philosophy, and in order for your organization to become a Lean organization, the philosophy needs to be embraced at the highest levels of your company. C-level employees need to have an equal and thorough understanding of what it means to be a Lean organization in order for your company to effectively become one. Lean beliefs and behaviors need to be exhibited on a daily basis by executive management, department management, supervisors, and associates alike.

Lean is certainly for everyone. An organization aspiring to become a Lean organization must therefore have a master plan to involve 100% of their employees in the transformation process. The concept is discussed in further detail in Chapter 5.

LEAN MANAGEMENT CONCEPTS

In order to effectively select and apply LSS tools to any process, the management team must have a basic understanding of Lean concepts as the driving force for a Lean operational philosophy. In the absence of this Lean operational philosophy, the random selection of process improvement tools will not yield the desired results that management hopes to achieve from its process improvement programs. In the remainder of this section, we discuss the dominant Lean management concepts that fundamentally guide effective LSS process improvement programs in your organization.

Waste

What is waste? Waste is typically defined as any activity that your customer is unwilling to pay for. It's usually described in terms of value-added (VA) activities versus no-value-added (NVA) activities (discussed below). Throughout the literature you'll find waste defined in terms of eight categories. These categories are overproduction, excess inventory, waiting, defects, extra processing, underutilized employees, motion, and transportation. All of our daily actions either add value from an external customer's standpoint or create one or more of these wastes in our operation.

The primary objective of the LSS practitioner is to learn how to identify where these wastes occur, how they occur, and what root causes led to the waste being manifested in your operation. Once we understand what these wastes are and how to identify them, we can use very specific Lean concepts and tools to eliminate these wastes permanently from our operations.

Value-Added Activities

A value-added (VA) activity is basically any activity that the employee conducts that the external customer is willing to pay for. These activities are usually comprised as the process steps required to convert some raw materials into a modified and useful product for the customer. VA activities in service industries typically refer to any series of events that enhance an external customer experience or assist them with things that they could not ordinarily do alone. In order to decide whether an activity is value-added or not, try putting yourself in the shoes of your external customer. If you can effectively say that your external customer would want to pay for the activity that you're about to conduct, then it's probably a VA activity.

No-Value-Added Activities

No-value-added (NVA) activities are activities that do not contribute to meeting external customer requirements and could be eliminated without degrading the product or service function or the business, i.e., inspecting parts, checking the accuracy of reports, reworking a unit, rewriting a report, etc. There are two kinds of NVA activities:

1. Activities that exist because the process is inadequately designed or the process is not functioning as designed. This includes movement, waiting, setting up for an activity, storing, and doing work over.

These activities would be unnecessary to produce the output of the process, but occur because of poor-process design. Such activities are often referred to as part of poor-quality cost.
2. Activities not required by the external customer or the process and activities that could be eliminated without affecting the output to the external customer, such as logging in a document.

It has been estimated that as much as 65% of all organizational activities and 95% of all lead time are consumed by employee NVA activities. This time and energy is consumed by an almost endless series of things that we build into our processes that the external customer has no use for. If what you're about to do does not appear to be something that you would be willing to pay for as an external customer, then the activity should be questioned. Ultimately, all NVA activities should be targeted for elimination from your processes.

Business-Value-Added Activities

Since the 1980s there has been dialogue concerning numerous activities that are NVA, but which are often required in order to deliver your product or service to your customer. We classify these as business-value-added (BVA). These could include a range of internal activities, such as accounting or order processing to external requirements that could include government regulations (e.g., Food and Drug Administration) or third-party stakeholders. In a pure Lean environment it could be argued that there are only two possible options for all activities: (1) value-added activities (i.e., anything that is required to deliver your product or service for the customer) or (2) no-value-added activities (anything that is not required to deliver your product or service for the customer). Introducing terms such as *business-value-added* (BVA) only serves to cloud the issue by promoting the concept that "some waste is allowable." How you choose to categorize and quantify VA and NVA activities is up to your organization. In most organizations it would be a major mistake to classify certain activities as NVA, such as activities related to the following: safety, personnel, accounts receivable, accounts payable, processing, payroll, environmental, legal, taxes, marketing, and many more of the activities required to run the business. Communicating to the people in these key processes within the organization that their work is NVA would destroy morale throughout the organization. We agree that the amount of effort and money spent in these areas should be minimized and waste should be removed from these important business-required processes, but the organization has to be aware of the legal requirements that are imposed upon the organization and of the commitments it has to its internal customers and its other stakeholders. Without living up to these commitments, the organization would not be able to meet its external customer requirements and expectations.

Waste Identification

Waste identification is the ability to see something that others cannot. In Lean we call this learning to see, a topic that was covered in detail in Chapter 3. It's not that Lean practitioners see any better or worse than we do; it's that they see differently and are viewing the organization with a different set of beliefs and preconditions. Waste identification is an ongoing process in an LSS organization. We are constantly learning to see waste and understand how waste is manifested. Our view of how waste negatively impacts our product quality or performance from a customer standpoint changes forever once we begin learning to see.

This process of waste identification using both Lean concepts and Lean tools can be used in two ways. First, waste can be identified using qualitative techniques, which means using a technique that doesn't necessarily require that we measure anything. This is the primary fundamental strength of the Lean portion of LSS. An example of this qualitative method is identifying the waste from a Lean management concept, such as point of use storage (POUS) or quality of the source. For example, once we understand the Lean concept of POUS, we can readily see areas in our organization where the concept may be applied for materials, tools, instrumentation, paperwork, or other necessary components used in the VA process. More importantly, you begin to see the management philosophy that resulted in poor-material storage to begin with.

Second, waste can be identified using quantitative data techniques. Quantitative techniques require that we use some form of measurement on a particular process to identify where the waste is occurring. This approach can be used on many waste identification activities, and it is a fundamental strength of the Six Sigma portion of LSS.

Waste Elimination

Once we have identified one or more of the wastes, we are in a position to begin to select Lean concepts or Lean tools to apply to that waste in order to eliminate it from our processes. At first glance, waste elimination appears to be a simple application of Lean tools in a given situation. However, it's not quite that easy. Much of the waste introduced into our organizational processes stems not from a simple flaw in the process itself; however, it stems from a management belief or operational philosophy regarding the control of all of our process inputs. As we will discuss later in this handbook, the key process inputs are materials, machinery, manpower, methods, and measurements. Our operational beliefs on managing these controlling-process inputs ultimately define how much waste there will be in our operation, where it will be located, what needs to be done to eliminate this waste, and what tools we will need to use to eliminate the wastes.

Developing a successful plan for waste elimination encompasses effective waste identification coupled with a Lean concept or tool selection and deployment. These activities are continuously conducted within the structure of a Lean management system where managers both support and encourage Lean practitioners to take the improvement actions.

There are three fundamental steps that must be followed in order to conduct a successful program using an LSS philosophy. These steps are so fundamental that each has been given its own chapter in this LSS handbook.

Value Stream

The value stream refers collectively to all those things required for your organization to produce perceived value for the customer. It includes materials, manpower, facilities, suppliers, or vendors; in essence, it includes everything that goes into creating an effective product or service that your customers are willing to pay for. The value stream contains all VA and NVA waste activities. A central activity for the Lean practitioner is to better understand all value stream components, both VA and NVA alike.

Visualizing the value stream is an integral complement of conducting any LSS process improvement program. There are several ways that a Lean practitioner can attempt to visualize the value stream. For example, to visualize a facility layout, the Lean practitioner can prepare a value stream map that identifies all the process steps, materials, equipment, facilities, employees, and activities required to add value for the customer. It can also include the measurements of time or output on those activities.

In a service industry environment defining the value stream is equally important. For example, the value stream in a service industry is probably composed of a series of steps that need to be effectively taken by employees.

They may or may not contain materials and equipment as is traditional with manufacturing operations. Much of the value stream can be composed of information management, documentation management, or activity management. Therefore, when trying to identify VA and NVA activities in the service environment, our challenge is to identify what activities need to be conducted, who needs to conduct those activities, and what accessories are required in order to provide a good service that the customer has defined.

When trying to define the value stream of a government agency or nonprofit organization, it can be even more complex than with the service industry organization or with the manufacturer. For example, nonprofit organizations are typically restricted by what kind of activities they can conduct, and rarely have the resources associated with the service industry organization or a manufacturing facility. As a consequence, the identification process of the value stream may be more cumbersome and not quite as defined.

Government agencies can have even more complex value streams. Whether it is a county, state, or federal government agency, there can often be a significant number of stakeholders that all have an influence on the operation of the value stream. Each of these stakeholders may have a completely different view of what value is from their perspective. As a consequence, the challenge for the Lean practitioner in identifying the value stream, of which the activities are VA and NVA, and defining the process improvement projects to improve the value stream becomes more complex.

In these types of situations it's often required that you break the value stream down into a number of connecting processes. In complex situations like this, whether at the service industry, government industry, or government agency, value stream management becomes a critical complement of any process improvement activity. Having said this, value stream identification and management have been effectively applied over the last several years in government agencies. For example, the Navy, Air Force, and Army LSS programs have been underway for the past 10 years. Having witnessed some of the significant cost reductions in these government agencies, the federal government is now considering trying to apply LSS to other federal government agencies.

Value Stream Management

Value stream management consists of a comprehensive approach to managing all aspects of value creation for the customer. A visual representation (value stream map(s)) of your entire value stream is created by identifying all activities required from the point that the customer initiates contact with your organization until the customer finally receives your product or service. This visual representation typically includes all the activities required for you to produce value from a customer standpoint. Throughout the visual representation the Lean practitioner endeavors to quantify how those activities are currently conducted in an effort to devise a methodology to improve those activities.

Consequently, how your organization defines your value stream is how you elect to develop partnerships and relationships with your vendors, the operational philosophy you have within your facility, and the methods with which you go about transferring your VA product or service to your customer, and how you measure performance; these are all defined as value stream management. An effective value stream management program, particularly for complex service industry organizations and government agencies or even simple manufacturing processes, typically requires the use of visual value stream mapping.

Continuous Flow

Establishing continuous flow is often not as easy as it might sound. The ultimate objective of establishing continuous flow is to link all of your VA steps seamlessly together, allowing no opportunities for downtime between steps. These steps could be automated manufacturing, manual assembly, general administrative tasks, warehousing, or shipping and distribution. In a service industry environment, the steps could include order processing, service delivery, interaction with the customer, or other VA activities.

Continuous flow has been referred to as the "Holy Grail" in manufacturing or service delivery. This is primarily because continuous flow refers to the state in which the VA entity flows from the point of inception to delivery to the customer. It is commonly accepted that once the organization achieves continuous flow, it achieves the highest level of product quality, productivity, and ultimately company profitability.

Continuous flow is important because once it has been established, it eliminates one of the primary wastes—the waste of waiting. Since up to 95% of all lead time can be waste, the importance of the concept of flow to reduce time to the customer cannot be overstated. The creation of flow eliminates the waste of waiting and ultimately decreases product cost.

Pull Systems

A pull system is one that is set up to respond directly to customer demand. The concept is that your organization will not expend any resources until a customer has placed demand on one or more of your VA products or services. For example, in a build-to-stock environment of a company that produces products A, B, and C, as the customer places orders and these products are consumed from your finished goods inventory, your production facility begins to replenish these orders based upon this external customer demand.

Similarly, within your organization sequential process steps can be described as “suppliers” and “customers.” As one process step (customer) consumes materials from the previous process step (supplier) this signals the supplier to replenish those materials that have been consumed. This process flows back through your organization to your incoming raw materials. In a true pull system all raw material purchases would be tied directly to customer demand. Pull systems work in conjunction with continuous flow to help organizations align all of their internal VA activities with customer expectations. Because this is such a powerful concept, some organizations start with focusing on only one thing—“making the product flow” at the demand of the customer.

Point of Use Storage

Point of use storage (POUS) is a term that we use to describe how we deal with our materials and tools that we use to add value for our customers. It is a concept that is exactly as it’s described—POUS means putting those materials or tools where they are used by your value-adding employees. Applying this concept, we can eliminate a number of wastes or NVA activities. For example, if we have materials that we need for a particular step in a process located in a warehouse several hundred feet from where they are used, we need to use resources in order to move those materials to the point at which they are used. This adds cost to your product or service but no value from a customer standpoint. Moreover, in this example, a material that is located a great distance away from where it is actually used introduces the wastes of waiting, motion, transportation, and extra processing into our process, just to retrieve it and move it to the point at which we choose to create value for our customer. The concept of POUS is a very powerful waste identification and elimination concept. It can be used in almost every environment in some shape or form.

Quality at the Source

Most every organization today uses some form of quality program to assure acceptable product quality for the customer. Typically this requires inspections or reviews of product or service quality. Unfortunately, one cannot inspect in quality; you can only identify that which you failed to make up to customer requirements. In the application of Lean concepts we strive to achieve "quality of the source," which is a term that's used to refer to producing quality at each individual VA step. The primary focus of quality at the source is to assure that we do not pass along a defect to the next step in a process. Similarly to the concept of POUS, quality of the source can be used at virtually every step in a process.

It is particularly important to use quality at the source when dealing with information management. Poor quality of information is a significant source of waste in many organizations today. Inaccurate information passed across an organization introduces virtually all nine wastes to varying degrees. Here are a few simple examples. Has order processing entered all of the specific information required to process an order effectively for your customer? Can your order be processed defect-free and delivered in 100% compliance with your customer specifications? If not, you could benefit from quality at the source in this process.

When considering a new process, process revision, or revised facility layout, quality at the source should be an integral part of that process. As with other Lean concepts, there is a Lean tool, mistake proofing, that supports this effort and is often used to achieve quality at the source.

Just-in-Time

Just-in-time (JIT) is a concept that's used to describe the just-in-time delivery of all services or materials to the next process in your VA process. The objective of JIT is to make sure that we minimize the amount of materials that we have in our possession at any point in time. Many organizations have business measures based upon inventory dollars or inventory turns. The concept of JIT was instituted to minimize the waste of excess inventory and the negative cash flows of inventory carrying costs.

In a JIT environment the universe of inventory is examined in detail with the ultimate target on minimizing inventory costs. Non-Lean organizations carry significant amounts of raw materials inventory, often measured in days', weeks', or even months' supplies. Typical world-class organizations carry much less inventory, with some inventory supplies being delivered to their operations in 2- to 4-hour increments.

How much inventory we have, how we purchased this inventory, how we receive and store this inventory, how we move our inventory around a facility—these all impact whether or not we are using JIT philosophy. All inventory represents cost for your organization. The greater your inventory levels, the greater your inventory carrying cost will be, and as a consequence, this cost will have to be reflected either in your product price or in decreased profitability.

Kaizen

The term *Kaizen* has been described several ways. *Kai* means "change," while *zen* means "for the better.' Another translation would be "change" plus "improve." Kaizen is most commonly described as continuous improvement. Kaizen is the foundation of all Lean or LSS process improvement initiatives. Kaizen can be conducted individually, as a part of a process improvement team, or as a response to process troubleshooting. These three fundamental applications of Kaizen are present and readily observable in all LSS organizations. A detailed description of the application of Kaizen concepts is presented and discussed in Chapter 5.

Regardless of whether you conduct Kaizen individually, as a team, or in process troubleshooting, applying Kaizen for process improvement requires several activities. First, one must be able to look at and take apart the current process. Second, you must be able to analyze all the elements of that process. Finally, based upon this analysis, you must be able to define an improved set of steps in the process that you were investigating.

While much of the literature today refers to Kaizen as a tool (and many organizations use Kaizen events as a sole tool for improvement activities), Kaizen is the single most important LSS operational philosophy. It is clearly the most underlying fundamental concept required for successful and sustainable process improvement. The ability to deploy Kaizen in any of its forms is a primary requisite to becoming an LSS manager. To this end, Lean practitioners actively engage all employees at all levels of the organization to practice Kaizen on a daily basis.

5M's—Materials, Machines, Manpower, Methods, and Measurements

The 5M's—materials, machines, manpower, methods, and measurements—represent categories of process input variables. In every organization each category is composed of several key process input variables.

The sum of all the key process input variables contained in each of these categories can be used to provide an accurate description of your value stream. Understanding the nature of each of these categories allows the Lean practitioner to accurately describe the component of your organizational processes and how each component influences the quality of product or service delivery for your customer.

Key Process Input Variables (KPIVs)

Fundamental key process input variables (KPIVs) are in essence composed of all the resources required to add value for your customers. More effective use of resources is typically a target of organizational process improvement programs. For example, under the category of materials, one can identify a series of questions that may require analysis in any process improvement project. Some of these questions may include:

- Which raw materials are needed?
- How much of each of these raw materials is needed?
- When were these materials purchased?
- How much of each material is purchased?
- How would these materials be packaged by my supplier?
- How often do I take delivery of each material?
- How are the materials transported to my facility?
- Where will the materials be stored within my facility?
- When and how will the materials be transported to the VA location?
- How are they consumed in our value-adding processes?

Key Process Output Variables (KPOVs)

Key process output variables (KPOVs) is basically another term used to describe output measures that you want from your processes. Usually KPOVs are measures or specifications that your customer has placed upon you for service or product delivery. For example, one customer may want a certain or specific on-time delivery of your product or service, say, within two business days of order. Another KPOV could be defined as some aspect of product specification.

KPOV can be either internally focused (designed to achieve an internal organizational measure) or externally focused (designed to meet some functional requirements set for you by the customer). LSS organizations attempt to focus KPOVs on attributes that are defined by the customer. Although a few internally focused measures may be desirable, great care must be taken not to have too many of these in your internal processes. Internally focused measures tend to drive organizations to take their eye off customer requirements. This is a common mistake of traditional organizations that typically develop a large number of KPOVs that are internally focused, and as a consequence, leads them to situations where they produce poor-quality product for their customers.

LEAN TOOLS

In this section we describe the most common Lean tools used in industry today and the fundamental nature of each tool. We discuss the purpose of the tool, when it is used, and some of the important aspects of why this tool is an essential component of your Lean transformation process. These sections are not meant to be comprehensive in nature. There are many Lean tool books that provide both in-depth descriptions and the methods of applying these tools.

As stated earlier, Lean is an operational philosophy, and in reality, there are a relatively small number of Lean tools. The real power in applying Lean tools is in coupling this effort with one of the previously described Lean concepts. One of the largest misconceptions in American management today is that Lean tools in and of themselves will produce significant productivity improvements. Applying Lean tools solely as tools does not yield the process improvement that many organizations are attempting to achieve or that are typically associated with true Lean initiatives. An overview of the prominent Lean tools is presented in the remainder of this section.

5S Workplace Organization and Standardization

Definition

5S Workplace Organization and Standardization is an LSS tool used to organize basic housekeeping activities and standardize materials, machinery, manpower, and methodologies used in your value-adding activities.

Just the Facts

The most fundamental of all the Lean tools is 5S Workplace Organization and Standardization. The concept of 5S Workplace Organization and Standardization was presented in the book entitled *Five Pillars of the Visual Workplace*, by Hiroyuki Hirano.* Workplace Organization and Standardization is a fundamental building block of any LSS organization. As described in the Lean concepts section, the Holy Grail in any process is continuous flow. If a process is poorly organized or not standardized, it is very difficult to establish continuous flow. This is one reason most organizations begin an implementation of Lean with the fundamental introduction to 5S Workplace Organization and Standardization. There are many case studies in the literature that cite the importance of 5S to their Lean initiatives. Moreover, an effective 5S program is a prerequisite to applying Lean tools, and achieving sustainable meaningful results.

The purpose of 5S is to arrive at a safe, neat, orderly workplace where everything required to perform for your customer is readily accessible by your employees. Implementing 5S Workplace Organization and Standardization results in a commonsense work area with an organized sequence of activities required in your value-added processes.

5S Means Action

5S Workplace Organization and Standardization programs are comprised of five phases of activities. Each of these five phases is required if your organization is to successfully implement a 5S program.

- **Sort:** The first S is sort. Sort means just that; you select a target area and sort through everything that is in that area. The objective is to eliminate anything that is not needed. That means you are trying to leave only the bare essentials that are required for the VA steps in that area.

  At first glance this may seem like a very easy thing to do. However, once you begin this process of sorting, you will quickly find that there are a large number of things in any given area that are wanted but probably not needed. Because this is common in most organizations, a procedure has been established to help with the sorting process. This procedure is referred to as red tagging.

* Hiroyuki Hirano, *Five Pillars of the Visual Workplace*, Productivity, Inc. Portland, OR, 1995.

Note: It is very important that the red tag is used to indicate that the item is "wanted but not needed" and is readily differentiated (size, shape, and shade of red) from the red tag used on rejected parts.

Red tagging is a concept whereby employees sort through *everything* that is located in the target area. When conducting red tagging, a few simple questions are used to decide whether or not what you've identified is needed or not. Is this item needed? If so, where should it be located? Also, what quantity of this item is needed? During this process you will run across a large number of items that you're not sure what to do with. These items should be "red tagged." A red tag is nothing more than a tag that identifies critical information about the item, such as what the item is, where it was found, what date you found it, reason it was tagged, and possible disposition for the item. Once red tagged, the item is moved to the red tag area, where it is reviewed by a supervisor or manager for proper disposition.

- **Set-in-order:** Once the sorting is completed, you are ready for set-in-order. The process of setting items in order of use is very simple in concept; however, it requires discipline on the part of the user to identify exactly where all VA materials should be stored and how they should be prepared for use in everyday activities. The act of setting items in order by its very nature requires that the Lean practitioner standardize the work area. Care should be taken during this process to consider all employee interactions. For example, one should consider the workforce, such as: Are the employees right-handed or left-handed? Are there possible disabled or handicapped employees, and how do we set-in-order for the employee that may be color blind? These are critical factors for set-in-order.

  Once completed, it should be visibly obvious to all where every tool, material, fixture, or other items used in the VA process are to be stored. This process of standardizing the workplace makes it easier for multiple employees to use the same work area and complete VA activities in a standardized fashion.
- **Shine:** With sort and set-in-order completed, the next objective is to make sure that the entire area is completely clean. Shine is another simple activity; it just requires that you clean an area and make it ready for use. Typically, there are two valuable components of the shine activity. First, make sure that the entire area is clean and swept; this includes

floors, aisles, and any areas in or around where VA activities need to be conducted. Second, owner's shine has to do with the removal of dust or dirt or grime from any of the floors, workbenches, tables, computer areas, and office equipment that are in use on a daily basis. Adding the discipline of shine to your 5S program requires that you really take a close look at everything that is used in the VA process. The result of this is that once completed, everything in or around the work area is in a state of readiness for use whenever it's needed.

The importance of shine cannot be overstated. Organizations that fail to do a thorough shine process inevitably regress to a less organized, less productive work space. The planning process should become a fundamental component of our everyday activities. In order to make the shine activity a standard part of your operational procedures, one may choose to look at it similarly to that of personal hygiene. Few, if any of us, would consider leaving for work in the morning without performing any one of a number of various activities required to make ourselves personally ready for our day, i.e., taking a shower, brushing our teeth, and fixing our hair. This kind of philosophy needs to be adopted during the shine campaign in your organization. It must be daily. It must be routine. It must be a part of who the organization is.

Another reason that the shine campaign is so important is that defects are often hidden in a clumsy, dirty, dusty, dark, or otherwise unclean environment. Many of the defects that we produce are often tied to an unclean work area. The cleaning process also affords the employee time to take a closer look at all of those items—materials, tools, computers etc.—that we use during the VA process. This type of inspection allows us to identify potential problems before they become too serious and result in poor-quality products or services for our customers.

- **Standardize:** The concept of standardize is a little different from the concepts of sort, set-in-order, and shine. Standardize requires a different thought process and is a technique that is used to define the consistent application of sort, set-in-order, and shine activities. Standardize is the difference between a sporadic, once-in-a-while approach to workplace organization and a systematic, continuous, and routine approach to maintaining your work space. Organizations that do not put the fourth S—standardize—in place and do not make standardize a daily habit revert back to old habits, and as a consequence, the workplace is never maintained in a state of readiness.

Standardize is the phase where companies begin to falter. The reason for this is that we have many activities that happen on a daily basis that we would consider not standard in our workplace. That means there are many things that happened only once in a while or only today or only this week. The result is that materials that are often used or excess supplies required for a one-time event begin to creep into our work spaces. When we apply standardize on a daily basis, it forces us to take action on these one-time events, that is, to change the uncontrolled "happen" event into the controlled or standardized "occur" event. How often have you heard yourself say: "I'll just put this here" or "This is just for today" or "We never really do this, but I have to do it this way this afternoon." Each of these thought processes results in the breakdown of standardize and the circumvention of sort, set-in-order, and shine philosophies.

Ultimately, the essence of standardize is to prevent any work area from returning to its original disorganized state. This is typically accomplished by using standardized tools, such as checklists or check sheets, cleaning schedules, and a visual map of what the area should look like. These tools typically identify all the activities required to effectively complete sort, set-in-order, shine, and standardize on a daily basis.

- **Sustain:** The fifth S is sustain. If your organization is to be successful at maintaining an effective 5S program, you must develop the discipline to keep sustain alive in your organization on a daily basis. This means allowing the time, energy, effort, and resources required to maintain your 5S program. Sustain is nothing more than making a habit of sort, set-in-order, shine, and standardize activities. It is basically demonstrating that your organization is committed to a company culture that accepts nothing less than a daily active 5S Workplace Organization and Standardization program. Many organizations make it through the first four S's, yet lack the commitment to make it a part of their core daily activities. As soon as the sustain efforts dwindle or falter, your organization will revert back to its previous disorganized environment.

Common Omissions When Implementing 5S

The 5S Workplace Organization and Standardization tool is for everyone everywhere in the organization all the time. It is both a concept and a tool encompassing an everyday way of life in your organization. The strength of the tool is that it can be started small; by that we mean that it can be used to improve small areas of your organization independently. Ultimately, the entire organization should be implementing 5S to the point where it is part of the organizational DNA. Unfortunately, efforts commonly falter at the standardize and sustain phases. In many organizations a casual walk-through exposes the remnants of past failed attempts to embrace and adopt this, the simplest of Lean tools. A list of 10 common omissions from 5S programs is listed below. These are implementation aspects to keep in mind when adopting a 5S Workplace Organization and Standardization program.

1. Not including the entire facility, inside and out. This includes every space—office, parking lot, hall, work space, closet, storage area, warehouse, break room, bathroom, copy center, call center, distribution facility, company vehicles, etc.
2. Not including the entire staff. If anywhere in your organization the mind set of “5S is for someone else” is present, your 5S program will not be successful.
3. Abandoning the 5S effort before all required 5S activities are in your organizational DNA and are a visible, daily, and cyclic set of commonly re-occurring activities.
4. Not allowing employee time for complete 5S understanding and implementation.
5. Not allowing employee time for regularly scheduled 5S activities to be conducted at certain points in the day or night, daily, weekly, and/or monthly.
6. Not investing in required housekeeping and standardization tools—cleaning materials, visual control indicators, or standard location markers.
7. Not documenting 5S activities. Many companies go through an initial 5S program, and because the concepts and activities are simple in nature, they do not document and standardize the processes. Checklists, cleaning schedules, visual maps, before and after photos, and visual controls are all essential components of 5S documentation.

8. Not having 5S standard work for all associates, managers, and executives. That's right—managers and executives. It is commonly recognized that today manager standard work is essential for 5S success.
9. Not setting up a 5S visual workplace display that shows the entire company's participation in the 5S program and is a living evolving display of ongoing 5S activities company-wide.
10. Not encouraging, supporting, acknowledging, and rewarding employee participation for ongoing creative 5S activities.

Overall Equipment Effectiveness

Definition

Overall equipment effectiveness (OEE) is an LSS tool used to monitor equipment effectiveness by measuring equipment availability, performance, and product quality of the equipment. In semiautomated or fully automated environments OEE becomes a foundation for Total Productive Maintenance programs and LSS process improvement projects.

Just the Facts

Overall equipment effectiveness is a tool that was developed predominantly for the manufacturing sector. It was created to measure equipment effectiveness in companies that use equipment to add value or create products for customers. This tool is perfectly suited for this outcome. However, the strength of this tool conceptually reaches far beyond equipment and can be applied in many nonmanufacturing environments. In addition to its application in a manufacturing environment, we will also show how to modify the OEE calculation to be used in nonmanufacturing environments.

The primary purpose of OEE is to measure how effective your VA equipment usage is. In a manufacturing environment OEE is also a secondary measure of productivity. OEE is often used as a foundation for evaluating the Total Productive Maintenance programs and deciding where to apply your maintenance dollars to receive the greatest improvement impact on equipment reliability. A well-run OEE measurement system can also be used as a foundation for assessing new capital equipment acquisitions.

Although OEE is an effective tool, as with all measures, there is one caveat here. The concept of optimizing equipment utilization at the expense of other key process inputs, such as materials or manpower, was a common mistake when OEE was first introduced, and is still present today. In an environment where salaries were low and equipment costs were high, it was easy to fall into the trap of requiring employees to be chained to equipment to achieve high productivity levels. An LSS environment mandates that we balance the proper amounts of materials (to reduce inventory), equipment (to meet customer demand), and utilization of the human factor (to be flexible to changing customer requirements). This synergy of process inputs demonstrates effective use of Lean concepts and tools and reinforces that we maintain a true process focus. It also provides a foundation that steers us away from the common misuses of LSS tools.

How to Use OEE

The use of OEE relies on three direct measurements of your equipment and the products produced by your equipment. These are equipment availability, equipment performance, and product quality. In essence, the objective of OEE is to identify and quantify equipment-related losses that decrease productivity. Once identified, these loss areas are targets for Kaizen.

- **Equipment availability:** Equipment availability is a measure of equipment readiness when your organization needs equipment to add value. This is basically composed of scheduled operating time minus any downtime losses that occur during that scheduled operating time. The result of the equipment availability analysis is the actual time that your equipment was running. This is often referred to as runtime or uptime.

  Availability = Uptime/Scheduled operating time

- **Equipment performance:** Once the runtime has been established, we can look at equipment performance. Equipment performance is defined as your target output for equipment running at maximum speed minus any speed losses that occur during operation. The result of equipment performance analysis is your actual output.

  Performance = Actual output/Target output

- **Product quality:** Product quality is a measure of your good product divided by actual (total) output. The result is your good product output.

Quality = Good products/Actual output

OEE (%) = Availability × Performance × Quality

Applying OEE in Nonmanufacturing Environments

The concept of overall equipment effectiveness can also be used in nonmanufacturing environments. To use the OEE calculation in nonmanufacturing environments all one needs to do is replace the middle term *equipment* with the term *value-added*. This gives us overall value-added effectiveness (OVAE). We can now define OEE in terms of OVAE.

- **Value-adding availability:** Every organization today conducts some form of VA activities for their customers. In order to complete the OVAE, your organization must define an element that adds value for your customers. This could be in the form of the number of process inputs that add value. For example, this could be the number of employees available in a service industry environment. In the health care industry this could be the number of nursing staff available in a specific wing of the hospital. In a call center this could be the number of call center representatives available to take incoming calls. The VA availability component of OVAE is therefore:

  VA availability = Working time/Scheduled time

- **Value-adding performance:** Value-adding performance can be defined as your actual service delivery output divided by your target service delivery output. Similarly to in an equipment environment, VA employees run into a number of speed-related delivery losses during their daily activities. The VA performance component of OVAE is therefore:

  VA performance = Actual service delivery output/Target service delivery output

- **Product or service quality:** Whether you are delivering a product or a service to your customer, the quality of that product or service can be measured. Service delivery defects can be described as actual service output divided by good service output. The service quality component of OVAE is therefore:

$$\text{Service quality} = \text{Good service output/Actual service output}$$

$$\text{OVAE (\%)} = \text{VA availability} \times \text{VA performance} \times \text{Service quality}$$

Mistake Proofing

Definition

Mistake proofing is an LSS tool used to minimize or eliminate mistakes (errors) that produce defects in a product, a process, or a service.

Just the Facts

Mistake proofing* is a tool that is used to minimize or eliminate errors and their subsequent defects in any process. No matter how hard we try, whenever we put together materials, machinery, manpower, and methods, we are bound to make some errors. These errors produce unwanted defects for our customers. The art of mistake proofing endeavors to eliminate mistakes where they occur by redefining activities and developing techniques to mitigate the defect before it happens.

So what is the purpose of conducting mistake proofing in our organization? Why would you want to eliminate mistakes from occurring in our organization? First, eliminating mistakes basically means that we are improving performance for customers, helping to generate customer satisfaction, and ultimately customer loyalty to purchase our product or service in the future. One way to look at mistake proofing is as a program to guarantee future income for your organization.

Another critical reason to conduct mistake proofing is that mistakes introduce additional cost to your product or service. Every time a mistake occurs, additional actions need to be taken before you can deliver your product or service to your customer, all of which add cost and no value from

---

* *Mistake-Proofing for Operators*, Shop Floor Series, Productivity Press, New York, 1996.

the customer standpoint. Many of these mistakes detract from company profitability but are not accounted for in everyday productivity monitoring systems.

Another factor is that every time a mistake occurs, resources are used. That means that lowering the number of mistakes in your organization will inherently be lowering how many resources are required by your organization at any point in time. This further decreases the overall cost of your product or service.

How to Use Mistake Proofing

Mistake proofing is typically applied using some variation of the Plan-Do-Check-Act (PDCA) cycle. It is primarily a tool to help your organization achieve the Lean concept of quality at the source. In its most basic form your objective is to isolate any process step or task during the Do-Check portion of the PDCA cycle. Achieving an effective check at the point of execution is critical to assuring that no error, mistake, or defect is produced at any individual process step and passed along to the next VA step.

One way to achieve this is by applying the concept of "negative analysis."* This technique is used to identify and define what can go wrong in a process, and subsequently designing a process that will not allow a mistake to occur. The analysis includes observations and investigations of the interactions between materials, manpower, and equipment during the process. Creative, "no-mistake" solutions are developed as a result of the negative analysis.

Mistake proofing can be used wherever there is an interaction between two entities during a process step. Some examples include interaction between two employees, an employee-customer or employee-supplier interaction, an employee and a piece of equipment, an employee-material activity, an employee-method step, or during any measurement activity. For example, an office environment example of mistake proofing may be something as simple as setting the fields of an order entry form so that an inaccurate piece of data could not be entered.

With the ever-increasing accuracy and speed of digital photography, even high-speed manufacturing processes (e.g., stamping) are able to inspect and record, via digital photograph, products that are produced at thousands of pieces per minute. As soon as a manufactured piece is out

---

* *Business Process Improvement: The Breakthrough Strategy for Total Quality, Productivity, and Competitiveness*, McGraw-Hill, 1991.

of a predefined visual specification range, the machine is stopped. Prior to this type of technology, thousands of pieces would've been stamped, creating a tremendous number of defects at a substantial cost to the manufacturer. This type of mistake proofing technology virtually eliminates the need in many instances for tedious and expensive quality control checks.

Cellular Manufacturing

Definition

Cellular manufacturing is an LSS tool used to organize VA activities into the most effective (highest productivity) and least resource consuming (lowest cost per unit) series of activities to deliver the perfect product or service to the customer.

Just the Facts

Cellular manufacturing is best described as a Lean tool that is used to make the best use of resources during your VA activities. These resources typically include raw materials, manpower, equipment, and facilities. Manufacturing cells are most effective when there is a known steady demand for your product or service and products can be grouped by common VA steps. With customer demand known, you are able to sequence the flow of materials in your facility and apply the necessary manpower and equipment to best deliver your product. Creating a cell typically requires that we tie together all manual and semiautomatic (equipment) VA steps with the ultimate goal of making all of the VA steps and consequently your product flow at the demand of the customer.

As with all manufacturing process improvement tools, the primary purpose of cellular manufacturing is to enhance customer satisfaction and improve organizational profitability. A well-designed manufacturing cell can provide several positive outputs for both company and customer:

- Decrease lead time to your customer
- Improve product quality
- Decrease total inventory required and inventory carrying costs
- Minimize labor content as a percent of product cost

Since the inception of cellular manufacturing and one-piece flow concepts, it has been generally accepted that manufacturing cells produce the highest-quality products, allow for the highest productivity, and ultimately produce the highest profitability when compared with traditional in-line manufacturing processes.

How to Create Manufacturing Cells

The application of manufacturing cells draws together the use of several Lean concepts previously discussed. These include point of use storage (POUS), quality at the source, just-in-time (JIT), Kanban, and facility layout (i.e., process flow layout, not a function department layout), to name a few. The objective in cell design is to identify and eliminate as much NVA time and activities as possible from the current process by organizing all VA activities in the best sequence.

There are five key steps to the successful design and implementation of a cell.

Step 1: Group products. The first step in cell design is to understand your product groupings. To do this, you must first construct a list of products and then identify all the process steps required by each product. It's usually easiest if you just make a matrix with product types on one axis and process steps on the other. After checking off which process steps are required by each product, it is easy to identify and group products by their respective process steps. After your products have been grouped, you can undergo the task of creating specific cells to manufacture or assemble all the products in each group.

Step 2: Measure demand (calculate Takt time). Once the products have been grouped, we must now calculate or measure the demand rate of each product. The demand rate is just the rate at which your customer requires that you produce your product or provide your service. The demand rate is the amount of working time available divided by the number of units sold and is typically reported as units per hour or activities per day. Understanding the demand rate of your customer is a critical prerequisite to designing a cell. Cells typically work best when the demand rate is somewhat constant. If your demand rate is erratic or unpredictable, you will need to consider adding a visual control supermarket along with your cell design.

Step 3: Chart current work sequence. For each product you must next chart your current work sequence. This is typically accomplished using a time observation chart to document each element of the work sequence and record the time to complete each work sequence element. This time observation process is an essential complement of deconstructing work activities and then re-assembling them back into a new and balanced continuous flow work sequence.

Step 4: Combine work and balance process. Once you have accurately recorded times for each element in the work sequence, you are now in a position to combine some work elements and balance the process to achieve the correct demand rate or output for the customer. This is accomplished by grouping elements to achieve the time that is less than or equal to the demand rate. For example, if the calculated demand rate is 10 minutes per unit for a specific product, then each individual element in the work sequence must be 10 minutes or less to complete. The closer you get each individual element to 10 minutes, the more continuous and smooth your work product flow will be. Work sequence balancing is not an exact science, but with some analysis of your time observation chart you will clearly be able to recognize significant disruptions in work flow associated with unbalanced work element times.

Step 5: Create new cell work sequence. After completion of steps 1 to 4, you are now in a position to create a completely new work cell sequence. During this step you complete a work flow layout that includes all materials and equipment manpower required to complete the work sequence. The primary objectives are to (1) simplify material flow by integrating process elements, (2) minimize material handling, and (3) make use of people for 100% of the demand rate time. In essence, your goal is to tie together and establish continuous flow of each work element. How you sequence these is dependent upon the actual work elements; however, it is best achieved by using one of the known successful cell configurations. The most common include the U-cell and S-cell configurations.

Kanban

Definition

Kanban means "signal." It is an LSS tool used to make visible the requirement for action on the part of employees.

Just the Facts

*Kanban** is the Japanese word for "card" or "sign," whose function is to relay information along with materials that tell employees exactly what to produce at any given point in your process. Most VA processes have a significant number of process steps. The beauty of Kanban is that it connects a series of process steps in a fashion that allows continuous flow. As we discussed earlier in this chapter, continuous flow is the Holy Grail of any process. Kanban is a critical tool to establishing flow in a process.

The purpose of using Kanban in a process is to regulate the flow of information and materials between employees by connecting sequential VA process steps. Kanban systems allow you to define the exact quantities of products that are required to meet your customer demand. The benefit of this system is that you produce only what the customer requested, therefore eliminating any tendency for overproduction, one of the nine wastes.

How to Use Kanban

Kanban is used as an information-inventory control system by two sequential steps in the process. It can be used by steps that are directly adjacent to each other, those that are separated by great distances, or between different types of equipment at varying stages of your VA process. The Kanban signal system answers the question of "what to do next" that is required by employees to achieve high levels of productivity. One caveat is that Kanban typically only works well within stable demand environments. With the relatively stable demand, order points and order quantities within successive steps of the process can be defined. These are the foundation of a Kanban system.

Typically, a Kanban card identifying product name, photo, requesting department, and quantity desired is shuttled between the consumer of the product and the producer of the product.

Kanban is also a very powerful supply chain tool. It can and is being effectively used to make the supplier-customer materials management process more effective up and down the supply chain. Kanban containers are common tools between world-class organizations managing materials to meet customer demand.

---

* *Kanban for the Shop Floor*, Shop Floor Series, Productivity Press, New York, 2002.

Another method of Kanban signals is with vendor-managed inventory scheduling. This typically has a supplier being granted secured access to the customers' information management system to monitor the real-time direct consumption of products. At prespecified points of consumption the supplier is signaled to prepare and ship product. This practice is becoming more common as companies strive to decrease lead times and inventories by taking advantage of global information sharing technologies. World-class companies are partnering through technology to create value for each other in ways not possible just 10 years ago.

Value Stream Mapping

Definition

Value stream mapping is an LSS tool used to map all activities (both VA and NVA) across your value stream. The tool allows for a visual representation or maps of resources allocation as you conduct business today (current state) as well as how you plan to add value in the future (future state).

Just the Facts

Value stream mapping (VSM) is a technique that's used to develop a visual representation of all the activities required for you to add value for your customer. It's typically conducted in a two-step process. The first step is to construct a current state map. In the current state you review every single activity that's currently being conducted in order to provide your product or service for your customers. The current state map is used to give a fairly accurate definition and description of what your organization currently does for your customer. This is critical for your organization to begin to understand the weaknesses of your current process and to identify what needs to be improved to improve performance for your customer. During the creation of the current state map, identify on the map as many of the nine wastes as possible, indicating where these wastes are present. Once a current state process map has been completed and significant waste throughout the process is identified, the second step of the VSM program can be completed, which is preparing a future state map.

The future state map defines and outlines a visual representation of how you want your organization to perform at some point in the future. It typically endeavors to describe an ideal state, that is, a state in which you identified and eliminated a significant amount of waste that existed in your current state map.

There are many software packages available today for VSM. Some are simple icon-based systems to help with preparation of current state and future state visual maps. Others are more complex and allow for the inclusion of many process variables, such as materials, employees, and cycle times and/or lead times. The most complex allow for computer process model development and sophisticated simulations of "what if" process change scenarios of selected process variables. Regardless of the complexity of your selected VSM solution, all options can allow you to significantly improve your processes using the VSM techniques.

Managing with Maps

Using VSMs to manage improvement activities is an effective way to organize, prioritize, deploy, and manage improvement activities. Using a technique called managing with maps is an effective approach to tie process performance acceleration and align organizational objectives from strategic plans to tactical improvement plans. In *The Organizational Alignment Handbook*, we describe the seven phases of the organizational alignment cycle.* The managing with maps technique would be developed and deployed in phases I, II, and VI. A brief overview of using managing with maps to align and manage organizational objectives is given below. Each view would require the preparation of current state and future state views at each level in the organization. These maps are used as the primary improvement management tools.

- **Strategic view:** The strategic view VSM shows a complete view of the value stream from customer contact to customer receipt of product or service. It identifies high-level process steps and includes key departments, high-level performance measures that all lower-level measures will roll up to. This map should allow all employees to see how their participation supports the organization's strategic objectives. It also identifies high-level improvement initiatives targeted for the map time period.

---

* H. James Harrington, Frank Voehl, *The Organizational Alignment Handbook*, CRC Press, Boca Raton, FL, 2012.

- **Department view:** The department view VSM shows a complete view of the value stream within a department. It identifies department-level process steps and assigns key department measures. These maps encompass resources coming into or being used within the department and all the VA products or services exiting the department. It identifies Kaizen activities or continuous improvement team activities.
- **Process view:** The process view VSM shows a complete view of each process within a department. It identifies detailed views of each process and assigns key department measures to each process. These maps encompass resources coming into or being used within each process and all the VA products or services exiting the process. It identifies Kaizen activities or continuous improvement team activities, current state performance, and future state targets for improvement.
- **Measure view:** The measure view VSM shows a complete view of all measures within a process. It identifies specific measures of various process steps and allows for monitoring and aligning multiple measures of a process. Using this technique allows for measurement management and eliminates conflicting measures within a process. This is particularly important in complex systems or regulated environments such as health care or medical products manufacturing.

Using the management with maps approach allows complete organizational alignment. It provides a mechanism for measurement roll-up from the most specific process step measurement to department objectives, and ultimately to strategic initiatives. These VSMs are a powerful visual management and communication tool.

Visual Controls

Definition

Visual control is an LSS tool used to describe "what to do next" for employees without using words. They are pictures or diagrams used to regulate employee activities, such as displaying activity instructions, identifying safety hazards, or restricting employee access, to name a few.

Just the Facts

Visual controls are simple signals that show at a glance what needs to be done. They are simplifications of systems that, when implemented effectively, require no communication between employees in order to signal what action should be taken. Think about that for a second—no communication! No e-mail, no information management system interaction, no phone interaction, no reading standard operating procedures or good manufacturing practices. These are all no-value-adding, time-robbing activities that the customer is unwilling to pay for.

Some visual control examples that describe necessary actions are where a material should be moved, what raw material or sub-assembly should be replenished, which orders need to be entered, which patients should be waited on first, etc.

How to Use Visual Controls

Visual controls are becoming more and more prevalent in many organizations. In multilingual environments visual controls can eliminate the need for translations. In organizations where employees may have handicaps, such as color blindness, for example, visual controls using geometric shapes can be incorporated into work instructions.

The visual control example in Table 4.3 will assist you in developing visual controls in virtually any environment. The table puts together control type, purpose, and several examples of common controls. This is a powerful tool for process improvement. Now that you have been introduced to them, look for aspects of your processes that would benefit from visual controls.

THE POWER OF LEAN CONCEPTS AND LEAN TOOLS

Throughout this chapter we have described a series of the most common but powerful Lean concepts and Lean tools used in process improvement activities. In this section we will present an example that demonstrates the proper application of the Lean operational philosophy, which combines the Lean concepts and tools described in this chapter. The synergy, strength, and power of coupling a Lean operational philosophy and Lean

TABLE 4.3

Visual Control Examples

Type	Purpose	Description
Items or parts	Identify the correct item or part	Signboards, photos, labels
Locations	Identify the correct location	Color coding, numbering, tape outlines
Quantities	Show the proper quantity	Min-max levels, container quantities
Methods	Describe the method	Standard procedures, visual work instructions, charts
Exception tags	Indicate special conditions or abnormalities	Read tags, repair tags, quarantine signs
Andon signals	Signal employee action	Visual flashing or rotating lights, bells, or buzzers
Kanban	Control materials movements	Card, containers, or empty spaces signaling production is required
Performance measurement displays	Visually show performance versus target	Safety, quality, or productivity performance measures
Defect displays	Make visible common problems	Board or table showing defective raw materials, tooling, or paperwork
Personnel boards	Show current availability or assignments	Availability (in/out), assignment department, or location

tools to effectively organize key input variables (5M's) are demonstrated in this example. Remember, only by managing inputs can we improve outputs. When looking to harness the power of LSS, endeavor to use as many of the concepts and tools as possible along your process.

Several Lean concepts or tools described in this chapter were used in unison in the composite U-cell case study described below. The term *composite U-cell* simply means we joined two U-cells to establish continuous flow for the production of the entire finished product. This example also shows both creativity and innovation that can be achieved to arrive at simple solutions that deliver powerful process improvements. The following 13 Lean concepts and tools were used in the composite U-cell case study.

1. Continuous flow
2. Kaizen

3. Pull systems
4. Cell design
5. 5S Workplace Organization and Standardization
6. Standardized work
7. Visual materials controls
8. Visual work instructions
9. Kanban
10. Mistake proofing
11. Point of use storage (POUS)
12. Quality at the source
13. Facility layout

Composite U-Cell Case Study

- Situation: A manufacturer of complex medical instrumentation was producing approximately 45 blood dialysis units/week operating two shifts in an enclosed electrostatic discharge (ESD) room. The company was regularly receiving orders for 60 units/week. There was no room for expansion, and the company was restricted to assembly in the current assembly area. The challenge was to implement process improvements and achieve 60 units/week with the existing staff, assembly area, and test equipment. Previous attempts to increase production beyond the 45 units/week level had resulted in significant quality problems and an actual reduction in output.
- Cell design activities: The Kaizen continuous improvement team created a "first of its kind" composite U-cell using the five-step cell design process. The composite cell is actually two cells operating in unison. Cell design for each cell included product grouping, Takt time, line balancing, work sequencing, and facility layout for a total of 38 workstations, over 50 assembly and test employees, and the assembly of over 4,000 parts.
- How the composite cell operates: Raw materials flow from the outside of each cell toward the inside of each cell via a series of sub-assembly workstations. Simultaneously, each major component, hydraulics

assembly and chassis assembly, flows in the U-shaped pattern on the inside of each cell, consuming sub-assemblies along the way. Final assembly of the hydraulics and chassis takes place at the work center between the cells as the last assembly step before transferring the final product to testing.

Lean Six Sigma Concepts and Tools Used

- Kaizen teams were used as the primary change management tool for cell design and implementation.
- The Kaizen and you method was used by individual employees for bench improvements along with 5S Workplace Organization and Standardization.
- During each cell design, the teams utilized continuous flow, pull systems, visual controls and instructions, standardized work, POUS, quality at the source, Kanban, mistake proofing, and facility layout.
- During cell "try-storms" workbench layouts were designated, and POUS and materials flows were established with replenishment occurring as materials were consumed using visual controls and Kanban cards and containers quantities.
- Quality @ the source and mistake proofing were incorporated into employee sub-assemblies activities. Two space visual controls located on the sub-assembly work center regulated materials flow within the cells. As the hydraulics assembly consumes each series of sub-assemblies, workers at each sub-assembly bench were signaled to replenish these parts.
- Well-defined Takt time assured continuous flow throughout each cell.
- Final assembly and testing proceeded without errors and sub-assembly rework that had previously hindered productivity.

Figure 4.1 shows details of the final composite U-cell layout, which includes raw materials' storage on the exterior of the cells, sub-assembly workstations, employee deployment, final assembly cart progression, and in-process and final test stations.

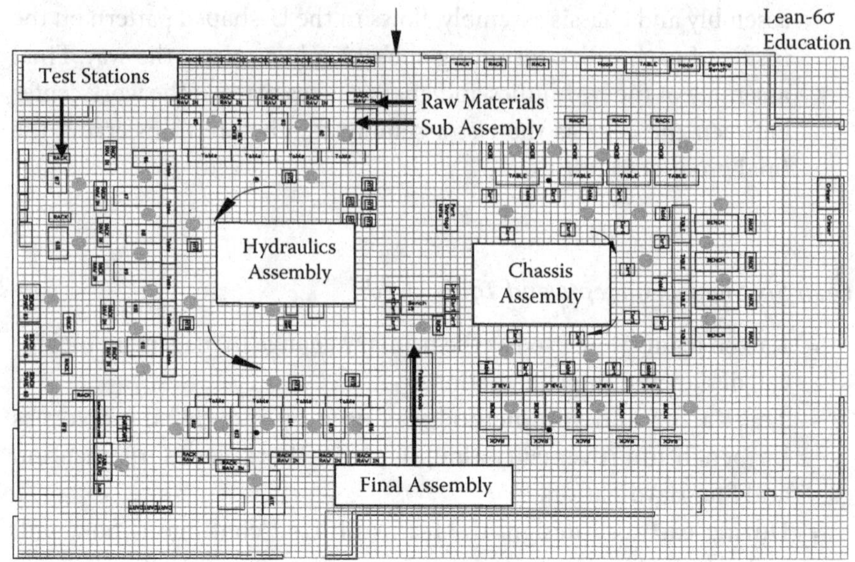

FIGURE 4.1
Composite U-cell layout. Economic impacts: Over a 60-day period the continuous improvement teams designed and implemented the integrated U-cells. The results were that the initial production rate of 45 units/week was successfully increased to 60 units/week with greatly improved product quality. This represents a 33% increase in productivity using the same floor space, number of employees, and existing test equipment.

SUMMARY

There are four key LSS learning points to take away from this chapter. The true LSS manager or practitioner integrates these four characteristics into his/her daily beliefs and activities to achieve performance improvement.

- Learn to "see" waste. Waste is almost everywhere in your organization. Increase your understanding of what causes each of the nine wastes. Wherever any of the 5M's interact, there is waste potential. Notice the activities revolving around VA processes. Focus on materials, machinery, methods, man/woman power, and measurements. These inputs are source points for all waste creation.
- Embrace the power of observation—stop and watch the process. Taiichi Ohno was famous for demanding observation of the process by his employees, so much so that he would draw a circle on the floor and

make shop floor managers stand in it and observe the process until they identified and understood where the waste was. These events could last hours with Ohno's intermittent checks to see if managers could identify the waste. His learning objective was a simple one: You cannot improve a process when you cannot see what to improve or where to improve.

- Learn the LSS concepts and apply the LSS tools. Begin to better understand the range of LSS concepts and how they interact with each other. Understand the warning signs that a traditional operational philosophy is at work and contributing to waste. These are typically starting points for process improvement projects. Apply the tools and monitor their effects on performance measures. Remember, LSS is a learning process. The more you use the tools, the better you will get at effectively improving processes.

Use the concepts and tools in groups to achieve enhanced performance improvement. Using LSS concepts and tools in groups can result in significant performance improvement. Revisit the composite U-cell case study as an example of grouping concepts and tools.

5

Three Faces of Change—Kaizen, Kaikaku, and Kakushin

Kaizen—Small change
Kaikaku—Transformation of mind
Kakushin—Innovation

It is not necessary to change. Survival is not mandatory.

—W. Edwards Deming

IN A NUTSHELL

Improvement means change. Today more than ever the ability of an organization to change quickly and adapt to changing customer requirements is mandatory. Almost every employee in every organization is faced with the challenge of changing how they do things today into how they want to do things tomorrow. For managers this can be a particularly daunting challenge. When discussing process improvement projects with managers, the LSS practitioner will often be faced with a barrage of "resistance to change challenges" across the organization. We hear statements such as:

- If only purchasing would do this or engineering would do that, our process could be improved.
- If this department would consider our needs, we would more effectively be able to deliver our product.
- If John's department could just deliver the materials that we need on time, our productivity would greatly improve.

These resistance-to-change scenarios are commonplace in many organizations today. This chapter is about gaining a better understanding of why there is so much resistance to change. How can successfully applying Kaizen to our change management activities be a fundamental approach to overcome this resistance? All LSS change management activities benefit from this approach, whether you are using a process change model of Define, Measure, Analyze, Improve, and Control (DMAIC) or a design changes model using Define, Measure, Analyze, Design, and Verify (DMADV). We discuss how individual and organizational beliefs and behaviors are responsible for some of the resistance to change that we encounter.

If we are to become true LSS practitioners, we need to gain a basic understanding of the ninth waste, behavior waste, and how we apply Kaizen to eliminate this and other wastes and facilitate change in our organization. The ninth waste, behavior waste, teaches us that change and improvement begin with us as individuals (personal waste) and collectively (people waste) as members of an organization. In this chapter we discuss the organizational path of change from a non-LSS environment to an LSS environment. The primary conduit of change management is Kaizen. Eliminating non-Lean beliefs and behaviors across the organization requires the sustained application of Kaizen.

Most organizations are familiar with Kaizen ("small continuous") changes; however, Kaikaku ("a transformation of mind"), which demonstrates large change or radical change, and Kakushin ("innovation") are equally important in your Lean transformation.

Kaizen* is not a new concept. It was coined in 1986 by Masaaki Imai in *Kaizen—The Key to Japan's Competitive Success*. This first book on Kaizen fully defines and describes the concept. One unique aspect of this book is that it gives us a look at how to apply Kaizen in a traditional organization. Coupled with his second book, *Gemba Kaizen*,† they represent powerful resources on change and process improvement available today. Concepts summarized and presented in this chapter are derived from these and other resources.

Kaizen is the lifeblood of a Lean organization and a Lean management system. It can be used individually, in teams, or for process troubleshooting. We show in this chapter that Kaizen is about taking positive action.

---

* Massaki Imai, *Kaizen—The Key to Japan's Competitive Success*, McGraw-Hill/Irwin, 1986.

† Massaki Imai, *Gemba Kaizen*, McGraw-Hill, New York, 1997.

Kaizen transitions into Kaikaku, a transformation of mind, and becomes the perfect conduit to deploy education, beliefs, and apply Lean tools across your organization.

The beauty and power of Kaizen, Kaikaku, and Kakushin is that they are easy concepts to learn, implement, and teach to others. Kaizen can become contagious (in a very positive sense) in your organization. You can always tell an organization that practices Kaizen within a few minutes of walking around the facility. The telltale signs are everywhere; there are visual controls and measurement systems. Kaikaku can be described as a series of Kaizen activities completed together, forming and exhibiting the presence of a Lean mindset. Kakushin or innovation is required for growth of all companies. Simply put, living the three faces of change—Kaizen, Kaikaku, and Kakushin—will show you how Lean can transform your organization.

INTRODUCTION

Kaizen means continuous improvement or change + improve. Many of us as managers want improvement, but often fear change. We embrace change only if we can completely control it. In many instances, management today wants to know the exact outcome of change before it allows any employee to take even the most fundamental steps to improve a bad situation. We have exposed individual and organizational beliefs as the root cause of this type of employee behavior, and now will present the change management required to eliminate it.

We present Kaizen, Kaikaku, and Kakushin here as the three fundamental tools to deploy change in your organization. What is the nature of change? How do we let go of old beliefs and behaviors? What are we changing into? How does Kaizen, Kaikaku, or Kakushin solve or address all these change questions?

It's time for individuals, teams, management at all levels, and employees in all functional areas of the organization to embrace change and to become the change that we want to see in the organization. In the words of Mahatma Gandhi: "We must become the change we want to see."

This means that we cannot look outward for change. We cannot stand and say that it's the outside world that needs to change—some other employee, department, function, policy, or procedure that needs altering.

Every individual in every organization can change aspects of many activities that touch their daily work environment. So let's start on our journey of Kaizen. We will never be the same, and neither will the organization.

Resistance to Change

Organizational change is often more difficult than it first appears. Why is individual change so difficult? What is it that makes organizational change so difficult? These are fundamental questions that must be at the forefront of any LSS transformation process.

There are three fundamental aspects that prevent, hinder, or inhibit employees from participating in change management. First, there is fear of the unknown. Second is the measurement system. Finally, there are individual beliefs and collective organizational beliefs that present resistance barriers.

Fear of the Unknown

When we say *fear*, we don't necessarily mean fear of retribution from a manager or another employee in the organization. What employees typically fear is the unknown associated with any change to their environment. The change from where they are now and what they are familiar with to where the organization is going can cause trepidation on the part of many employees. Often changes are not clearly or completely defined. When there is uncertainty, people may be hesitant to walk into the unknown. Addressing this employee fear is a component of any successful LSS organization.

Measurement Systems

It has often been said that measures drive behavior and bad measures drive bad behavior, or non-Lean measures drive non-Lean behavior. In either event, the measurement system can be a significant source of resistance among employees, departments, or divisions of the organization. When embarking on any LSS project, understanding the nature of the measurement system that's in place is a critical factor to the success of the project. A review of measures may help in the formulation of your LSS project. Some types of measures that should be considered include:

- Individual employee performance measures
- Department measures

- Division measures
- Bonus performance measures
- Process measures
- Productivity measures
- Customer-related performance measures
- Cost measures
- Corporate measures
- Stakeholder measures
- Regulatory agency measures
- Government regulation measures
- Risk management measures
- Liability measures
- Contract measures

Whether the ability to change these measures is real or just perceived, the above list can have a substantial influence on the resistance to change that you face on your LSS project. Clearly, many of those described can deter an employee on an LSS team from taking action.

Beliefs

Human beings have a fundamental desire to feel that the actions they are taking every day are correct. These actions are derived from each individual's beliefs. We normally don't look at ourselves as individuals who are deliberately doing something that is wrong. We generally believe that our behaviors and actions are "the right thing to do." Because of this, people tend to hold on to their beliefs and behaviors tightly, making it difficult to change a behavior pattern that is not producing effective performance of the organization.

OVERCOMING RESISTANCE TO CHANGE

Overcoming resistance to change is rarely easy and can range from a difficult task to a nearly impossible one. People tend to hold on to their beliefs fairly tightly. Many LSS efforts are stalled or fail outright because of the inability to overcome resistance to change.

There is a wide range of excellent literature on the subject of change management. Some focus on managerial psychology, while others use various tools and techniques, and still others present a combination of the two approaches. In this section we introduce the three basic phases that employees typically go through during the implementation of LSS concepts and tools for effective results.

Leaving Old Beliefs Behind

There are three distinct phases required to rewrite our beliefs in successfully transitioning to a new set of daily beliefs and behaviors. The first is to adopt the ability to "let go"* of our old beliefs to which we hold on to so tightly. The second is to open up our mind to the possibility of new ways of thinking and new ways of looking at the daily activities that we conduct. During this period of receptive activity to new ideas, it is incumbent that we try to implement these ideas in new ways in our organization. Over time, these new ideas become new patterns of behavior that become dominant; this third phase is identified as the emergence of an LSS environment.

So how do we leave our old beliefs behind? What is the first step? One of the fundamental characteristics of operating in an LSS environment is the ability of its employees to challenge current beliefs and practices. These may be individual beliefs or organizational beliefs. Initially, this must start with the individual, who on a daily basis must become proficient at questioning whether or not the activities that he or she will be conducting today are creating value for the customer. When this practice becomes commonplace with individuals within our organization, we will be well on our way to developing the discipline to change old non-Lean beliefs.

Only when we have begun to adopt the concept of changing old beliefs can we even consider the possibilities of adopting a different way of doing things. Many companies today run into significant problems with adopting LSS concepts and tools simply because they are unwilling to change old beliefs. LSS is not a concept that can simply be slapped on top of an old belief system, deployed in a vacuum, and be successful. Trying to adopt LSS without letting go of old beliefs is a mistake made by many organizations today, and one of the leading reasons why many LSS initiatives do not produce the results anticipated and are unsustainable.

---

* William Bridges, *Managing Transitions*, 3rd ed., Da Capo Press, Cambridge, MA, 2009.

Considering New Possibilities

Considering new possibilities is a fundamental activity of Kaizen and the essence of this chapter. When we consider new possibilities, we open ourselves up to trying new things and begin to examine the way we think, our point of view, and the methods we use in daily activities. Developing an organizational personality that considers new possibilities is the essence of LSS in that it allows a "learning environment" to emerge.

The identification and development of new possibilities is neither quick nor easy. It requires dedication at all levels of the organization and a commitment to practice Kaizen on a daily basis. Employees must be allowed the elbow room to experiment with try-storms to gauge the relative success of possible new improvements. They must be allowed the opportunity to make mistakes, course correct, and define better paths to higher productivity.

Management's insistence on a "results-only focus" can stifle the creativity of employees to try new process focus concepts. In this respect, Lean management becomes critical in the development of new processes. The rewriting of our beliefs requires an acknowledgment by management that our current belief system is ineffective and necessitates a deliberate management approach that supports the elimination of old beliefs followed by the consideration of new LSS beliefs.

Emergence of LSS

The continued practice of these two fundamental phases—leaving old beliefs behind and considering new possibilities—results in the emergence of an environment where individuals, groups, and your organization as a whole are on a journey to an LSS environment.

When considering change, start small and follow these simple steps. First, identify a waste in your organization and the old belief that contributed to the production of that waste. Second, put the old belief aside and consider new possibilities. Follow the three faces of change to become an LSS organization.

THREE FACES OF CHANGE

There are three faces of change that all improvement activities can be classified as: Kaizen, Kaikaku, and Kakushin. The fundamentals of each are presented in this section. True LSS organizations have employees well versed in all three, and activities of all three types of change management are visible across the organization.

Kaizen—Continuous Improvement

What is Kaizen? Kaizen has been defined many ways. Several definitions follow to give you a range of interpretations of the term *Kaizen*.

1. Kaizen: Change + improve.
2. Kai: To take apart and make new. Zen: To think or become enlightened.
3. Kaizen: Continuous improvement using small incremental changes.
4. Kaizen: When applied to the workplace means continuous improvement involving everyone, managers and workers alike.

> It is not the strongest of the species that survive, nor the most intelligent, but the one most responsive to change.
>
> **—Charles Darwin, English naturalist (1809–1882)**

In Lean organizations, management has two roles—maintenance and improvement.* *Maintenance* means to standardize and sustain a process, while *improvement* means to move the process forward to higher levels of performance. Virtually all process improvement programs start with Standardize-Do-Check-Act (SDCA) and move to Plan-Do-Check-Act (PDCA). Figure 5.1 shows this progression.

Kaizen process means to establish and live the PDCA cycle (see Figure 5.2):

- **Plan:** Establish a target for improvement.
- **Do:** Implement the plan.
- **Check:** Determine whether implementation has brought planned improvement.
- **Act:** Perform and standardize the new procedures to prevent recurrence of the original problem.

* Imai, *Gemba Kaizen*. McGraw-Hill, N.Y., 1997.

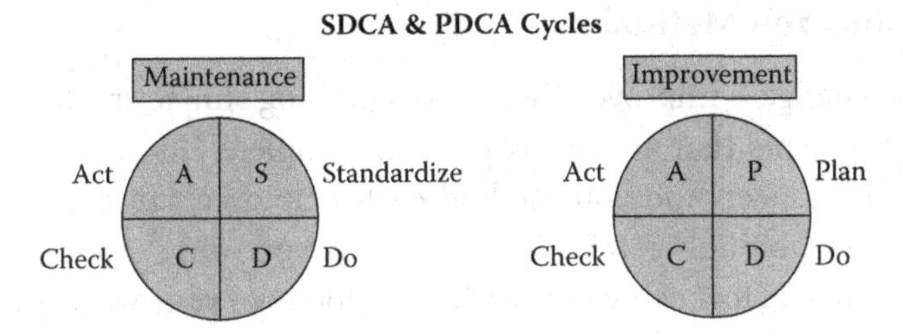

FIGURE 5.1
SDCA and PDCA cycles.

Where to start Kaizen? If you are not sure where to start Kaizen in your organization, use the four K's of Kaizen; observe your organization, and start with one of the following:

- **Kusai:** Things that smell bad.
- **Kitsui:** Things that are hard to do or are in dark areas.
- **Kitanai:** Things that are dirty.
- **Kiken:** Things that are dangerous.

How do you use Kaizen? There are three basic approaches to using Kaizen effectively: Kaizen and you method, Kaizen for process troubleshooting, and Kaizen teams. Each is briefly described below.

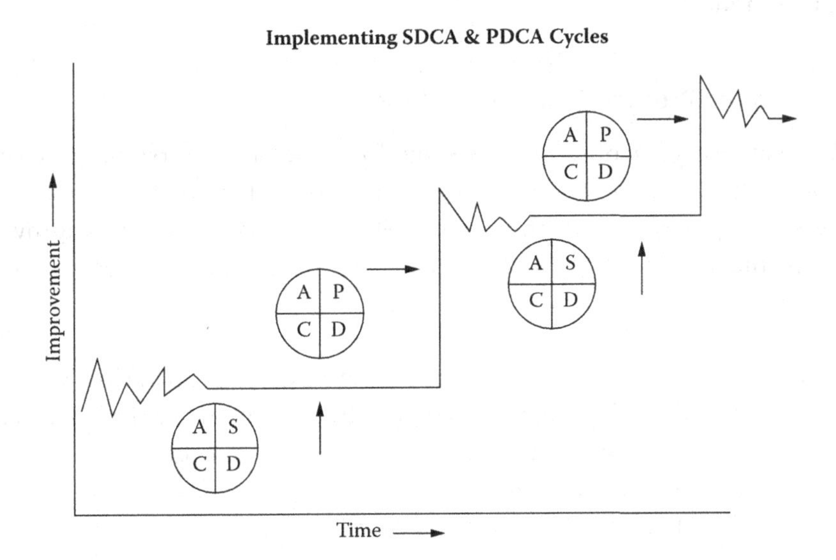

FIGURE 5.2
Implementing SDCA and PDCA cycles.

Kaizen and You Method

Kaizen is change + improve. That means making simple small incremental improvements that any employee can complete. That's all there is to it. With the Kaizen and you method each employee can begin today to "change your point of view," "change the way you work," and "change the way you think." How can you start Kaizen today in your work area?

- **Stop** doing unnecessary things.
- **Reduce:** If you can't stop, reduce them somehow.
- **Change:** Try another way.

There are three laws that make the simple concept of Kaizen work, but you absolutely have to do all three or you will *never* be good at Kaizen.

1. **Surface:** Write the idea down.
2. **Implement:** You make the change.
3. **Share:** Post it, review it, and talk about it.

With Kaizen you can change yourself and you can change your workplace. Kaizen should be done to benefit you. The person who gains the most from the Kaizen is the person who does the Kaizen. Start doing Kaizen today!

Kaizen for Process Troubleshooting

Everyone in your organization should completely understand and be able to successfully complete Kaizen using the following five-step process.* Employees who cannot complete this process cannot improve their individual areas, and consequently, you cannot improve the organization.

- Step 1: When a problem (abnormality) arises, first, go to *Gemba*.
- Step 2 : Check the *Gembutsu* (relevant objects surrounding the problem).
- Step 3: Take temporary countermeasures "on the spot."
- Step 4: Find the root cause(s).
- Step 5: Standardize to prevent recurrence.

* Ibid.

Step 1: Go to Gemba

Gemba is the most important place in your company. It is where *all* of the value is created for your customers. This concept was so important that Soichiro Honda, founder of Honda Motor Company, did not have a president's office. He was always found somewhere in Gemba. Gemba means the "real place," or where the "action" is. When a problem arises, go to Gemba first and look to solve specific problems.

Step 2: Conduct Gembutsu

Gembutsu means to assess all the relevant information in Gemba that surrounds the problem. Interview several employees; ask questions of what was happening when the problem occurred. Seek information in a nonthreatening way; the employees in the area want the problem to be resolved as much as you do. Don't place blame, insinuate wrongdoing, or belittle an employee's work performance. This is a search for the facts of what happened. Remember to gather information regarding all 5M's: materials, machinery, manpower, measurements, and methods.

Step 3: Take Temporary Countermeasures "on the Spot"

There is nothing more reassuring to an employee than knowing that management will support employees when problems arise. This is best demonstrated by a manager that takes sound action immediately. Your ability to remain calm in a critical situation and gather relevant information, understand the situation, and take action on the spot is one sign of good leadership and will be respected by all employees. Most importantly, it should be recognized by all, especially management, that these are temporary measures. The most common mistake is that organizations stop here at step 3; they never find the root causes, and consequently the organization lives with many "band aid" solutions that never get resolved and result in constant nagging poor-quality and poor-performance issues.

Step 4: Find Root Causes

Once temporary measures are in place, root cause analysis must be conducted. It is imperative that the root causes are found and eliminated. These can be conducted using techniques such as the five whys, cause-and-effect analysis, or failure mode and effects analysis (FMEA). This is a critical step, for if the root cause(s) is not found, the organization is doomed to repeatedly revisit the problem.

Step 5: Standardize to Prevent Recurrence

Standardize means to put in a control system to prevent the problem from appearing again. Depending on the nature of the problem, this often requires management tools such as a revised maintenance schedule, revised standard operation procedures or visual work instructions, or process control charts. These preventative measures must be reviewed regularly to assure that the problem has been eliminated. During standardize you must:

- Eliminate root causes
- Implement a permanent solution
- Confirm the effectiveness of the permanent solution
- Standardize the use of the new procedure

Kaizen Teams

Kaizen events that are conducted with cross-functional teams are powerful tools for process improvement. They allow the creativity of staff from across the organization to take a fresh objective look at where the organization stands. These teams often readily identify dozens of roadblocks and opportunities for improvement (OFIs). Teams working together inevitably generate ideas that do not arise in the normal daily operation.

> The best way to have a good idea is to have lots of ideas.
>
> **—Linus Pauling**

A startling statistic that brings home the difference between Lean organizations and traditional organizations is the concept of idea generation. Typical American companies generate only 0.5 ideas per worker per year.* Typical Japanese companies generate nine ideas per employee per year.† Toyota generates 70 improvement ideas per worker per year!‡

* Reference for "Small Business," *Encyclopedia of Business*, 2nd ed.
† Ibid.
‡ Bunji Tozawa, Norman Bodek, *The Idea Generator—Quick and Easy Kaizen*.

Possible Target Areas for Kaizen Teams

- Customer service—can be improved
- Quality—can be improved
- Costs—can be lowered
- Schedule—improve delivery and production time
- Cycle time, setup time—can be reduced
- Inventory—reduce the unnecessary stock
- Safety—reduce possible accidents
- People—improve workers' skills and knowledge
- Equipment—improve downtime and efficiency
- Environment—improve air quality, reduce odors
- Visual—use colors, clean up, find things easier
- Location—reduce unnecessary motion or facilitate necessary interaction, etc.

Preparing for Kaizen

Preparing for a Kaizen event is almost more important than the event itself. Many events have failed to achieve results anticipated due to poor preparation. This can lead both management and staff to lose faith in an incredibly valuable tool. The basics of preparing for Kaizen include:

- **Management selects target area:** This should not be too broad or too narrowly focused. The examples above give some great topics for Kaizen.
- **Set time:** Two to five-day Kaizen events are typical. These can be followed by 6 to 8 weeks of continuous improvement team meetings to complete solutions implementations before disbanding the team and re-forming to attack another topic.
- **Set scope or project boundaries:** The team needs to know what's included and what's off limits. This helps to keep focus and achieve better results.
- **Select Kaizen team members** (4 to 6 cross-functional): These should include 1 to 2 employees from the target area, suppliers and customers of the target area. For example, a Kaizen in manufacturing may want to include someone from order processing (supplier) and distribution (customer). Teams should also include at least two outsiders from different departments or functional areas.
- **Objective "eyes" assigned to project:** This is typically an external consultant, but can be an internal source if there is a highly

trained LSS practitioner on staff. One of the many advantages of using an external source is the wealth of project implementations he or she typically brings to your organization. Another is mentoring of team members during the event. Perhaps the most important role of an external consultant is the ability of the external person to expose sensitive weaknesses that employees may be reluctant to point out.

Team Member Roles in Kaizen

Each team member's role is to work synergistically with other team members to improve the target area. When conducted correctly, Kaizen becomes contagious. It is the constant doing and sharing of Kaizen results that stimulates others to do Kaizen! Team members should remember to:

- Participate
- Study the process in the target area
- Use creativity before capital
- Use Kaizen
- Share ideas
- Ask questions
- Experiment with changes

Overcoming Obstacles during Kaizen

The "we can't" syndrome is the common outcry of employees in traditional organizations. The list of reasons the authors have personally been given why Kaizen won't work in a particular organization is staggering. Once you begin your journey to becoming an LSS practitioner, you immediately realize how ridiculous all the following assertions are:

- We produce such a variety of products.
- Our products are custom; we can't standardize.
- The change is too rapid.
- We are not in a mass production situation.
- Our people are too busy.

Some reasons for not conducting process improvement reach a level of absurdity that is difficult to comprehend. Let us share a true story with you that one of the authors experienced. Several years ago we were conducting a process improvement program in a traditional organization with a management philosophy similar to the Theory X philosophy. The management team was tremendously cost driven in every aspect of their organization. One of our opportunities for improvement surfaced the need for additional employee training and cross-training. It was readily apparent to the entire team that employees at this organization were in dire need of training and cross-training. However, the very thought of spending money, time, energy, and effort on training and cross-training was out of the question for senior management. Management's response to the team was: "If we train them, they will leave the company."

From our perspective this was understandable since it was blatantly obvious that senior managers were living one of the traditional belief philosophies of "low pay, high turnover." So as to not be too confrontational, we posed a question to management. "Is there anything that you can think of that would be worse than trained employees leaving your company?" There was a puzzled look on some faces as they were struggling to comprehend the question; clearly, it hadn't dawned on them that anything could be worse. After a few moments of silence, my response was, "The only thing worse than trained employees leaving the company is *untrained* employees that stay with the company."

Needless to say, process improvement programs didn't get very far in that organization. Senior management held closely to their belief and completed nothing but the bare essentials of employee training. Even in the face of poor product quality and poor customer satisfaction they were unwilling to change their beliefs and prepare their staff to deliver what the customer wanted. This is a message to all managers in all organizations that "beliefs drive behavior." Work daily on developing your LSS belief system.

So, whenever you are faced with the employee response "we can't" change for any reason, use the following response: "I know we can't do this because …, but if we could, how would we …?"

People sometimes fear change. I have been asked on several occasions, What if the Kaizen doesn't work? The answer is simple: Do Kaizen again! Yeah, but what if it still doesn't work? If it still doesn't work, do Kaizen again! What if a problem develops from a Kaizen? Do another Kaizen until the problem goes away! You need to conduct Kaizen until you reach a point where you have faith and have lost your fear of taking improvement

10 Attributes of Kaizen

1. Kaizen involves everyone in the organization—owners, presidents, senior managers, department managers, team leaders, and supervisors!
2. Kaizen fosters process-oriented thinking (i.e., processes must be improved for results to improve).
3. Kaizen focuses human efforts on what we can do, not on what we can't do.
4. Kaizen defines a new role for management.
5. Kaizen is process-oriented rather than results-oriented.
6. Kaizen follows Standardize-Do-Check-Act (SDCA) for process maintenance activities and the Plan-Do-Check-Act (PDCA) cycle for improvement activities.
7. Kaizen puts quality first all the time.
8. Kaizen puts the customer first all the time, even before the boss.
9. Kaizen defines the "next process," either internal or external process, as the customer.
10. Kaizen is not a home run. It's a single!

FIGURE 5.3
Ten attributes of Kaizen.

actions. That is when you know that you are a change agent, a true LSS practitioner. (See Figure 5.3.)

KAIKAKU—TRANSFORMATION OF MIND

Similar to Kaizen, Kaikaku has been defined in many ways. Several definitions follow to give you a range of interpretations of the term *Kaikaku*.*

1. Kaikaku: Change + radical.
2. Kai: To take apart and make new. Kaku: Radically alter.
3. Kaikaku: Transformation of mind.
4. Kaikaku: Can also mean innovation, although the authors will discuss innovation later in this chapter as Kakushin.

How Do We Recognize Kaikaku (Transformation of Mind)?

Kaikaku is the result of successive Lean learning and Lean doing until Lean becomes a part of you. Looking back, I am not even sure when it first

* Norman Bodek, *"Kaikaku," The Power and Magic of Lean; A Study in Knowledge Transfer.*

occurred in me. One day I was training in an organization when it became clear to me that my mind was completely rewired with Lean beliefs and behaviors. Since that day I have been on a mission to help employees at all levels transition from a traditional belief system to a Lean belief system.

In Kaizen we talked about small incremental changes. Many of the tools presented in this handbook are used in Kaizen. Some, however, are tailored to Kaikaku in that they require a more complex set of activities, a more comprehensive set of Lean beliefs and behaviors, or simply put, a "transformation of mind" has taken place in order to properly deploy the Lean tool. This entire handbook is about trying to bring about this transformation of mind in all your employees.

The misconception of management in most Western companies is that Lean is just a set of tools. Management and employees learn a few Lean tools, conduct some Kaizen, and believe that their organization is Lean. Until the transformation of mind has occurred in employees at all levels, your organization cannot reap the true benefits of Lean.

It is only when an employee reaches a significant point in the Lean journey that he/she begins to recognize Kaikaku in himself or herself. Their own personal transformation of mind from traditional beliefs to Lean beliefs is occurring on a daily basis. Those in your organization still practicing traditional beliefs will neither recognize your efforts nor consider them to have any value. Do not let this deter you. You must continue on your journey and help them to see and live Lean beliefs and behaviors. Two examples of Lean tools that can reflect the presence of Kaikaku are cell design and facility layout.

Kaikaku in Cell Design

Let's consider the five steps of cell design and some of the changes that have to occur in order to successfully implement a cell. Many changes to traditional beliefs about materials, people deployment, equipment, and methods of production need to be considered and include:

- Grouping similar products
- Looking at customer demand and understanding how it changes
- Calculating customer demand
- Redeploying people in nontraditional roles (to meet customer demand, not fill a functional job description)
- Changing the number of people as needed

- Placing equipment to allow continuous materials flow during the value-add process
- Connecting the signals that initiate flow of materials to customer demand
- Using POUS for materials and equipment whenever possible
- Using quality at the source
- Creating visual work instructions

Individually each of these attribute changes may not appear too significant; however, collectively they represent a radical change or transformation of mind. The Kaikaku case summary of a composite U-cell in Figure 5.4 (discussed in detail in Chapter 4) demonstrates how this transformation of mind occurs collectively for a large number of employees working together in a cell.

Kaikaku in Facility Layouts

Over the years we've heard many managers touting that they were implementing a Lean layout. However, if we take a closer look at these facilities, we find that few Lean tools had been successfully incorporated into the process. A true Lean facility layout requires that Kaikaku is present. Even more so than a single cell, an entire Lean facility layout is comprised of hundreds of considerations. The most important aspect is to begin with customer demand and configure your facility to respond to what your customer wants. Some of the major considerations are:

- Assessing build to order and build to stock products
- Creating a mixed-model pull system
- Order processing
- Purchasing practices
- Inventory control
- All production activities
- Equipment layout
- Materials placement and POUS
- Employee deployment
- Distribution

One approach when considering a new facility layout is to complete a walk-through Lean layout. Figure 5.5 shows the results of a Lean facility

Kaikaku Case Study Summary: Composite U-Cell
Blood Dialysis Instrumentation Assembly

Situation: The company was producing approximately 45 blood dialysis units/week, operating two shifts in an enclosed electrostatic discharge (ESD) room. There was no room for expansion, and the company was restricted to assembly in the current assembly area. The company also had orders for 60 units/week. The challenge was to implement process improvements and achieve 60 units/week with the existing staff, assembly area, and test equipment. Previous attempts to increase production levels had resulted in significant quality problems and an actual reduction in output.

Kaizen Events—Continuous Improvement Teams: Our first actions were to assemble a cross-functional team and conduct Kaizen events in the dialysis instrument assembly room. The team worked together and separately to attack both the hydraulics and chassis assembly areas. During the Kaizen events, each team uncovered approximately 30 different opportunities for improvement.

Cell Design Activities

The Kaizen teams completed the five-step cell design process in both the hydraulics and chassis areas and linked them to produce a composite U-cell. This included Takt time, line balancing, and facility layout for 38 workstations, over 50 assembly and test employees (green dots), and the assembly of over 4,000 parts.

Economic impacts: Over a 60-day period these Kaizen activities produced a Kaikaku (transformation of mind) with the newly implemented cell. The resulting economic impacts were realized from the team effort.

Initial production rate was 45 units/week, and the final production was 60 units/week. This is a 33% increase in production using the same floor space, employees, and equipment. This translated to a significant revenue increase and a dramatic increase in profitability, since all fixed costs remained the same.

FIGURE 5.4A
Kaikaku case study summary.

layout project. Using only blue painter's tape and wood pallets, the team created a functional facility layout. The objective of the project was to produce a Lean layout design that allowed senior management to "walk through" the facility prior to any equipment or material delivery and physically see how the facility would operate in the new Lean mixed-model pull system. Locations for everything were identified on the facility floor. This included raw materials POUS, all equipment, materials motion through the facility, employee deployment, production personnel offices,

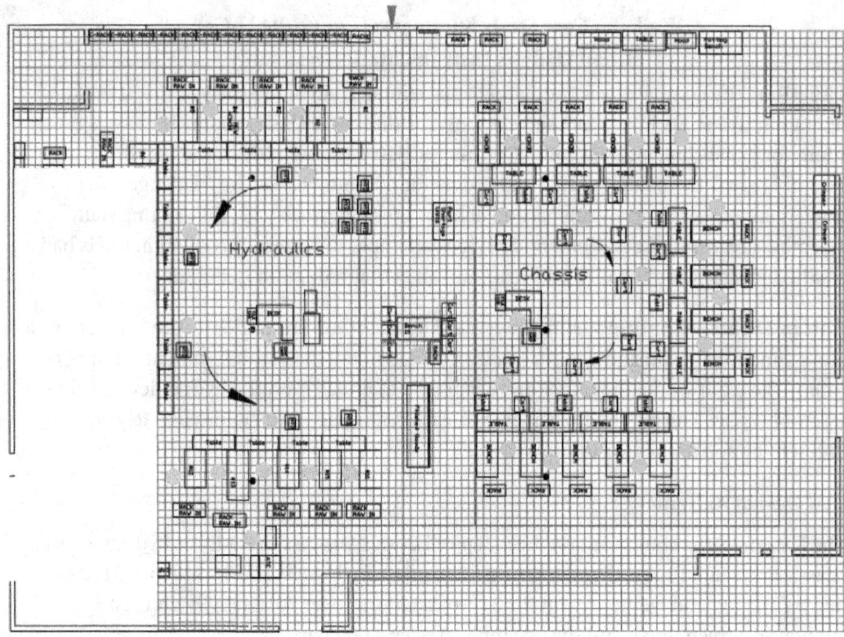

FIGURE 5.4B
Proposed hydraulics and chassis U-cell layout.

FIGURE 5.5
New facility layout—walk-through design process.

finished goods inventory, and ground, less than truckload and truckload distribution. From this simple layout we could show management how materials would be received, where POUS storage would be, how orders would be processed on the floor, how employees would be deployed, electrical and other utilities considerations, and how order fulfillment would occur for distribution.

Both of the above examples require a transformation of Mind to complete effectively. How we think about materials handling, employee deployment, equipment utilization, and adding value for the customer all must be radically changed from those of a traditional organization. They require the adoption of so many small changes that at this point the organization has undergone a radical change in beliefs and about how value is created for the customer. At this point the mental transformation has occurred and you are operating as a Lean organization.

KAKUSHIN (INNOVATION)

Virtually all process improvement requires some form of change that could be described as innovation, or at a minimum, creativity. In this section we look at innovation as a process in and of itself. As such, it can be defined, described, and managed like any other process. In the remainder of this section we present the 20-20 innovation process.

The 20-20 Innovation Process

Organizations are focusing on innovation as the necessary skill for revenue growth in that most corporations have pretty much exhausted the opportunities for restructuring and re-engineering. The new strategic mantra is revenue growth resulting from four primary strategies: (1) geographic expansion; (2) alliances, acquisitions, and mergers; (3) greater market penetration; and (4) product development and enhancement. And product development and enhancement ultimately depend upon product development as their foundation. Market penetration literally depends upon marketing innovation for its success, with product development and enhancement, along with cost advantages, contributing greatly. To

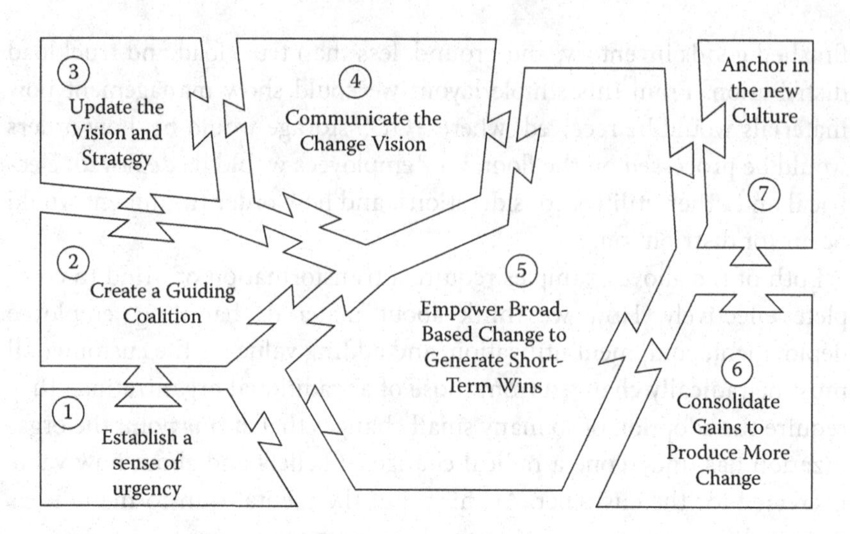

FIGURE 5.6
The 20-20 innovation process model.

obtain cost advantages, an organization also depends upon innovation, as process design and continuous improvement are process innovation activities. Finally, the related process of management innovation is an untapped resource that can help organizations improve the value that they add to the value chain of key organizational processes. (See Figure 5.6.)

To compete, organizations either attempt to differentiate themselves from their competition or attempt to achieve some relative low-cost position, and in both cases, innovation is the key. The journey from novice to expert in any field begins by understanding these essentials, practicing them, mastering them at one level, and then moving on toward the limits of your potential. At some point in the process, the best innovators rise above their profession in a multidisciplinary manner.

Each of the six essentials of the 20-20 innovation process represents a bundle of habits, skills, and knowledge that come together in innovation-driven problem-solving personalities, and each personality draws its strengths from a variety of specialties. A select number have been identified and mapped against the six essentials as a basic guide to the interdisciplinary skill set, as shown in Table 5.1.

These six personalities serve as an effective way to assess yourself and your organization, so that you can determine what your strengths and weaknesses are and how to assemble a complete innovation and problem-solving capability. This approach should be considered a guide, not a

TABLE 5.1

Six Essentials of the 20-20 Innovation Process

Six Essentials of Innovation	Essential Description	Related Innovation Personality	Personality Description (likely professional source of knowledge and skill)
1. Generate the mindset	Creating organizations and teams that are innovative, creative, and decisive	The innovator	Visionary, designer, entrepreneur, artist, counselor, spiritual leader, poet
2. Know the territory	Acquiring and absorbing strategic knowledge on a continuous basis	The discoverer	Historian, scientist, researcher, journalist, investigator, teacher, accountant
3. Build the relationships	Building loyalty and trust by exchanging and delivering value	The communicator	Politician, social worker, civil servant or leader, legislator, publicist, salesperson, agent
4. Manage the journeys	Picking the right projects to work on, for the right reasons, at the right time, with the right tools	The playmaker	Commander, executive, physician, consultant, judge, coach
5. Create the solutions	Designing the best end-to-end solutions that are complete and well supported	The creator	Builder, architect, engineer, inventor, investor, author, trader
6. Deliver the results	Implementing solutions that are effective for complex and competitive situations	The performer	Athlete, attorney, entertainer, musician, nurse, customer representative

magic formula, in that each individual and organization has a unique mixture of these skill sets that we use to feed into our "Knowledge Wizard." Some are intensively focused on one or two of these skill sets, and others have a broader blend—the greater the blend, the higher the probability of success. In other words, the better you get to know them, the better innovator and problem solver you will become. Great innovators know their strengths and weaknesses; they build teams to compensate for the weaknesses and create wholes that are synergistic in that they are greater than the sum of their individual parts.

At its basic fundamental level, all competition in some way relies upon innovation and creativity. There are six essential innovation and problem-solving skills that apply to any type of innovation or opportunity. The difference between the best and the worst innovators and problem solvers lies in how many of these skills they can marshal by themselves and with others, and how deeply the skills are understood, both individually and collectively. Poor innovators understand them incompletely, and therefore cannot develop an effective and complete capability. Great innovators and problem solvers know them well enough to pull together and manage all six, or exhibit one in great depth, as part of a team. The following outlines in more detail the six essentials of the innovation process:

- Essential 1: Generate the innovation mindset. Identify opportunities and potent ideas and attitudes for success and create a strategy for various alternative points of view, while defining the program expectations and metrics. This will improve your organization's effectiveness in moving creatively through the process and related problem-solving effort.

  The innovator's mindset and central idea is that the best innovation comes from taking alternative points of view. They have a unique and flexible way of seeing things, and thus the ability to both develop original viewpoints and incorporate those of others around them, which are the roots of both creativity and objectivity. The innovator's journey is the sum of many different points of view. Tapping into these alternate perspectives of both ourselves and others, then testing them and choosing the most valuable, requires sustaining a flexible viewpoint through further inquiry. Our innovation quotient questionnaire helps to integrate these perspectives in new ways, which constitute the powerful mindset of the Innovator.

The components of this first phase of *developing the mindset* are: (1) developing potent ideas and attitudes using a project scope sheet, (2) turning problems into opportunities using the IQ assessment, (3) committing to the challenge, (4) creating strength out of weakness and vulnerability, (5) not settling for half-measures, and (6) transcending the limits.

- Essential 2: Know the territory. This stage concentrates on moving from innovation to insight by asking the right questions and obtaining sound and timely information and data. It also involves the collection and analysis of the system's past (using the Knowledge Wizard).* Better knowledge helps organizations define the opportunities and problems more effectively, choose the better pathways, and clearly identify what's at stake. The discoverer's skill set helps to answer the following:
  - At this stage, we simply don't know enough to define the opportunity or problem well, so why don't we step back and answer some key unknowns before we proceed any further?
  - Although we may have a seemingly well-structured proposal, what have we really learned from our past mistakes in that area?
  - How can we avoid an unworkable journey or one that creates unnecessary risk?

  The discovery process and knowledge of the territory brings insight and understanding, which often reveal the most likely problems and opportunities in the context of higher relief. By performing more investigation, the implications become more apparent as a foundation for action. Without this essential, the innovator may choose an unworkable journey or one that creates unnecessary risks. We need to remember that knowledge is the key asset of our day and age. Knowing the territory, the work of the discoverer using the Knowledge Wizard (KW) deals with acquiring the right information about the critical elements of the environment that we are solving problems in.
- Essential 3: Build the relationships. This stage involves the basics of how to move from isight to relationship building by cultivating quality communications and interaction, so that we can create an ever-expanding circle based upon service, identity, and loyalty. It gives your organization the support and human context that is needed

* Knowledge Wizard® software, Indention International, Inc.

to effectively create and foster innovation and implement change. Good communicators probe the functions and contradictions to ask whether ideas and opportunities are really worth it:

- What will our customers and stakeholders think about this? Will our efforts add value and build loyalty?
- Are these opportunities going to threaten some of our best business relationships in the process, and are they really worth it?
- How will the people be affected by the opportunities?

Through the mastery of relationship building, good communicators connect potential journeys to their actual implications and contradictions for real people, in order to help determine whether the innovation or problem solving is worthwhile, for whom, and why. Then they generate a core group that will tackle the journey, along with a virtual network that will support the effort. The support network is key, as this is not a one-person brainstorm. Without this essential, there may be no compelling reason to innovate and solve the problem at all. Or worse yet, there may be a reason, but no one convinced enough to take part.

The components of this third phase of building the relationships are: (1) cultivate quality communication and interaction, (2) create meaningful communication, (3) move from communication to give-and-take relationships, (4) advance from relationships to a core team, (5) the core team leads to a network, and (6) the network advances to a living community.

- Essential 4: Manage the journeys. This stage focuses on moving from building the community to giving that core community a sense of direction and clear priorities by choosing destinations and strategies. The focus is on alignment and commitment to projects and chartering, along with defining how to pursue the selected directions and revealing evolutionary resources, i.e., the technologies, materials, products, processes, skills, knowledge, etc., that could be utilized in the development using the innovation knowledge base (which could include patterns describing the evolution of technological and social systems). Fostering the understanding of the stages of the innovation journey helps the organization set goals, define success, and develop effective plans. Playmakers are those who ask about alternatives and eliminations with the following questions:
  - While the ideas may be interesting, are we reaching high enough here?
  - What is our real aim and purpose?

- The technology or tools may be great, but are we solving the right problem?

  The components of this fourth phase of managing *the journeys* are: (1) choose destinations and set directions, (2) solve the right problems at the right time by moving from disorientation to selection, (3) choose where you want to go, what to eliminate, and the paths to get there, (4) define success, (5) plan for the unexpected (chance) with alternative operators, and (6) lead the way.
- Essential 5: Create the solutions. This stage involves moving from leadership to power by designing, building, and maintaining optimal solutions. It helps organizations to bring the best people and technology together, along with the necessary tools and software, in order to generate complete and flexible solutions for the innovations and problems that you are trying to solve. A good creator uses modeling to ask the following:
  - Even if we have come up with a great idea, do we have the right people and the capability skill sets behind it?
  - Do we really understand the purpose of our designs?
  - Even if it is the right innovation area or problem to work on, are we willing to invest in the needed technology that it may take to deliver it?
  - Do we really understand if the economics are viable and doable?

  A key to this stage is revealing and solving problems using modeling, resolving contradictions, overcoming limitations, etc., that can hinder the achievement of set goals using various instruments for problem solving, including the most recent ideation DMADV methodology. It could also involve some objectives: deployment, customer needs, customer experience, measurements, and management reviews. The components of this fifth phase of creating the solutions are: (1) design, build, and maintain optimal solutions using DMADV, (2) find the right people for the right work, (3) get the right tools to do the job in order to move from team to capability, (4) learn how to conserve scarce resources, (5) get the right information to the right people, (6) design solutions that evolve, and (7) shift the balance of power to create the dominant solution.
- Essential 6: Deliver the results. This stage concentrates on moving from power to sustainable advantage through the stages of intuitive and disciplined implementation, which allows the organization to continually exceed expectations, and conquer complexity, scale,

and friction. The key is performance, which is accomplished with simplicity, discipline, and a competitive advantage. This stage is all about performance, which uses the Knowledge Wizard as the following questions illustrate:

- We may have a brilliant strategy, but can be execute and implement it?
- Are the timeframes and resources unrealistic?
- What are the biggest risks, and how can we manage them up front?

The components of this sixth phase of delivering the results are: (1) practice intuitive, disciplined execution using the Knowledge Wizard, (2) simplify and specify, (3) set the pace and pilot the course, (4) make the right decisions at the right time, (5) optimize risk and return, (6) learn to fail small and early on to win big later on, and (7) maintain your leading cutting edge.

SUMMARY

One of the few things that we know for certain is that change is constant in organizations. This change can come in many forms and can include positive or negative attributes for individuals, departments, and the company at large. This chapter was structured to give you approaches that drive structured positive change and provides a fundamental foundation for successful LSS process improvement projects.

First, we discussed the three primary obstacles or resistance to change factors that can inhibit effective LSS process improvements. The most common resistance to change factors are fear of the unknown, beliefs, and measurements systems. The LSS practitioner should be prepared to recognize and understand these obstacles and use LSS concepts and tools to overcome them.

The essence of this chapter is the concepts of Kaizen, Kaikaku, and Kakushin, which are the three fundamental tools or conduits to deploy change in your organization. With the application of Kaizen in each of its common forms—Kaizen and you, Kaizen for process troubleshooting, and Kaizen teams—the LSS practitioner can become a driver of positive change with virtually all employee engagements. This can be assisting employees with small process improvement projects that each individual can complete on his/her own. The most widely known and used form is the Kaizen event or Kaizen continuous improvement team. Kaizen teams typically attack projects that are larger in scope than any one person can handle. These projects can be 2 to 4 days in length or can include team activities that can last several weeks.

The second fundamental concept and milestone is the transformation of mind concept, which reflects the LSS practitioners' advanced knowledge as exhibited by the use of multiple LSS tools in unison. The ability to think and act across the entire value stream has become a part of your organization's daily activities.

The third fundamental concept in this chapter is the systematic and standardized approach to change by innovation. The 20-20 innovation process demonstrates the six essentials of innovation that can be applied to LSS organizational improvements.

When employees begin to deploy the change management concepts presented in this chapter, a powerful positive set of changes emerge and the organization will begin systematically eliminating waste, improving processes, and expanding LSS knowledge for employees. These are the building blocks to improved product quality, increased employee productivity, and enhanced company profitability.

Section 3

SSBB Overview

So far in this book we have concentrated on the Lean side of the LSSBB methodology. This section provides a short overview of the SSBB side of the LSSBB methodology. The term *Six Sigma* was originated by Bill Smith of Motorola when he convinced William J. Weisz, then COO of Motorola, that the standard of performance for all business activities should be 3.4 defects per million opportunities. This was statistically equivalent to plus-or-minus six sigma when long-term drift was considered. Previously, most process control standards were set at plus or minus three sigma, which is equivalent to 66.8 defects per million opportunities. To meet this new, more rigid requirement, Motorola implemented an extensive program, training people on how to solve problems. This program was called Six Sigma; it was the start of a worldwide movement to produce products and services that were near perfection.

The Six Sigma methodology is based upon two concepts:

- Concept 1: The more variation is reduced around the midpoint of a specification, the less chance there is to create defects or errors. (See Figure S3.1.)

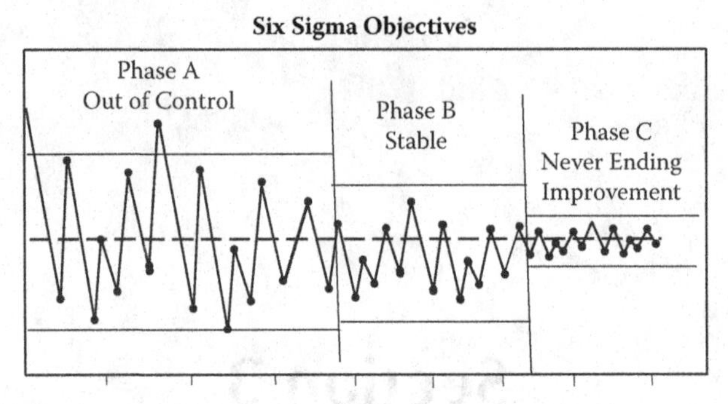

FIGURE S3.1
Reduced variation around the midpoint.

This is a concept that was well accepted in the 1940s and popularized in the 1970s by Taguchi. As early as the 1920s, Shewhart set the standard of performance in his control charts threat plus or minus three sigma. Motorola felt that Phil Crosby's zero defects performance criterion was unattainable (and they were right), so they set a specification to a target variation of performance at six sigma. This is known as a process capability of 2.0 (Cp). Motorola felt that over time the process would drift plus-or-minus 1.5 sigma, so the actual long-term process capability target was set for 1.5 (Cpk), or spec to an actual variation target of 4.5 sigma. (See Figure S3.2.)

The change in target variation from three sigma to six sigma is where Six Sigma got its name. At the six sigma level, the long-term

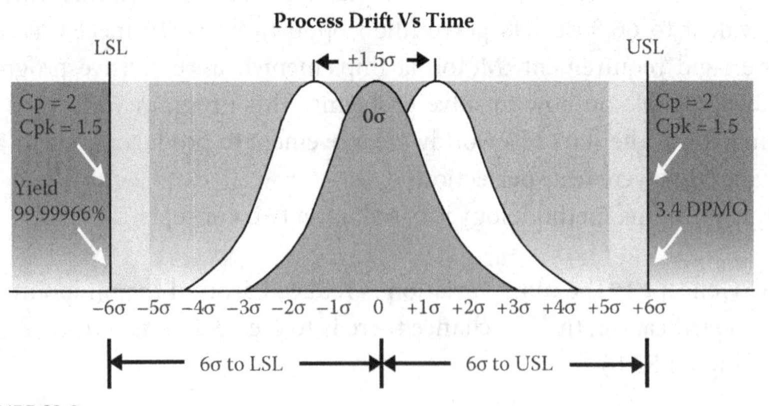

FIGURE S3.2
Process drift versus time.

process capability results in 3.4 errors per million opportunities. (Note: This is not 3.4 errors per million items/transactions.) This is near to perfection without requiring perfection. It is a goal to seek, not a performance standard that has to be met.

Another important point is that it is related to opportunities, not to the total output. For example, a car could have a million opportunities for failure due to its complexity. In this case, at the six sigma level there would be an average of 3.4 defects per car produced. This is the reason that every car that rolls off the Toyota assembly line still has its breaks checked, to be sure they work.

- Concept 2: The lower the costs, cycle time, and error level, the higher the profits, market share, and customer satisfaction. This approach focuses on streamlining the process, driving out waste and no-value-added activities. It was based upon the business process improvement concepts that were made popular in the early 1980s. (See Figure S3.3.)

  In this concept, reducing the mean, not reducing variation, is the objective. For example, it is much better to be able to manufacture a part for $100 ± $10 than to manufacture the same part for $500 ± $1. In fact, in some cases variation is welcomed as long as it is on the lower side of the midpoint. Often, investigating these lower levels of variation can lead to additional reductions in cost, cycle time, and error rate.

Another part of the Six Sigma concept is called *Lean*. This methodology was developed by Henry Ford Sr. in the 1910s where in-process stock was almost eliminated, continuous flow lines were designed, and

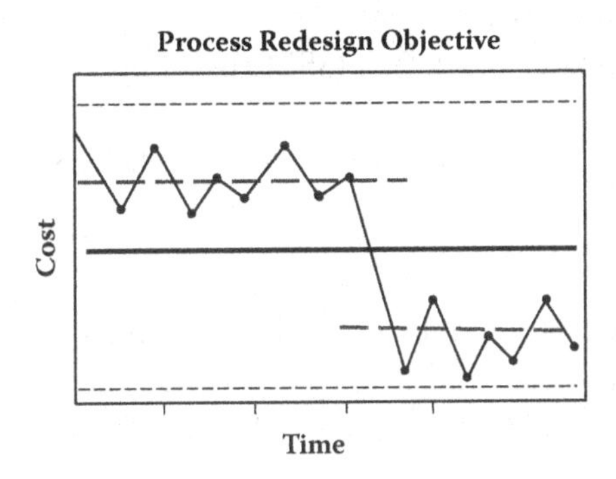

FIGURE S3.3
Reduced error levels.

manufacturing equipment was error proofed. These Lean concepts were further refined by Toyota to develop the most effective auto assembly process in the world. (The Lean concepts were discussed in detail in Section 2 of this book.) Using these two primary concepts as an objective, Motorola needed to define ways to implement them. It turned to Shewhart's Plan-Do-Check-Act (PDCA) cycle, which was developed in the 1930s and was extensively used in the Total Quality Management (TQM) methodology that was popular in the 1980s and 1990s. In this case, Dr. Mikel Harry from Motorola modified it to RDMAICSI (Recognize-Define-Measure-Analyze-Improve-Control-Standardize-Integrate). Later it was further defined to Define, Measure, Analyze, Improve, and Control (DMAIC). Bill Smith, who became the leader of Motorola's Six Sigma efforts, felt that the complexity of the problems Motorola was facing required much more sophisticated techniques to solve them. Unfortunately, many of the professional quality engineers were gone from Motorola, and along with them were the needed statistical analysis tools that were part of the quality engineering profession. As a result, Motorola's Six Sigma program was heavily focused on statistical training and analysis. This resulted in a cadre of special problem-solving people who became proficient in statistical tools like design of experiments. They were called Six Sigma Black Belts (SSBBs).

The Six Sigma methodology has created a special term for people within the organization who have been trained in how to help the organization solve its most difficult problems or take advantage of new improvement opportunities. The key levels are:

- Six Sigma Yellow Belt (SSYB)—Level 1
- Six Sigma Green Belt (SSGB)—Level 2
- Six Sigma Black Belt (SSBB)—Level 3
- Six Sigma Master Black Belt (SSMBB)—Level 4

Six Sigma Black Belts are the work horses of the Six Sigma System.

—H. James Harrington

Six Sigma Black Belts (SSBBs) are highly skilled individuals who are effective problem solvers and who have a good understanding of the most frequently used statistical tools required to support the Six Sigma system. Their responsibilities are to lead Six Sigma problem-solving teams and to define and develop the right people to coordinate and lead the simple Six

Sigma projects (Green Belts). Candidates for SSBBs should be experienced professionals who are already highly respected throughout the organization. They should have experience as a change agent and be very creative. SSBBs should generate a minimum of US$1 million in savings per year as a result of their direct activities. SSBBs are not coaches; they are specialists who solve problems and support the SSGBs and SSYBs. They are used as Six Sigma problem-solving team managers/leaders of complex and important projects. The position of SSBB is a full-time job; they are assigned to train, lead, and support the Six Sigma problem-solving teams. They serve as internal consultants and instructors. They normally will work with two to four problem-solving teams at a time. The average SSBB will complete a minimum of eight projects per year, which are led by the SSBB himself/herself or by the SSGBs who they are supporting. The SSBB assignment usually lasts for 2 years. It is recommended that the organization have one SSBB for every 100 employees. A typical SSBB spends his/her time as follows:

- 35%—running projects that he/she is assigned to lead
- 20%—helping SSGBs that are assigned to lead projects
- 20%—teaching either formally or informally
- 15%—doing analytical work
- 10%—defining additional projects

The SSBB must be skilled in six areas:

- Project management
- Leadership
- Analytical thinking
- Adult learning
- Organizational change management
- Statistical analysis

Based upon our experience, most of the SSBB training has been directed at analytical skills. Even the selection of the SSBB is based upon his or her analytical interests. This is wrong. Traits to look for in selecting a SSBB are:

- Trusted leader
- Self-starter
- Good listener

- Excellent communicator
- Politically savvy
- Has a detailed knowledge of the business
- Highly respected
- Understands processes
- Customer focused
- Passionate
- Excellent planner
- Holds to schedules
- Motivating
- Gets projects done on schedule and at cost
- Understands the organization's strategy
- Excellent negotiation skills
- Embraces change

SSBBs should be specialists, not just coaches. It is important to build a cadre of highly skilled SSBBs. However, they must not be placed in charge of management of the improvement process. SSBBs are sometimes responsible for managing individual projects, but not directing the overall improvement process—that should be the job of management. Don't just keep management engaged in the process—keep them in charge of the process.

SSBBs are required to have mastered the two Six Sigma improvement methodologies. They are:

1. Design, Measure, Analyze, Improve, and Control (DMAIC)
2. Design, Measure, Analyze, Design, and Verify (DMADV)

In this book we will only be providing you with an overview of the Six Sigma methodology to understand the basic foundation of this methodology. We suggest you read the book *The Six Sigma Green Belt Handbook* (Harrington, Gupta, and Voehl; Paton Press, Chico, CA, 2008), which defines in detail the basic tools used in these two methodologies. (See Sections 4 and 5 of this book for a detailed review of the advanced statistical and nonstatistical SSBB tools used in this methodology.)

6

On Integrating LSS and DMAIC with DMADV

IN A NUTSHELL

Lean and Six Sigma DMAIC and DMADV are two fundamentally different but compatible approaches to process improvement. They are both very effective methodologies to be applied when the circumstances call for either one. Lean's objective is to identify and eliminate waste in all processes, while DMAIC/DMADVs' objectives are to identify and eliminate variation in a process. While leaders do not have to work the mechanics of the LSS tools, they need considerable savvy in the reading and interpretation of each tool's outputs. Those who prepare well in this area are able to expect teams to use certain tools to answer certain kinds of questions (thereby driving the proper use of tools). Their understanding helps them to challenge some team findings, to coach teams that may get stuck or off track, and to anticipate the tools and data that will be useful in an upcoming review. Integrating Lean with DMAIC/DMADV is the subject of this chapter and offers the best of both worlds for process improvement practitioners that understand the strengths of each and integrate them seamlessly where applicable.

OVERVIEW

The identifying and eliminating waste concepts of Lean and the identifying and eliminating variation concepts of DMAIC/DMADV have been difficult for the process improvement community to merge together in a seamless fashion, if at all. In fact, even today there remains poor understanding of how to make both of these excellent tools work in unison. Lean and DMAIC/DMADV show the individuality of each approach, and the complementary nature of these two unique and specific approaches together makes a powerful process improvement philosophy come alive. The goals of Lean can and do use some quantification; however, the Lean philosophy is predominantly qualitative in nature, and this overarching qualitative approach puts a few core beliefs above all activities. In fact, all activities should be traceable back to one of these qualitative business drivers. These are all activities driven by customer demand, establishing continuous flow of products or services to meet these customer needs, and driving waste out of the process until the customer receives perfection as a result of your efforts. The goals of DMAIC/DMADV are rooted in quantitative analysis and the scientific method. In their book *Science for All Americans* (Oxford Press, New York, 1990), Rutherford and Ahlgren note: "Scientific habits of mind can help people in every walk of life to deal with problems that often involve evidence, quantitative considerations, logical arguments, and uncertainty … involving four key values: curiosity, openness to new ideas, skepticism, and critical thinking. Curiosity means being filled with questions, seeking answers, and verifying how good the answers are. Openness means being discovery-oriented, even if the ideas are at odds to what is currently believed. Skepticism means accepting new ideas only when they are borne out by the evidence and logically consistent. Critical thinking means not being swayed by weak arguments. Collectively these 4 key values represent the foundation for Scientific Thinking.

Points to Remember

Some of the difficulties that arise in applying the scientific thinking principles are:

- Overcoming resistance to change even when an innovative change is suggested, as it is difficult to get people to try the change and adapt themselves to the new situation
- Satisfying a diversity of viewpoints, as different team members may have varying viewpoints as to what constitutes an improvement
- Thinking that any change using the DMAIC/DMADV process would be an improvement in its own right
- Taking the time to meet the objectives of the problem-solving process once the change is agreed to
- Recognizing when a change is an improvement through proper testing and follow-through

While the above difficulties are real, they can be overcome by using the DMAIC/DMADV problem-solving process outlined in this application and by helping people overcome the roadblocks, which is the true focus of the science and art of this approach. Because all products, services, and outcomes result from a complex system of interaction of people, equipment, and processes, it is crucial to understand the properties of such systems. Appreciation of the DMADV design and verification process helps us to understand the interdependencies and interrelationships among all of the components of a system, and thereby increases the accuracy of prediction and the impact of recommended changes throughout the system. However, a person can use the methods described in this application without knowing the theory behind them, just as a person can learn to drive a car without knowing how it moves. Dr. Edwards Deming once said that one need not be an expert in any part of scientific thinking in order to understand and apply it.

GOALS OF LEAN DMADV

The goals of Lean DMADV have their roots in Lean Manufacturing, a comprehensive term referring to manufacturing methodologies based on maximizing value and minimizing waste in the manufacturing process. Lean Manufacturing evolved in Japan from its beginning in Ford Motor when Toyota built on Henry Ford's concepts for the elimination of waste and just-in-time stocking to create the Toyota Production System (TPS) in Japan. Many of the most recognizable phrases, including Kaizen and Kanban, are Japanese terms that have become standard in Lean Manufacturing and in the past 10 years or so have rapidly spread to the service, government, military, and not-for-profit sectors.

Lean Design

In today's globally competitive environment, speed is the currency. Global competition can respond to a market need virtually overnight, and in various and sundry product categories, the ultimate market winners and losers are being decided in a matter of weeks rather than years. To survive in this maelstrom, design teams need to be fast, efficient, and highly effective. True excellence in time-to-market, however, requires cooperation among several functional areas (at a minimum, marketing, design engineering, and manufacturing), along with something that is often in short supply in industry today: discipline. Lean design and development is a process and, as such, can only work if the process is actually used. During the past 20 years, we have worked with countless firms that have a well-defined product development process—on paper. Somewhere on a dusty cobwebbed shelf there is a thick, formal document that describes how the development cycle should proceed. However, this formal process is not consistently applied, critical activities are often waived, and in some cases, the process is abandoned entirely in the interest of "getting the product out the door." Why this happens is no mystery. Most organizations have adopted a canned development process that either: (1) worked somewhere else and was borrowed, (2) has been bestowed with the title of "best practice" but as history has proven, no practice stays as a best practice for long, or (3) was implemented by an outside consultant that initially made a great sales pitch and many promises, but failed to recognize the unique nature of the firm's industry, culture, and customers. The result is often an overblown, cobbled-together process that is so cumbersome and restrictive that it just begs to be circumvented.

The solution is a set of Lean design methods that can take you from slow and steady to quick and agile. Each method addresses a different aspect of the product development process: the harvesting of initial customer inputs, the planning of a development project, resource allocation and prioritization, time and workflow management, and several practical techniques for improved organization, communication, and execution. All methods are intuitive, team-friendly, and designed for flexibility. Furthermore, all of the best-known Lean product design methods are intended to guide you toward what we consider to be a waste-free ideal.

So where do you start? Naturally, your highest priorities for improvement will greatly depend on the nature of your specific market situation, but in general, cost reduction (the dimension we are calling Lean 3P) is the most logical starting point. Why? Depending on your business environment, it might be that slashing time-to-market or driving toward higher levels of innovation will give you greater overall benefit. However, reducing manufacturing cost is the fastest and surest way to achieve a measurable increase in profits. Speeding up the development process often requires disruptive changes in how a firm operates, and those changes may impact virtually everyone in the company. Moreover, the benefits won't be felt for months or years, depending on your typical development cycle time. Cost reduction, on the other hand, can be applied to both new product ideas and existing successful products, requires minimal organizational change, and can yield immediate bottom-line results. Therefore, slashing costs has always been considered a great place to begin your journey toward Lean product design excellence.

There are numerous opportunities to slash manufacturing cost during the Lean design cycle, including:

- Reduce direct material cost: Common parts, common raw materials, parts-count reduction, design simplification, reduction of scrap and quality defects, elimination of batch processes, etc.
- Reduce direct labor cost: Design simplification, design for Lean manufacture and assembly, parts-count reduction, matching product tolerances to process capabilities, standardizing processes, etc.
- Reduce operational overhead: Minimize impact on factory layout, capture cross-product-line synergies (e.g., a modular design/mass customization strategy), improve utilization of shared capital equipment, etc.

- Minimize nonrecurring design: Cost platform design strategies, parts standardization, Lean QFD/voice of the customer, Six Sigma methods, design of experiments, value engineering, production preparation (3P) process, etc.
- Minimize product-specific capital investment: Production preparation (3P) process, matching product tolerances to process capabilities, value engineering/design simplification, design for one-piece flow, standardization of parts, etc.

To positively impact these five critical factors in product cost, we've outlined Lean design tools that address all aspects of cost reduction, from capturing early voice-of-the-customer inputs to ensuring a smooth and successful transition to a Lean Manufacturing environment. The Lean design approach offers tremendous flexibility, allowing firms to easily create their own customized cost reduction strategy. Lean design helps its adopters design for profit at the source. A product's design determines most of a company's costs and rewards.

Focusing on the design, Lean design helps its adopters realize tremendous increases in profits and market share. Lean design calculates life cycle costs, including the costs of quality, and eventually calculates a sigma number. Design trade-offs can be quickly analyzed to see the effects on total accounted costs. The Lean design approach is becoming a proven methodology that ultimately builds business cases that guide their SSBBs and sponsors to the areas that show the highest returns on investment.

Points to Remember

At the heart of Lean design methods is the determination of value. Value is defined as an item or feature for which a customer is willing to pay. All other aspects of the manufacturing or service delivery process are deemed waste. Lean design is used as a tool to focus resources and energies on designing and producing the value-added features while identifying and eliminating no-value-added activities at the beginning, not years after the prototype is in operation. For the purposes of this chapter, a Lean design is an approach toward a determination of a desired future an organization wishes to achieve. It describes what the organization is trying to accomplish and bring to market.

GOALS OF DMAIC/DMADV

The goals for both DMAIC and DMADV are based upon systematic quantitative approaches to process improvement and process design, respectively. The basic structure and goals of each of these approaches is focused on identification and elimination of variation in both process design and process improvement.

The following is a clarification of the DMAIC process:

- Define: Select an appropriate project and define the problem, especially in terms of customer-critical demands.
- Measure: Assemble measurable data about process performance and develop a quantitative problem statement.
- Analyze: Analyze the causes of the problem and verify suspected root cause(s).
- Improve: Identify actions to reduce defects and variation caused by root cause(s) and implement selected actions, while evaluating the measurable improvement (if not evident, return to step 1, Define).
- Control: Control the process to ensure continued, improved performance and determine if improvements can be transferred elsewhere. Identify lessons learned and next steps.

Overview of How DMAIC Works

The tools of process improvement are most often applied within a simple performance improvement model known as Define, Measure, Analyze, Improve, and Control (DMAIC). DMAIC is summarized in Table 6.1. DMAIC is used when a project's goal can be accomplished by improving an existing product, process, or service.

DMAIC is such an integral part of both the Lean and Six Sigma approaches that it has been used by the authors to organize the material for a major part of our previously published books, *Six Sigma Green Belt Handbook* and *Six Sigma Yellow Belt Handbook*. It provides a useful framework for conducting Six Sigma projects and is often used to create a "gated process" for project control.

TABLE 6.1

Defining How DMAIC Works

D	*Define* the goals of the improvement activity. The most important goals are obtained from customers. At the top level the goals will be the strategic objectives of the organization, such as greater customer loyalty, a higher return on investment (ROI) or increased market share, or greater employee satisfaction. At the operation's level, a goal might be to increase the throughput of a production department. At the project level, goals might be to reduce the defect level and increase throughput for a particular process. Obtain goals from direct communication with customers, shareholders, and employees.
M	*Measure* the existing system. Establish valid and reliable metrics to help monitor progress toward the goal(s) defined at the previous step. Begin by determining the current baseline. Use exploratory and descriptive data analysis to help you understand the data.
A	*Analyze* the system to identify ways to eliminate the gap between the current performance of the system or process and the desired goal. Use statistical tools to guide the analysis.
I	*Improve* the system. Be creative in finding new ways to do things better, cheaper, or faster. Use project management and other planning and management tools to implement the new approach. Use statistical methods to validate the improvement.
C	*Control* the new system. Institutionalize the improved system by modifying compensation and incentive systems, policies, procedures, MRP, budgets, operating instructions, and other management systems. You may wish to utilize standardization such as ISO 9000 to assure that documentation is correct. Use statistical tools to monitor stability of the new systems.

Overview of How DMADV Works

The following is a clarification of the DMADV process:

- Define: Define design goals that are consistent with customer demands.
- Measure: Identify and measure product characteristics that are critical to quality (CTQ).
- Analyze: Analyze to develop and design alternatives, create a high-level design, and evaluate design capability to select the best design.
- Design: Complete design details, optimize the design, and plan for design verification.
- Verify: Verify the design, set up pilot runs, implement the production process, and hand it over to the process owners.

One strategic objective of any organization is the continual improvement of its processes in order to gain competitive advantage, enhance its

TABLE 6.2

Defining How DMADV Works

D	*Define* the goals of the design activity. What is being designed? Why? Use QFD or the analytic hierarchical process to assure that the goals are consistent with customer demands and enterprise strategy.
M	*Measure.* Determine critical to stakeholder metrics. Translate customer requirements into project goals.
A	*Analyze* the options available for meeting the goals. Determine the performance of similar best-in-class designs.
D	*Design* the new product, service, or process. Use predictive models, simulation, prototypes, pilot runs, etc., to validate the design concept's effectiveness in meeting goals.
V	*Verify* the design's effectiveness in the real world.

performance, and benefit interested parties, such as customers, employees, and shareholders. In many situations, however, improving a process is not a sound business option. Rather, a complete process redesign is required. DMADV is the Six Sigma methodology that focuses on process design/redesign. In Define, the project purpose and scope are established. In Measure, voice-of-the-customer data are translated into critical to quality characteristics (i.e., design measurements) that the design must meet. The project team then generates innovative design concepts, evaluates, and selects the best concept for the design (Analyze). High-level designs are then developed and tested (Design). Verification against design requirements and validation against intended use are followed by transitioning the new design to process owners for rollout, implementation, and control, completing the DMADV methodology. (See Table 6.2.) Another approach, used when the goal is the development of a new or radically redesigned product, process, or service, is DMADV (Define, Measure, Analyze, Design, and Verify). DMADV is part of the Design for Lean Six Sigma (DFSS) toolkit.

COMPARING DMAIC AND DMADV

On one hand, DMAIC is a process improvement tool that is used to modify a process that already exists and does not always provide the performance that is desired, and provides a foundation for a systematic and structured examination of any process. On the other hand, DMADV is a process definition and creation tool that outlines and uses a systematic approach to define, create, and execute a new process for a situation where no process currently exists. Although DMAIC and DMADV have some similar characteristics, DMADV holds the upper hand where a completely clean starting point to design is needed, instead of a continuous improvement building upon the existing process/system. DMADV promotes creative and innovative thinking to create the best process possible from the customer standpoint. It allows process designers the freedom to brainstorm with the approach "If I had a genie who gave me three wishes and a magic wand that could create the perfect process, then what would it look like?"

Here is an important point to remember: Use DMAIC to improve processes and use DMADV to design new ones.

No matter how you approach deploying improvement teams in your organization, they will all need to know what is expected of them. That is where having a standard improvement model, such as DMAIC, is extremely helpful. It provides teams with a roadmap. DMAIC is a structured, disciplined, rigorous approach to process improvement consisting of the five phases mentioned, where each phase is linked logically to the previous phase as well as to the next phase.

INTEGRATING LEAN WITH DMAIC/DMADV

The worlds of Lean and DMAIC/DMADV (Six Sigma) have been seemingly at odds for many years. This is primarily due to the difficulty of practitioners grappling with the two fundamentally different concepts—identifying and eliminating waste versus identifying and eliminating variation. As different approaches to process improvement, when applied in the proper situation, Lean and DMAIC/DMADV can integrate to form a more powerful tool than either can be standing alone.

Virtually all Lean concepts integrate well with DMAIC and DMADV. Our purpose in this section of the handbook is to focus on special considerations for using the Lean concepts integrated with the DMAIC/DMADV process in any environment, including both methods and tools that are particularly helpful as well as hints on how to model the people side of each phase.

Lean thinking supports two basic disciplines for speeding up the knowledge creation process—short, frequent learning cycles and delayed commitment. Short, frequent learning cycles are the antithesis of thorough front-end planning, but they are the best approach to processes that have the word *development* in their title. In product development, for example, the typical process is to define requirements, choose a solution, refine the solution, and implement the solution. Although this process is common, it is not the best way to generate knowledge. Conversely, the delayed commitment approach of Toyota is much faster and delivers products of superior quality that consistently outsell the competition. Toyota builds sets of possibilities to satisfy customer needs, and then, through a series of combining and narrowing, the new product emerges. The combining and narrowing process is paced by milestones that define stages of the narrowing process. Milestones are always met, despite the fact that there are no task breakouts or tracking. Decisions are delayed as long as possible, so that they can be based on the maximum amount of information.

However, Lean thinking sometimes has a big blind spot for the most powerful aspect of Six Sigma—its ability to connect business leaders and key project teams in a potent two-way, fact-based dialogue. To see and understand that an exploration of Six Sigma roadmap architecture is needed, Lean tools can be dovetailed nicely into virtually any DMAIC or DMADV project, regardless of the size and scope involved. They can be used to accentuate DMAIC or DMADV concepts and tools, or as some basic stand-alone techniques to produce a vital component of the DMAIC/DMADV project. Table 6.3 illustrates how and where Lean tools can be integrated with DMAIC/DMADV Project Framework.

As demonstrated in Table 6.3, the application of Lean tools and techniques fits neatly into the structured approach of Six Sigma. Whether you are conducting a DMAIC or DMADV project, Lean can play a significant role in either process design or process improvement. Both concepts fit nicely in the tollgate format, and both align with a structured approach, and have at their core the use of root cause analysis (RCA) as a means for driving effective and permanent process improvement.

Root Cause Analysis and Lean

The common crossover point between Lean and DMAIC/DMDV is in the root cause analysis (RCA). The primary difference between Lean and DMAIC/DMADV is that Lean projects can use both qualitative and quantitative RCA analysis, such as the five whys, cause-and-effect diagrams, and failure mode and effects analysis (FMEA), to name a few that are common to both.

TABLE 6.3

Lean Six Sigma Tools for DMAIC and DMADV

Project Phase	Lean Six Sigma Tools
Define	Project charter Define customer requirements Identify Lean measures Lean Kaizen plan Lean value stream map (VSM) Voice-of-the-customer matrices SIPOC-RM KJ tree diagram
Measure	Lean benchmarking or scorecard Current state charts (handoff, spaghetti, process maps, process flows) Lean quantitative measurements, zero defects/waste Lean qualitative measures assessment (quality @ source, POUS, continuous flow, nine waste checklists) Descriptive statistics and data mining Six Sigma quantitative analysis—Pareto analysis (process defects) 5 whys
Analyze	Current state charts analysis Lean qualitative measures RCA analysis Lean quantitative measures RCA analysis Cause-and-effect diagrams and 5 whys Tree diagrams and matrices Process maps/value stream maps Design of experiments (DoE) Hypothesis testing (enumerative statistics) FMEA Inferential statistics Simulations and modeling
Improve (design)	Apply the "old" Lean tools: 5S, quality @ source, POUS, standardized work, cells, Total Productive Maintenance (TPM), facility layout, single-minute exchange of dies (SMED), batch reduction, Kanban, visual controls, VSM, and Kaizen blitzes Create and deploy future state charts (handoff, spaghetti, process maps, process flows) Create future state qualitative measures Create future state quantitative measures
Control (verify)	Lean sustainability—Kaizen action plans Visual measures deployment Lean educational plan deployment Lean communications plan Basic and advanced SPC Process FMEA Process monitoring system

The idea is to become more acquainted with the Lean RCA toolbox and apply the appropriate tools and technique in order to address a serious Lean workplace situation. Further, since problem solving is an integral part of the Lean management continuous improvement process, RCA is viewed as one of the core building blocks of the Lean organization. In itself, RCA will not produce any results, as it must be made part of the larger problem-solving effort, i.e., part of the conscious attitude that embraces a relentless pursuit of improvements at every level in every department or business process of the organization. In short, RCA is a highly versatile analysis approach that needs some structure in order to be successful. The sheer number of groups of tools available can be enough to dissuade anyone from embarking on analysis.

Groups of Root Cause Analysis Tools

See Figures 6.1 to 6.6.

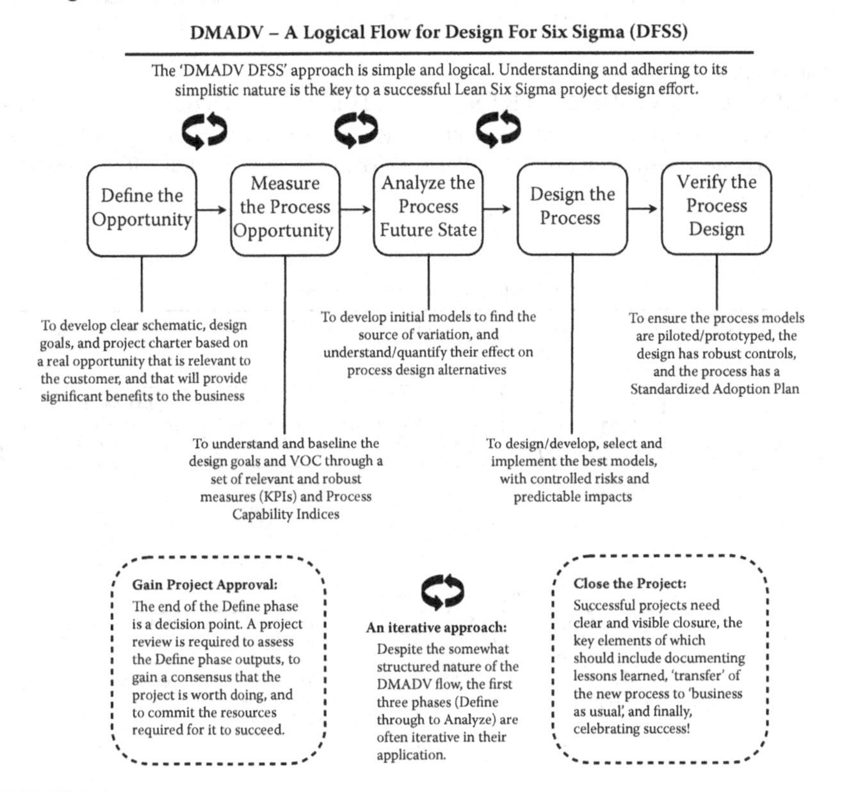

FIGURE 6.1
DMADV/DFSS approach. This figure is based substantially on the *DMAIC structure proposed by Quentin Brook in the publication, Lean Six Sigma and Minitab: The Complete toolbox Guide for All Lean Six Sigma Practitioners,* published by OPEX Resources Ltd. 2004, 2007, 2010.

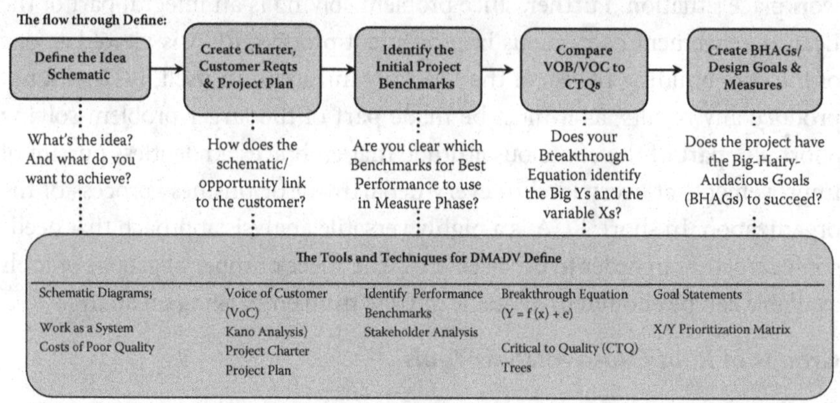

FIGURE 6.2
DMADV: Overview of Define. This figure is based substantially on the *DMAIC structure proposed by Quentin Brook in the publication, Lean Six Sigma and Minitab: The Complete toolbox Guide for All Lean Six Sigma Practitioners,* published by OPEX Resources Ltd. 2004, 2007, 2010.

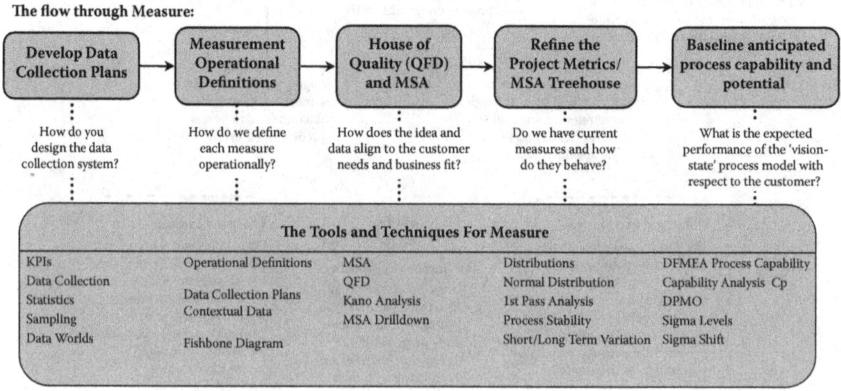

FIGURE 6.3
DMADV: Overview of Measure. This figure is based substantially on the *DMAIC structure proposed by Quentin Brook in the publication, Lean Six Sigma and Minitab: The Complete toolbox Guide for All Lean Six Sigma Practitioners,* published by OPEX Resources Ltd. 2004, 2007, 2010.

DMADV: Overview of Analyze

The Analyze phase aims to develop initial models to find the source of variation, and understand/quantify their effect on process design alternatives.

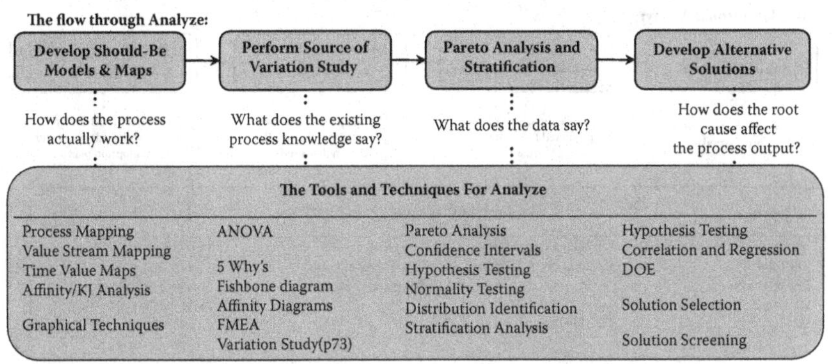

The Process Door Roadmap: The first two steps of Analyze are also referred to as the 'Process Door' because they aim to understand and gain clues directly from the process itself. The tools focus on gaining an in-depth understanding of how the process really works, and so most of them involve the people who know the process best – those who 'make it happen'.

The Data Door Roadmap: The last two steps are also referred to as the 'Data Door' because they focus on gaining clues and understanding from the data itself. These tools include a range of graphical and statistical tools to analyze the data.

FIGURE 6.4
DMADV: Overview of Analyze. This figure is based substantially on the *DMAIC structure proposed by Quentin Brook in the publication, Lean Six Sigma and Minitab: The Complete toolbox Guide for All Lean Six Sigma Practitioners,* published by OPEX Resources Ltd. 2004, 2007, 2010.

DMADV: Overview of Design

The Design phase aims to design/develop, select and implement the best models, with contrôlled risks and predictable impacts.

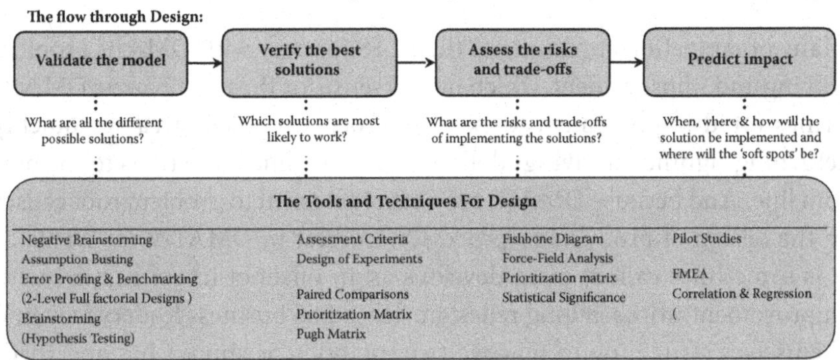

How is the success of the Improve phase measured? During the Analyze and Improve phases, the data collection systems developed in Measure should remain in place, with the KPI charts being updated regularly and reviewed at the beginning of each project team meeting. The success of the improve phase is **not** based upon the successful implementation of the selected solutions, but instead when the process measurement (KPIs) have improved and this has been validated with appropriate statistical techniques (graphs, hypothesis testing, etc.).

FIGURE 6.5
DMADV: Overview of Design. This figure is based substantially on the *DMAIC structure proposed by Quentin Brook in the publication, Lean Six Sigma and Minitab: The Complete toolbox Guide for All Lean Six Sigma Practitioners,* published by OPEX Resources Ltd. 2004, 2007, 2010.

DMADV: Overview of Verify

The Verify phase aims to ensure that the process motifs that have been validated and refined become 'operationalized,' so that the process design will be sustained and has a standardized adoption plan.

The flow through Verify:

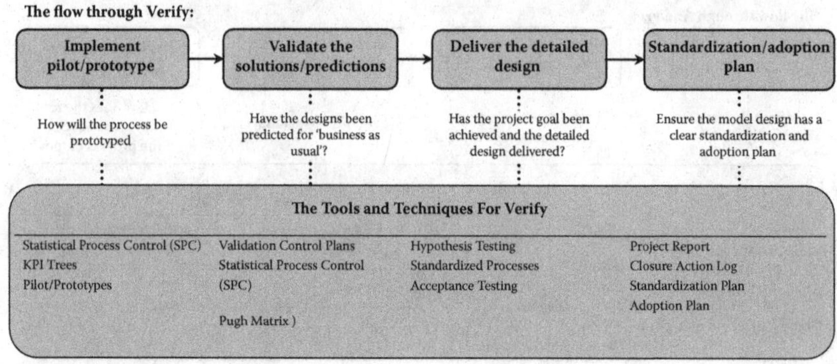

Ongoing measurement of the process: Processes must have rigorous data collection systems in place before a project can be closed. This involves defining who is responsible for collecting and reviewing the data, as well as ensuring that the measurements have been integrated in to the organization's KPI trees/dashboards.

Closing projects is a critical element of a successful Six Sigma program. There are often opportunities to apply the lessons learned from projects to different areas of the business, which requires clear action plans, and a 'knowledge management' approach to documenting projects.

FIGURE 6.6
DMADV: Overview of Verify. This figure is based substantially on the *DMAIC structure proposed by Quentin Brook in the publication, Lean Six Sigma and Minitab: The Complete toolbox Guide for All Lean Six Sigma Practitioners,* published by OPEX Resources Ltd. 2004, 2007, 2010.

SUMMARY

Many organizations first initiate their LSS journey with DMAIC problem solving and improvement, which makes sense for the most part as DMAIC brings rapid improvement to existing process problem areas by quickly returning significant savings dollars and cycle time reductions to the bottom line. And because DMAIC projects often point to problem root causes in the design of products or processes, interest in DMADV (or DFSS, as it is sometimes called) often develops as an offshoot in connection with improvement work needing redesign. Thus, LSS business leaders may find themselves struggling to manage two methods or approaches and roadmaps instead of a unified one. However, the good news is that in a world where "innovate and design" are naturally separated from "improvement work," a two-roadmap system may work just fine.

In many of our client-related engagements, many of which are profiled in this handbook, design is often interwoven with existing products and processes, and improvement often means revisiting the core fundamental design. In those situations, the SSTs, belts, and champions can waste a ton of energy worrying, "Is this a DMAIC or DMADV project, or both?" When an organization starts to feel that too much time is being spent worrying about the distinctions between DMAIC and DMADV (or is it DFSS?), it may be time to move forward and integrate and simplify things. Experienced LSS practitioners may notice that the thought processes have some parallels, especially in the DMAIC Define and Control phases. While it is tempting, and in some cases even possible, to integrate the approaches in new and creative ways, it is best to use some common approaches with an eye toward caution, as effective integration of the roadmaps requires special attention be paid to the subtle ways in which they are different.

Table 6.4 is a major step in guiding a team toward the correct methodology, no matter where on the roadmap the team was when the project's nature became clear.

The concluding section of this chapter was summarized in Table 6.4, which depicted a synthesis and integrated view of what happens if the DMAIC and DMADV thought processes are distilled to a high enough level that one map might be laid out instead of two. The "Operational Definition" columns showed the various distinctions that might be important to a particular team, depending on whether its current project was DMAIC or DFSS (or somewhere in the middle). The intended takeaway from this view is that DMA is where many of the DMAIC and DMADV distinctions lie. While they can be overlaid, it is important for each project team member to understand as early as possible where they are on the DMAIC-DMADV continuum. Some projects start out "thinking they are DMAIC" only to find (somewhere in D, M, or A) that they are really more DMADV—or vice versa.

All practitioners appreciate that roadmaps are needed to guide the work in Six Sigma, and that there is an understandable need to simplify and integrate when their complexity starts to get in the way. While considering a "branched" and a "parallel" approach to integrating DMAIC and DMADV, SSBBSs must be armed with as much insight as possible before deciding what is best in their particular environment.

TABLE 6.4

Summary Analysis of DMAIC versus DMADV Usages

	Operational Definition	
Lean Six Sigma Critical Thinking Process	**DMAIC (Improvement)**	**DMADV (Design/Innovation)**
Define		
What are the project goals?	Removing waste or reducing a problem in an existing process or work product (e.g., defects, rework, waste, and delays)	Identifying and capitalizing on a new design opportunity as in a new design patent or next-generation product or service
The business case?	Reducing costs of poor quality (e.g., rework, scrap, waste) Returning savings to the bottom line	Increasing business net value through new product sales, profitability, market share Bringing increased revenue to the top line
Project scope?	Defined and bounded by the problem (focused and narrow)	Defined by potential opportunity for new program or product (broad at the outset)
Customers/ stakeholders?	Those involved in or impacted by the problem; already familiar players	Internal or external potential "markets" connected with the opportunity; could be new players
As-is process?	Studied to reveal: clues about the problem, measurement points, and things not to break in the course of improvement	Studied to reveal: compensatory behavior, lead user "aha's," and future trends
What are key input and output requirements?	Must-be needs vs. wants, and satisfier requirements—to be sure the solution improves the primary *Y* while maintaining or improving performance across the board	"Must be" and "satisfiers" as a base, but special attention toward identifying latent requirements; VOC data gathering more widely exploratory

TABLE 6.4 (CONTINUED)

Summary Analysis of DMAIC versus DMADV Usages

	Operational Definition	
Lean Six Sigma Critical Thinking Process	**DMAIC (Improvement)**	**DMADV (Design/Innovation)**
Define		
	The focus and curiosity brought to VOC data gathering is that of a detective looking for what's important, but also clues about the problem, its implications, and its location(s)	Problems are interesting, but additionally, quirks in how things are done and future trends are pursued for their value in uncovering latent requirements and robust design clues
	Identifying and capitalizing on an opportunity (i.e., a new or next-generation product or service)	
Measure		
What are the most important measures and their drivers?	*XY* prioritization, operational definitions, and measurement systems analysis (MSA) are useful in all projects	*XY* prioritization, operational definitions, and measurement systems analysis (MSA) are useful in all projects
	Segmentation based on the location of the problem and its symptoms	Segmentation based on potential locations of the opportunity
What measures to collect? (where, how much, etc.)	Data collected, using the as-is process as the source of facts, to shed light on the root cause drivers (for this project's focused problem or problems)	Additional measures used to model "prospective value" (e.g., conjoint analysis); data collected, with or without an as-is process, to characterize and prioritize requirements (including prospective latent requirements); as appropriate, data collected or developed through modeling to shed light on design drivers

(*continued*)

TABLE 6.4 (CONTINUED)

Summary Analysis of DMAIC versus DMADV Usages

	Operational Definition	
Lean Six Sigma Critical Thinking Process	**DMAIC (Improvement)**	**DMADV (Design/Innovation)**
Analyze		
What has the team learned about current performance/capability?	This will apply directly, as there is a great deal more current performance to document and analyze for a typical DMAIC-type project	There may not be a current process for the work under consideration, but generally the performance of a relevant as-is process helps document the state of the art or leading edge
What can the team learn from patterns and statistical contrasts in the data?	"Peeling the onion" to get down to root causes and fundamental drivers in order to remove those causes and fix the problem at its core—removing or reducing the waste in the as-is process; the historic/current process data are the key source of data and insight	Understanding the design drivers in order to guide upcoming decisions about which factors will be included in and adjusted to optimize business net value; models, prototypes, and industry benchmarks are key sources of data and insight
Can the team verify the root cause or driving impacts of the *X*'s on the *Y*'s?	In DMAIC the *X*-*Y* connections are part of the known, existing system; the work here focuses on quantifying that relationship	In DMADV, the *XY* connection may be new and may depend on the solution to be selected; the work at this step may involve prototyping or modeling the *XY* connection in order to estimate the nature of the relationship
Improve vs. Design	**Improve**	**Design**
What is the best solution/choice?	Selecting among solution options using force field analysis	Some basic design and modeling of solution options may have been started as part of the *XY* relationship assessment in analyze
How should the solution be detailed for best practical implementation?	Some use of modeling, depending on the case	For software, the "coding/construction" happens here—realizing the design

TABLE 6.4 (CONTINUED)

Summary Analysis of DMAIC versus DMADV Usages

	Operational Definition	
Lean Six Sigma Critical Thinking Process	**DMAIC (Improvement)**	**DMADV (Design/Innovation)**
How will it work?	Use of SCAMPER, FMEA, and piloting	Continued use of modeling, prototyping/alpha release to predict and/or verify performance and reduce risk
Control and Verify		
What factors are important to control over the life of the improvement?	Similar considerations for both DMAIC and DMADV; however, in DMADV, the new designs need to be verified and validated; DMAIC uses DoE and DMADV both DoE and Taguchi loss experiments	
How will ongoing operation be monitored at all levels, with control signals at every level?	Process management and monitoring Statistical process control KPI dashboards	

Section 4

LSSBB Advanced Nonstatistical Tools

INTRODUCTION

Having the right tool makes the difference between success and failure.

—H. James Harrington

In this section we present the key body of knowledge for advanced *nonstatistical* tools that all LSSBBs will be using the most often. In Section 5 we present the advanced *statistical* tools that are used most often by LSSBBs. In these two sections we provide you with the tools to solve your most difficult problems and take advantage of the many opportunities to improve your processes. The tools in these two chapters are in addition to the tools that a potential LSSBB should already have mastered as an SSGB. As an LSSBB, it is essential that you are competent at using these tools in order to lead your LSS project teams. We liken using just the basic Green Belt tools to improve complex opportunities for improvement to trying to build a house using just a yardstick, hammer, and handsaw. It can be done, but the quality of the results will be poor, it will take a long time to do it, and the costs will be high. Today a professional builder wouldn't think about building a house without using power saws, power hammers, power screwdrivers, and laser measurement tools. If you use these power tools to build a house, the result will be that the construction will be done faster, with higher quality and lower costs. Similarly, these additional tools provide the LSSBB with the means and power to define the very best solution to complex situations.

The tools in Section 4 are presented in alphabetical order. Most of these tools are first defined, and then you are instructed on how to use them. Next, you are provided with examples of how they have been used before. The content of the chapters that make up this section include

- Chapter 7: "Black Belt Nonstatistical Tools (A through M)"
  - 5S
  - Benchmarking
  - Bureaucracy elimination
  - Conflict resolution
  - Critical to quality
  - Cycle time analysis and reduction
  - Fast-action solution technique (FAST)
  - Foundation of Six Sigma
  - Just-in-time
  - Matrix diagrams/decision matrix
  - Measurement in Six Sigma
- Chapter 8: "Black Belt Nonstatistical Tools (O through Q)"
  - Organizational change management
  - Pareto diagrams
  - Prioritization matrix
  - Project management
  - Quality function deployment
- Chapter 9: "Black Belt Nonstatistical Tools (R through Z)"
  - Reliability management systems
  - Root cause analysis
  - Scatter diagrams

- Selection matrix/decision matrix
- SIPOC
- SWOT
- Takt time
- Theory of constraints
- Tree diagram
- Value stream mapping

OVERVIEW

There are three main objectives for learning the LSSBB methodology:

- To master two systematic approaches to problem solving that have a very good track record of producing excellent results—DMAIC (Define, Measure, Analyze, Improve, and Control) and DMADV (Define, Measure, Analyze, Design, and Verify)
- To master a number of very effective tools that can be used to solve problems
- To master the way to apply LSSBB methodology to bring about transformation within an organization

To become an SSGB, you learned how to use the DMAIC approach to problem solving that was developed to minimize process variation. In addition, you were introduced to the following basic problem-solving tools:

- Affinity diagrams
- Brainstorming
- Cause-and-effect diagrams
- Check sheets
- Failure mode and effects analysis
- Flowcharts
- Force field analysis
- Graphs and charts
- Histograms
- Kano model
- Nominal group techniques

- Pareto diagrams
- Root cause analysis
- Scatter diagrams
- SIPOC diagrams

Note: In addition, an SSGB was introduced to many statistical tools.

As an active SSGB, we hope you have had a chance to use most of these tools in solving problems within your organization. The tools that were presented in the SSGB classes are the most commonly used tools, and they are designed to solve most of the problems that are assigned to a Six Sigma Team.

As an LSSBB, you will be required to have expanded skills that will prepare you to solve the most complex problems. You will also learn how to effectively use the DMADV problem-solving approach that focuses on designing a total new process or product. DMADV is the Six Sigma approach to adapting process re-engineering to the Six Sigma philosophy. DMADV is used when the DMAIC approach will not produce the required level of improvement. It is designed to bring about a drastic improvement in performance. While DMAIC focuses on reducing variation, DMADV focuses on changing the performance level of the product or process. Figures S4.1 and S4.2 will help you understand the difference between the results of decreasing variation and improving the total process.

In the case of decreasing variation, the centerline remains the same but the amount of variation above the centerline is greatly reduced. In the case of total process improvement, the focus is not on decreasing variation but changing the position of the centerline.

As an LSSBB, you will need to understand the following 26 advanced nonstatistical tools. (In Section 5 we will discuss the additional advanced

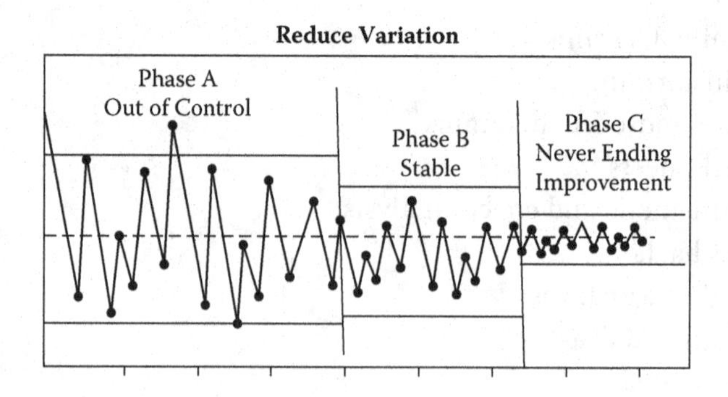

FIGURE S4.1
Reduced process variation.

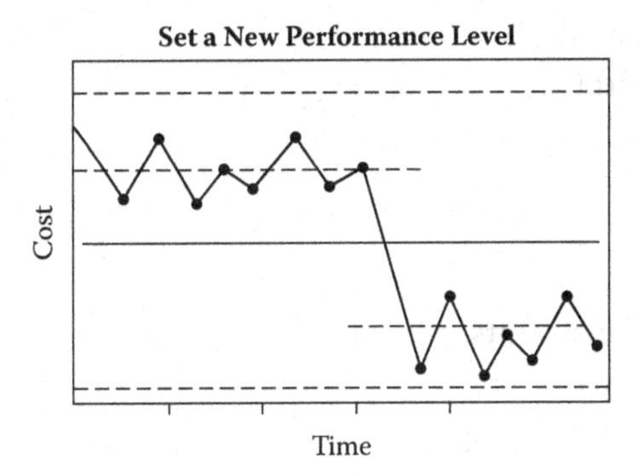

FIGURE S4.2
Setting a new level of performance.

statistical tools that an LSSBB needs to understand and be able to use.) These 26 tools can and should be applied to both the DMADV and DMAIC problem-solving approaches. In fact, an LSSBB is often required to teach these tools to the members of a Six Sigma Team when they are needed to solve a problem. So it is imperative that as an LSSBB you understand these tools thoroughly. It is for this very reason that we have set aside a large part of this book to discuss these 26 tools. You will find all of these tools useful in working with both the DMAIC and DMADV methodologies.

1. 5S
2. Benchmarking of processes
3. Bureaucracy elimination methods
4. Conflict resolution
5. Critical to quality
6. Cycle time analysis
7. Fast-action solution technique
8. Foundations of Six Sigma
9. Just-in-time
10. Matrix diagrams/decision matrices
11. Measurement in Six Sigma
12. Organizational change management
13. Pareto diagrams
14. Prioritization matrix
15. Project management
16. Quality function deployment
17. Reliability management systems

18. Root cause analysis
19. Scatter diagrams
20. Selection matrix
21. SIPOC
22. SWOT
23. Takt time
24. Theory of constraints
25. Tree diagrams
26. Value stream mapping

For ease of reference, the tools are presented in alphabetical order rather than in the order they would be used in solving a problem.

WHERE ARE THESE TOOLS USED?

Table S4.1 defines what phase in the two Six Sigma methodologies that each GB and LSSBB tool is most apt to be used in. This does not indicate that the tool should be used each time a project goes through the indicated phase; in fact, some of the tools may not be used in 1 in 10 projects that go through the indicated phase.

SUPPORTING SOFTWARE

The tools that we present in the next chapters use many different types of graphs, tables, charts, diagrams, and statistical formulas that are time-consuming to use and prepare. To aid you in using these tools, a number of software packages are available, and newly upgraded ones are coming out all the time.

The two most frequently used software applications for statistical analysis are Minitab and JPM/SAS. These software applications give thorough statistical analysis results. They are both advanced tools, and users should have more than an introduction to statistics understanding in order to avoid making invalid decisions.

Software applications have been written for many Six Sigma tools. Software has been written to help use the tools listed in Table S4.2.

TABLE S4.1

Basic Tools

Tool	D	M	A	I	C	D	M	A	D	V
Affinity diagrams	X					X				
Benchmarking of processes			X					X		
Brainstorming				X					X	
Bureaucracy elimination methods	X					X				
Cause-and-effect analysis			X					X	X	X
Check sheets	X					X				X
Conflict resolution	X	X	X	X	X	X	X	X	X	X
Cycle time analysis			X					X	X	X
Failure mode and effects analysis			X	X				X	X	X
Fast-action solution technique (FAST)	X	X	X	X	X	X	X	X		
Five S's	X	X		X	X	X				
Flowcharts	X	X	X	X		X	X	X	X	X
Foundations of Six Sigma	X					X			X	X
Force field analysis	X					X				
Graphs and charts		X	X	X	X		X	X	X	X
Histograms		X	X		X		X	X	X	X
Kano model	X					X			X	
Matrix diagrams/decision matrices	X					X				
Measurement in Six Sigma		X					X		X	X
Nominal group techniques	X			X		X			X	
Organizational change management	X			X		X			X	
Pareto analysis		X	X		X		X	X		X
Plan-Do-Check-Act	X	X	X	X	X	X	X	X		
Prioritization matrix	X					X			X	
Project management	X	X	X	X	X	X	X	X	X	X
Quality function deployment	X					X				
Reliability management systems		X	X		X		X	X	X	X
Root cause analysis	X	X	X			X	X	X	X	X
Scatter diagrams		X	X		X		X	X	X	X

(Continued)

TABLE S4.1 (CONTINUED)

Basic Tools

Tool	D	M	A	I	C		D	M	A	D	V
Selection matrix	X						X			X	
SIPOC (Supplier, Input, Process, Output, Customer)	X						X				
SWOT			X						X		
Theory of constraints		X	X	X				X	X	X	X
Value stream mapping	X	X	X	X			X	X	X	X	X
5W's	X	X	X	X			X	X	X	X	
5W's and 2H's	X	X	X	X			X	X	X	X	

Many applications cover more than one of these categories. While many different companies offer specialized and global applications, there is no one program in a category that is best. The choice of a program is dependent on the individual and/or company's preferences. The applications listed here are not meant to be all-inclusive, and they are only offered as a reference for what is available at the writing of this book.

TABLE S4.2

Tools with Software Programs Available

Affinity diagram/KJ analysis	Pareto
Analysis of variance (ANOVA)	Poka-yoke
Brainstorming	Process mapping
Capability indices/process capability	Project charter
Cause and effect	QFD/House of Quality
Control charts	Regression
Design of experiments (DoE)	Risk management
FMEA	Sampling/data
Graphical analysis charts	SIPOC/COPIS
Hypothesis testing	Statistical analysis
Kano analysis	Value stream mapping
Measurement systems analysis (MSA)/gauge R&R	Variation
Normality	Templates

While it is impossible to describe all of the software applications that are available, the following list is a sample of functions for which applications have been written. The software in each category is listed in alphabetical order:

- Process mapping (flowcharting)
    - Autoflowchart
    - Crystal Flow
    - Cute Flowchart
    - Diagram Designer
    - Diagram Studio
    - Edraw
    - Flowbreeze
    - Flowchart.com
    - Gliffy
    - Microsoft Office
        - Excel
        - PowerPoint
    - Paqico
    - Reflow
    - SmartDraw
    - Vizio
    - Wizflow flowcharter
- Statistical analysis
    - Amos
    - JMP/SAS
    - Mathematica
    - MATLAB*
    - Minitab
    - NVivp
    - SAS
    - SPSS (formerly PASW)
    - Systat
- Design of experiments
    - Anova-TM (Advanced Systems and Designs)
    - DC-Pro (PIster Group)

* MATLAB™ is a registered trademark of The MathWorks, Inc.

- Design-Expert (Stat-Ease)
- DOE Kiss (Air Academy Associates)
- DOEpack (PQ Systems)
- DOE-PC (Quality America)
- DOE Wisdom (Launsby Consulting)
- ECHIP – (ECHIP)
- Experimental Design Module (Domain Manufacturing Corp.)
- JMP/SAS (SAS Institute)
- Minitab (MINITAB)
- Quantek-4 (Nutek)
- Statgraphics Plus for Windows (Manugistics)
- Statistica with Statistica Industrial Systems (StatSoft)
- Statistical Sample Planner (Dyancomp)
- Statit (Statware)
- Strategy (Experiment Strategies)
- Trial Run (SOSS)

- Brainstorming
  - Aibase (aibase-cs.com)
  - BrainStorm (Marck Pearlstone & David Tebbutt; Brainstormsw.com)
  - Curio (sengobi.com/products/curio)
  - Edraw Max (Edreawsoft.com)
  - eXpertSystem (store.richcontent.com/exv7.html)
  - Incubator (mindcad.com)
  - MindManager (mindjet.com/us/products)
  - MindView (Innovationtools.com)
  - ParaMind (Paramind.net)
  - Personal Brain (thebrain.com)
  - SmartDraw (smartdraw.com/exp/mim)
  - ThoughtOffice (thoughtrod.com and brainstormingsoftware.org)
- Affinity diagrams
  - Affinity Diagram for Excel (freebizfiles.com)
  - PathMaker (skymark.com)
  - SmartDraw (smartdraw.com)
- Cause and effect (fishbone, Ishikawa chart)
  - Fishbone (Ishikawa) Diagram Software (sigmaflow.com)
  - QI-Macros (qimacros.com)

  - SigmaXL (sigmaxl.com)
  - SmartDraw (smartdraw.com)
  - Statgraphics Centurion (Manugistics)

- FMEA
  - APIS IQ-Software (apis.de/en)
  - Byteworx FMEA Software (byteworx.com)
  - FMEA Software (ReliaSoft.org)
  - FMEA-Pro (diadem.com)
  - SFMEA (softrel.com)
  - XFMEA (reliasoft.com)
- Kano
  - Online services
    - Isixsigma.com
    - Kanosurvey.com
    - Sourceforge.net
    - 12manage.com
- Measurement system analysis (MSA)
  - QI Macros (qimacros.com)
  - MSA Pro (omnex.com)
  - SigmaXL (sigmaxl.com)
- QFD (House of Quality)
  - QFD Capture (qfdcapture.com)
  - PathMaker (noweco.com)

7

Black Belt Nonstatistical Tools (A through M)

Without sound data you can't make sound decisions.

—H. James Harrington

INTRODUCTION

In this chapter a number of very important tools are presented. Read them carefully and take time to think about how you can apply each of the tools to one of the improvement opportunities that are available within your organization and/or the problems that you solve as an SSGB or LSSBB.

In many organizations the LSSBB will be required to teach these tools and the Green Belt tools to Six Sigma Team members. Take time to think about how you would teach each tool to the people who would make up a Six Sigma Team that you would be leading. We recognize that you may feel that you don't have the experience or the skills to teach all of these and the SSGB tools after just reading about them or just attending a class on them. That is why before a person can function fully and effectively as an LSSBB, he or she needs to have experience working on projects. It is best if the potential LSSBB has already been introduced to some of these tools when he or she was working on a project led by an LSSBB.

The tools that are presented in this chapter include the following:

- 5S
- Benchmarking
- Bureaucracy elimination
- Conflict resolution
- Critical to quality
- Cycle time analysis and reduction
- Fast-action solution technique
- Foundation of Six Sigma
- Just-in-time
- Matrix diagrams/decision matrix
- Measurement in Six Sigma

5S

> We often spend more time looking for the things to do the job, than doing the job.
>
> **—H. James Harrington**

Definition

Created by Ohno Shingo, *5S* is a workplace organization methodology that uses a list of five Japanese words: *seiri*, *seiton*, *seiso*, *seiketsu*, and *shitsuke*. Translated into English, they all start with the letter *S*. They are:

1. Sorting (seiri)
2. Straightening or setting-in-order (seiton)
3. Sweeping or shining or cleanliness/systematic cleaning (seiso)
4. Standardizing (seiketsu)
5. Sustaining the discipline or self-discipline (shitsuke)

Just the Facts

5S is a system to optimize productivity by reducing waste and maintaining an orderly workplace. These actions are geared to achieving better and more efficient operational results. This is usually the first method implemented when starting a Lean process. It is intended to clean up and organize the workplace to make it more effective and efficient.

Example

Organizing the placement of tools and replacing tools in their specified location allows for more efficiency in the use of the tools. A simple look at the storage area will immediately show which tools are missing or in use. (See Figure 7.1.)

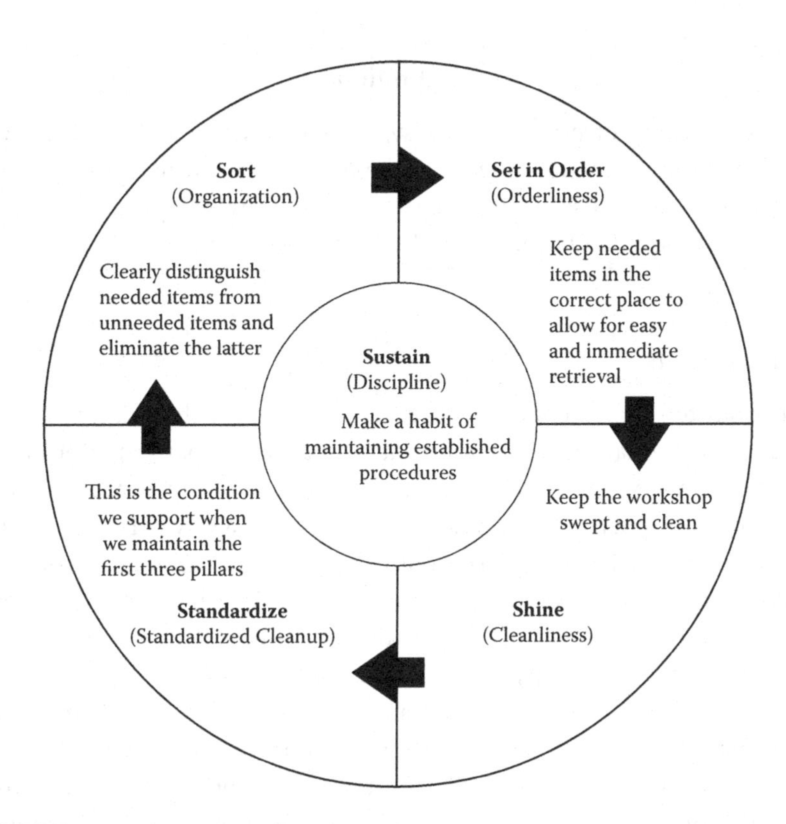

FIGURE 7.1
The 5S approach.

Additional Reading

Picard, Daniel (ed.). *The Black Belt Memory Jogger* (GOAL/QPC and Six Sigma Academy, Salem, NH, 2002).
Wortman, Bill. *The Certified Six Sigma Black Belt Primer* (Quality Council of Indiana, West Terre Haute, 2001).

BENCHMARKING OF PROCESSES

Benchmark to know how much you can improve and how to do it.

Definition

Benchmarking of processes is a systematic way to identify superior processes and practices that are adopted or adapted to a process in order to reduce cost, decrease cycle time, cut inventory, and provide greater satisfaction to the internal and external customers.

Just the Facts

Xerox started the recent enthusiasm for process benchmarking when it credited benchmarking as the tool that caused the company's performance to turn around in the mid-1980s. As John Cooney, one of Xerox's marketing managers, puts it, "Benchmarking has become a way of life at Xerox."

Benchmarking helps organizations bring on new ideas. It makes organizations reexamine the way they are doing things. When U.S. organizations dominated the world's market, they didn't have outside pressures driving their improvement efforts, so they could take their own sweet time. Today U.S. organizations no longer dominate the world's market, so their benchmarking activities must take on an international focus.

Don't misunderstand. The United States is not second best in everything. In many industries, it is still the world leader. But there is a lot of competition out there, hungry to take away customers. Foreign competitors are just as creative as American organizations; for example, foreign organizations accounted for over 50% of new U.S. patents in 1994. And they can still remember what it is like to go hungry. In my view, there is a direct correlation between the last time a worker went hungry and his or her work ethic.

IBC, based in Houston, Texas, conducted a survey of 76 organizations and found that:

- Benchmarking was considered to be a necessary tool for survival.
- Most firms did not know how to systematically conduct a benchmarking project.
- 95% of the organizations surveyed felt that most firms do not know how to benchmark.
- 79% of the organizations believed that top management must be very actively involved if the benchmarking process is going to be successful.

In a survey of 770 organizations in Europe, the Benchmarking Centre (UK) found that:

- 89% of the organizations rated "finding competent benchmarking partners" as their most important requirement.
- 70% of the UK organizations were doing benchmarking.
- 95% of the organizations were willing to share information with a benchmarking center.

During World War II a recruitment poster read: "Join the Navy and see the world." Too many managers and professionals today view benchmarking in the same light. They have changed the slogan to read: "Join a benchmarking team and see the world." This is far from the truth. Studies show that for every hour spent visiting a benchmarking partner's site, over 200 hours are required in planning, collecting, and analyzing data and implementing changes. Benchmarking is not site visits. It is creative analysis of thousands of pieces of information to bring about a change in the benchmark item—a change that is often as drastic as the caterpillar's transformation into the monarch butterfly.

When all is said and done, the most difficult part of the benchmarking process is awakening to the fact that your operation does not know it all. Most U.S organizations are just too egotistical to admit that they can learn anything from another organization. The benchmarking process starts when you admit that your counterparts in Mainland China, Portugal, Malaysia, Sweden, the United Kingdom, France, and elsewhere may have an idea that is better than yours.

David T. Kearns, chief executive officer of Xerox Corporation, defines benchmarking as "the continuous process of measuring products, service, and practices against the toughest competition or those companies recognized as industrial leaders." Benchmarking is a never-ending discovery and learning experience that identifies and evaluates best processes and performance in order to integrate them into an organization's present process to increase its effectiveness, efficiency, and adaptability. It provides a systematic way to identify superior products, services, processes, and practices that can be adopted or adapted to your environment to reduce costs, decrease cycle time, cut inventory, and provide greater satisfaction to your internal and external customers.

Benchmarking is often viewed as simply purchasing competitive products to compare them with the ones manufactured by the organization. This is called *product benchmarking* or *reverse engineering* and is not the one that will be discussed in this approach.

Process benchmarking is a lot like a detective story, and the person doing the benchmarking operates a lot like a detective. He or she must search through the many clues available in the public domain to find leads and then follow up on these leads to identify and understand the truly world-class processes. It can be an exciting and enlightening adventure.

What Will Benchmarking Do for You?

Benchmarking requires a lot of work, and once you start, it is an ongoing process to keep it updated. Why, then, should any organization even begin? It is because the benefits of benchmarking far outweigh the effort and expense. Benchmarking:

- Provides you with a way to improve customer satisfaction
- Defines best applicable processes

- Improves your process
- Helps eliminate the "not invented here" syndrome
- Identifies your competitive position
- Increases the effectiveness, efficiency, and adaptability of your processes
- Transforms complacency into an urgent desire to improve
- Helps set attainable but aggressive targets
- Increases the desire to change
- Allows you to project future trends in your industry
- Prioritizes improvement activities
- Provides your organization with a competitive advantage
- Creates a continuous improvement culture
- Improves relationships and understanding between benchmarking partners

History of Benchmarking

During the 1960s, IBM realized that its costs could be reduced significantly, and the quality of its process-sensitive products improved, if its worldwide locations adopted the best existing practices. As a result, a corporate procedure was written requiring all process-sensitive products to be manufactured using compatible processes. This launched a corporate-wide effort to have common practices at all locations or, if that proved impractical, to have compatibility between common processes. The determination to identify the best manufacturing processes gave IBM a significant international competitive advantage.

In the late 1970s, in a similar move, Xerox decided to compare its U.S. products with those of its Japanese affiliate, Fuji-Xerox. Xerox was shocked to discover that Fuji was selling copiers at a price equivalent to what it cost U.S. Xerox to manufacture the copiers. This discovery spearheaded a successful program to reduce costs in the U.S. manufacturing process. Based on the success of this pilot program, Xerox management later incorporated benchmarking as a key element in its corporate-wide improvement efforts. The formal initiation of the program began around 1983. Through this program, benchmarking took on new dimensions, and benchmarking techniques were applied to support processes as well as product processes.

Today many organizations use benchmarking to help drive their continuous improvement efforts. For example, Motorola cites benchmarking as one of the major tools powering its improvement process, for which it was awarded the Malcolm Baldrige Award in 1988.

Types of Benchmarking

Essentially, there are two generic types of benchmarking:

- Internal
- External

Internal Benchmarking

Internal benchmarking compares practices within one part of an organization to those of a more advanced unit within the same organization. This type of benchmarking is the easiest to conduct because there are no security and confidentiality problems to overcome.

The 14 Steps to Do Internal Benchmarking

There are 14 basic steps in an internal benchmarking effort:

1. Identify what to benchmark. Identify the products, processes, and/or activities that should be benchmarked. Plan to develop a database of benchmarking information.
2. Obtain management support. Obtain management support for the benchmark activities. This support must include project approval and approved human and financial resources. It may be beneficial to form a corporate benchmark committee, both to build a base of support and to review the data after they are collected.
3. Develop benchmark measurements. Benchmark measurements use both qualitative and quantitative data.
4. Determine how to collect data. Four commonly used data collection methods are:
   - Exchange of process data, procedures, and flowcharts
   - Telephone interviews and surveys
   - Joint committees
   - Site visits

   In most cases, all four will be used.

5. Review plans with location experts. Search out people who understand the process being evaluated and ask them to:
    - Review the data plan
    - Recommend other locations that are doing the same or similar activities well
    - Suggest contacts for information
6. Select locations. Select the locations to be benchmarked using the updated data collection plan.
7. Exchange data. Contact key process people within your location or from other locations and explain what you are trying to accomplish. Ask them to become partners in the benchmarking effort. Offer to send them copies of your measurement data, procedures, and process flowcharts. Request that they review these data and send back their data in a similar format, along with any comments they have.
8. Conduct telephone interviews and surveys. After carefully reviewing the other locations' data, call your contacts and discuss the data you received from them to clarify key activities. In complex studies, a written survey should be used instead of a telephone interview because it allows the individual to acquire exact data and to document the response.
9. Conduct location visits. The benchmarking team or corporate benchmarking committee should conduct a detailed tour of the process being benchmarked. Each member should then write process review reports defining strengths and weaknesses observed.
10. Analyze the data. Construct a process flow diagram including the best processes and practices from all locations. Then estimate the expected performance if the optimum process is implemented.
11. Establish a process-change plan. Based on the process diagram developed in step 10, the team or committee should conduct a change-impact analysis to prioritize the implementation activities. In some cases, organizations may decide not to make any process changes. The decision to implement the change now or later will depend on the projected gains resulting from the change.
12. Implement one change at a time. Implement the high-priority changes one at a time. By implementing one change at a time, it is easier to assess the impact of each change on the total process. It is also far less disruptive to the overall operation.

13. Measure changes. Develop a system that measures the impact of each change on the total process. Many changes can have both positive and negative effects. Any measurement system should evaluate the impact of a change on effectiveness and efficiency as it relates to the total process.
14. Report on an ongoing basis. Establish a measurement report comparing performance by location. Issue this report every six months.

External Benchmarking

External benchmarking applies the benchmarking approach to comparable organizations outside your own. External benchmarking may be conducted with competitors or with world-class noncompetitors.

External benchmarking is divided into 3 sub-classifications. They are competitive, world-class operations, and activity-type benchmarking. For most processes, both internal and external benchmarking are used.

- External competitive benchmarking, also known as reverse engineering, is a form of external benchmarking that requires investigating a competitor's products, services, and processes. The most common way to do this is to purchase competitive products and services and analyze them to identify competitive advantages.
- External world-class operations benchmarking is a form of external benchmarking that extends the benchmarking approach outside the organization's direct competition to involve dissimilar industries. Many business processes are generic in nature and application (e.g., warehousing, supplier relations, service parts logistics, advertising, customer relations, employee hiring) and can provide meaningful insights despite the fact that they are being used in an unrelated industry. Benchmarking dissimilar industries enables you to discover innovative processes not currently used in your particular product types that will allow your processes to become best in class. It is usually advisable to establish some criteria when selecting organizations to be benchmarked. Some of these criteria might include:
  - Requirements of the external organization's customers: High quality and high reliability or low quality and one-time usage.
  - The characteristics of the external organization's product: Size, shape, weight, environment, etc.

- Type of industry: Grocery industry, office products industry, electronics industry, transportation industry, etc. In this case, consider broad industrial categories, not the specific products.
- After the external organization is selected, the benchmarking approach can be applied to a product, a process, or even an individual activity.

The 14 Steps for External Benchmarking

There are 14 basic steps in an external benchmarking effort; the first 5 steps are the same as for internal benchmarking.

1. Identify what to benchmark. Identify the products, processes, and/or activities that should be benchmarked. Plan to develop a database of benchmarking information.
2. Obtain management support. Obtain management support for the benchmark activities. This support must include project approval and approved human and financial resources. It may be beneficial to form a corporate benchmark committee, both to build a base of support and to review the data after they are collected.
3. Develop benchmark measurements. Benchmark measurements use both qualitative and quantitative data.
4. Determine how to collect data. Four commonly used data collection methods are:
   - Exchange of process data, procedures, and flowcharts
   - Telephone interviews and surveys
   - Joint committees
   - Site visits

   In most cases, all four will be used.
5. Review plans with location experts. Search out people who understand the process being evaluated and ask them to:
   - Review the data plan
   - Recommend other locations that are doing the same or similar activities well
   - Suggest contacts for information
6. Develop a preliminary list of the best external organizations. Using information generated by internal benchmarking efforts or by direction of the corporate benchmarking committee, list organizations with a world-class reputation. (This preliminary list will later be refined.)

7. Develop a data collection plan. There are many ways to collect information on external organizations. Some of the most common are:
   - Searches of literature
   - Use of information developed by professional and trade associations
   - Use of external experts

   A great deal of data about the competition, and about world-class organizations in general, exists in the public domain. Devote the time and effort necessary to collect these data. The more you know about an organization before contacting it, the more likely you are to gain real insights into its processes.
8. Make a final selection of benchmark organizations. Review the data you have collected to date and update your database. Identify any voids in the collected data. After a detailed analysis of these data, pinpoint the organizations to target for benchmarking and identify key contacts there.

   At this point, you should have reduced your potential benchmark organizations to between three and five. Some organizations that looked good at first may have to be dropped from the list because of:
   - Unwillingness to share data
   - Lack of data
   - Existence of better candidates
   - Reputation as not the best performer
   - Processes not comparable to yours
   - Communications problems
   - Travel costs
9. Gather data. There are a number of ways to collect data from external benchmark organizations. Among these are:
   - Anonymous surveys administered by a third party
   - Focus groups organized and facilitated by a third party
   - Site visits

   It is generally a good idea to conduct surveys before site visits. When conducting site visits, limit the visit team to two to eight people and identify the role of each in advance. Immediately after the visit (during the same day, if possible) hold a meeting of the team to consolidate thinking and to record observations.
10. Establish and update the benchmark database. The previous activities generate vast amounts of data that must be captured and analyzed.

The best way to do this is to constantly update the benchmark database as each activity is performed.

11. Analyze the data. Two types of data are collected and used in the benchmarking approach. The first type is qualitative data (word descriptions) and the second is quantitative data (numbers, ratios, etc.). For quantitative data, design a data matrix that highlights the parts of the process requiring additional data and study. Some effective ways to present and analyze qualitative data are:
    - Word charts
    - Written step-by-step procedures
    - Flowcharts
12. Establish a process-change plan. Conduct a change-impact analysis to prioritize the implementation activities. In some cases, organizations may decide not to make any process changes. The decision to implement a change now or later will depend on the projected gains resulting from the change.
13. Implement one change at a time. Implement the high-priority changes one at a time. If you implement one change at a time, it is easier to assess the impact of each change on the total process. It is also far less disruptive to the overall operation.
14. Measure changes. Develop a system to measure the impact of each change on the total process. Many changes can have both positive and negative effects. The measurement system should evaluate the total impact of the change on the effectiveness and efficiency of the process as a whole.

Guidelines and Tips

Use the *Encyclopedia of Trade Associations* to develop an initial list of target trade and professional associations.

- Qualify each potential association for:
    - Its membership
    - The types of conferences offered
    - Prominent industry figures who are members
    - Periodicals or journals published
- Develop a final list of leading associations in the industry or field.
- Contact the representatives of each association.

- Make sure you fully understand the product, process, or activity to be benchmarked before attempting to generate data.
- Make a concerted effort to establish a broad base of organizational support for the benchmarking effort. This is part of the change process and is essential to using the data.
- When conducting competitive benchmarking, be careful not to expose yourself to risk by engaging in practices that could be considered illegal or unethical. Wherever possible, use information about competitors from the public domain.

What Are the Primary Reasons for Using Process Benchmarking?

The two primary reasons for using process benchmarking are goal setting and process development. Every person, process, and organization needs goals to strive for. Without them, life is unrewarding, and we drift on a sea of confusion. We all want to improve. No one likes to be average. In the past, goals usually were based on the organization's and/or the process's past performance. There was little correlation between our goals and the ultimate standard of excellence. Occasionally, our goals exceeded what is achievable, but more commonly they fell far below what had been, or could be, achieved. As a result of setting low goals for ourselves, we enjoyed a false sense of accomplishment. We stopped trying to improve because we so easily met the low standards we set for ourselves. This prevented many individuals, processes, and organizations from maturing to their full potential. Because it provides a means for setting challenging targets and attainable goals, benchmarking is the antidote to this self-imposed mediocrity.

Even more important, process benchmarking provides a way to discover and understand methods that can be applied to your process to effect major improvements. This is the unique value of process benchmarking; it not only tells you how good you need to and can be, but it also tells you how to change your process to get there.

When setting up your measurement system for process benchmarking, you need to plan for two things—the what and the how. Both of these functions are important. What good is knowing how good world class is if you do not know how to improve your process to obtain it? Knowing that you are bad, but not being able to improve it, just discourages everyone. Similarly, what good is a new process idea if you do not know whether it will have a positive impact on your process? You need to design your process benchmarking to provide both the what and the how.

The What

Measurements are absolutely crucial. If you cannot measure, you cannot control. If you cannot control, you cannot manage. It's as simple as that. Quantitative data are absolutely essential ingredients in becoming and staying world class.

As critical as measurements are, it would seem that everyone would know just what needs to be measured. Unfortunately, this is not the case. In fact, in most instances it is just the opposite, and this is particularly true when you talk about business processes. Process benchmarking should measure things such as:

- How much
- How fast
- How good
- When
- Where
- How long
- Size, shape, form, and fit

Although most product-type measurements are physical in nature, most process measurements are effort, cost, and time related. Consequently, it often is best to establish ratio measurements (e.g., return on investments, returns per year, unit costs, productivity rates) rather than actual values. The use of ratios allows data exchange without disclosing absolute values or production rates. This encourages free exchange of information between organizations.

The How

Another real advantage of process benchmarking is that it provides you with insights into how others have become the best. This aspect focuses on discovering how world-class organizations developed their processes and systems to ensure superior performance. At this juncture, we seek and analyze the how-to, the knowledge, the ways, the processes, and the methods responsible for making an organization, a process, or an activity the best of its kind. We then apply this knowledge to our process, adapting it to meet the unique requirements of our products, employees, customers, and organization's personality.

The Five Phases of Internal and External Combined Benchmarking Process

- Phase I: Planning the benchmarking process and characterization of the item(s)
- Phase II: Internal data collection and analysis
- Phase III: External data collection and analysis
- Phase IV: Improvement of the item's performance
- Phase V: Continuous improvement

These benchmarking phases comprise a total of 20 activities. (See Table 7.1.) Each activity is sub-divided into a number of specific tasks.

TABLE 7.1

The 5 Phases and 20 Activities of the Internal and External Combined Benchmarking Process

Benchmarking Phase	Related Activities
Phase I: Planning the benchmarking process and characterization of the item(s)	Identify what to benchmark Obtain top management support Develop the measurement plan Develop the data collection plan Review the plans with location experts Characterize the benchmark item
Phase II: Internal data collection and analysis	Collect and analyze internal published information Select potential internal benchmarking sites Collect internal original research information Conduct interviews and surveys Form an internal benchmarking committee Conduct internal site visits
Phase III: External data collection and analysis	Collect external published information Collect external original research information
Phase IV: Improvement of the item's performance	Identify corrective actions Develop an implementation plan Gain top management approval of the future state solution Implement the future state solution and measure its impact
Phase V: Continuous improvement	Maintain the benchmarking database Implement continuous performance improvement

> If you can't meet a world standard of quality at the best price, you're not even in the game.
>
> **—Jack Welch, CEO of General Electric**

Examples

See Figures 7.2 to 7.5.

> Benchmarking is creating better solutions based upon a firm knowledge base. It is not copying the best.

Additional Reading

American Productivity and Quality Center. *The Benchmarking Management* (Portland, OR: Productivity Press, 1993).

Bendell, Tony, Louise Boulter, and John Kelly. *Benchmarking for Competitive Advantage* (New York: McGraw-Hill, 1993).

Camp, Robert C. *Business Process Benchmarking: Finding and Implementing Best Practices* (Milwaukee, WI: ASQ Quality Press, 1995).

Harrington, H. James. *The Complete Benchmarking Implementation Guide* (New York: McGraw-Hill, 1996).

Harrington, H. James, and James S. Harrington. *High Performance Benchmarking* (New York: McGraw-Hill, 1996).

International Benchmarking Clearinghouse. *Planning, Organizing, and Managing Benchmarking Activities: A User's Guide* (Houston, TX: 1992).

Xerox Quality Solutions. *Benchmarking for Process Improvement* (Rochester, NY: Xerox Quality Solutions, 1994).

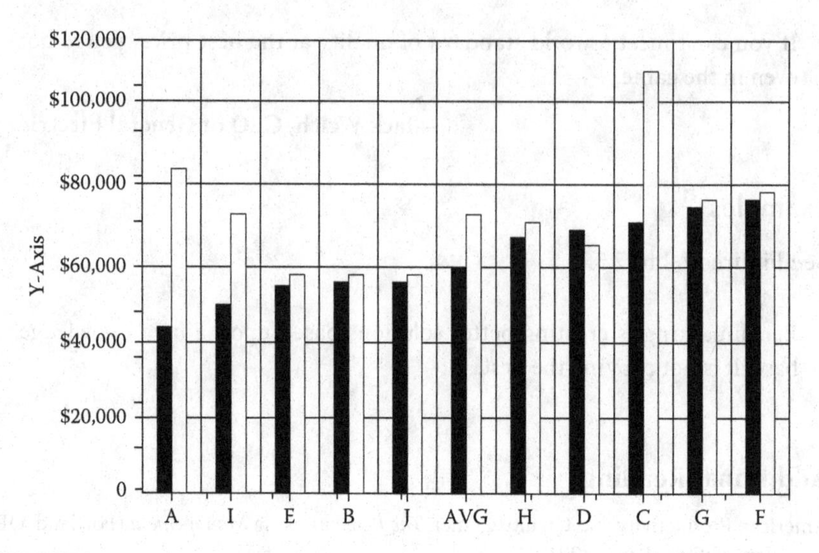

FIGURE 7.2
Compensation and benefits per employee.

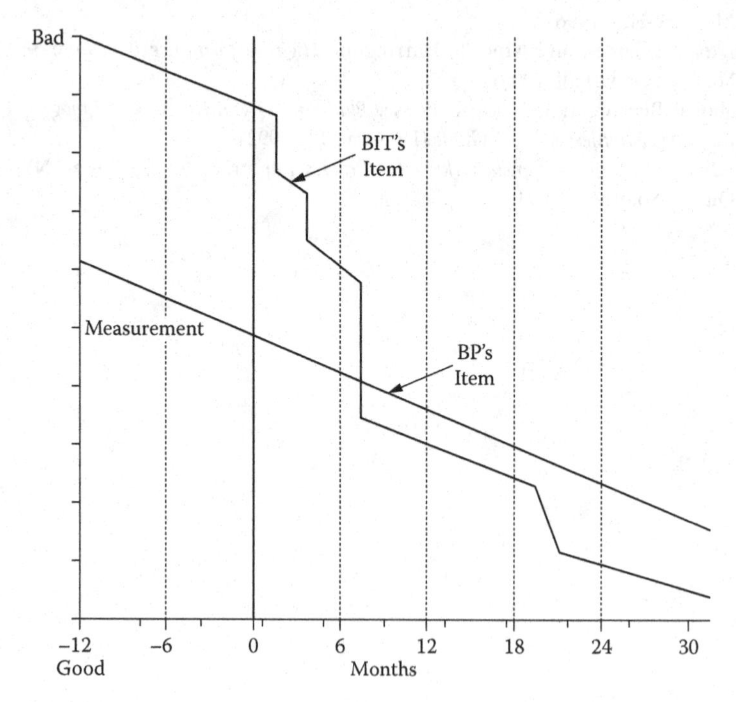

FIGURE 7.3
Process benchmarking.

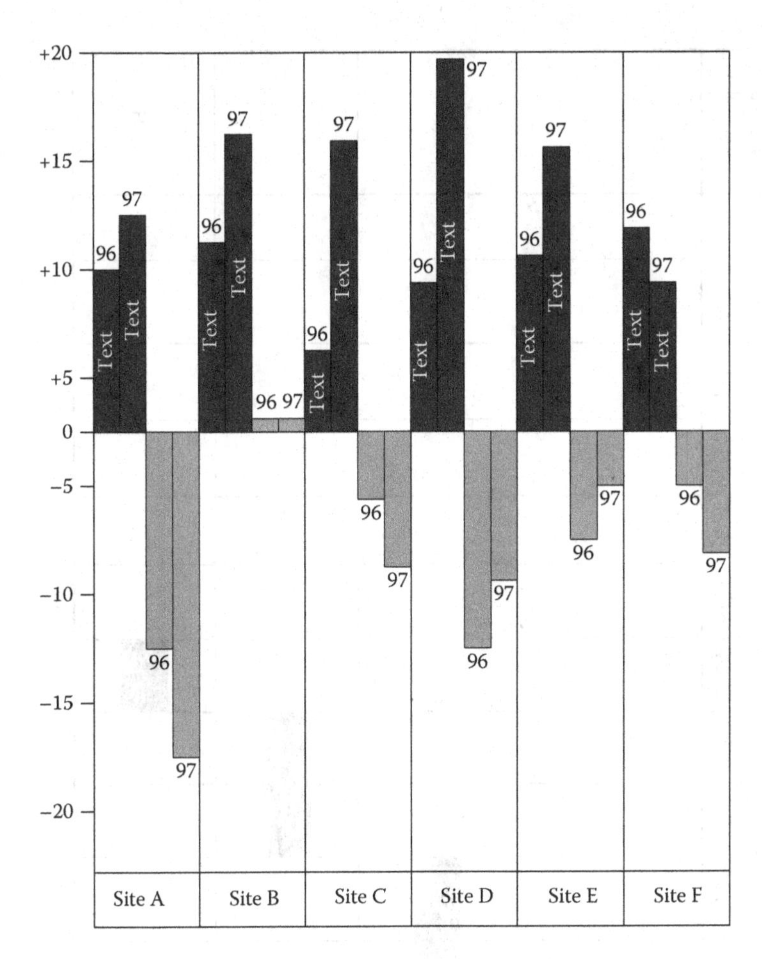

FIGURE 7.4
Changes in two measurements over 2 years for six sites.

Activity	2/3	2/10	2/17	2/24	3/3	3/10	3/17	3/24	3/31	4/7	4/14	4/21	4/28	5/5	Persons Responsible
1. Order new software	■														LH
2. Pilot new software				■	■										SS
3. Write new procedures						■	■								SS
4. Train employees								■							LH

FIGURE 7.5
Implementation of the Gantt chart (*continued*).

5. Install Improvement #17									■	■	■	■	■	■	SS
6. Hold feedback meeting with employees									X		X			X	JSH
7. Measure improvement									■	■	■	■	■	■	JSH
8. Report results														X	SS

FIGURE 7.5 (***continued***)
Implementation of the Gantt chart.

TABLE 7.2

Finance Job Profile

	A	B	C	D	F	G	H	I	Avg.
Vice president	3%	2%	1%	2%	1%	1%	2%	1%	2%
Director	3%	5%	3%	3%	6%	4%	5%	58%	4%
Senior manager	7%	18%	7%	2%	8%	5%	1%	8%	7%
Manager/ supervisor	14%	9%	18%	25%	18%	10%	17%	13%	15%
Subtotal: Managers	27%	34%	29%	32%	33%	20%	25%	27%	28%
Financial analyst	11%	40%	33%	29%	27%	38%	6%	28%	27%
Accountant	36%	21%	19%	23%	23%	12%	38%	28%	25%
Clerical	26%	5%	19%	16%	17%	30%	31%	17%	20%
Subtotal: Nonmanagers	73%	66%	71%	68%	67%	80%	75%	73%	72%
Total	100%	100%	100%	100%	100%	100%	100%	100%	100%

Note: Company E omitted due to incomplete data.

TABLE 7.3

Common Areas of Focus

	A	B	C	D	E	F	G	H	I	J
Systems improvement	X	X	X	X	X	X	X	X	X	X
Overall efficiency			X	X	X		X	X	X	X
Financial close/reporting			X	X	X	X	X	X		X
Budget and forecast	X	X	X		X					
Revenue and A/R		X		X	X		X	X		
General ledger			X	X			X			
Cost accounting				X		X	X	X		X
Asset/cash management	X						X	X		
EDI		X	X					X		
Purchasing and A/P				X						
Organization structure				X						
Business model	X					X		X	X	

TABLE 7.4

Number of Domestic Systems by Function

	A	B	C	D	E	F	G	H	I	J	Avg.
General ledger	1	1	3	2	1	1	1	2	4	2	1.8
Accounts receivable	1	1	3	2	1	3	1	2	1	1	1.6
Accounts payable	1	1	3	2	1	1	1	2	3	2	1.7
Fixed assets	1	1	3	2	1	1	1	2	3	2	1.7
Cost accounting	1	1	1	2	1	4	1	1	1	3	1.6
Purchasing	1	1	3	1	2	2	1	2	3	1	1.7
Order entry	1	1	3	2	1	3	1	2	6	1	2.1
Financial reporting if different than G/L	1	2	1	2	1	1	0	2	3	1	1.4

TABLE 7.5

Measurement Interaction Chart

Corrective Actions	**Measurements**							
	1	**2**	**3**	**4**	**5**	**6**	**7**	**8**
A	0	+	0	0	+	–	+	0
B	+	0	0	0	0	0	0	0
C	+	–	+	0	+	0	0	–
D	–	0	0	+	0	–	0	+
E	0	0	0	–	0	+	+	0
F	0	+	+	–	0	+	+	0
G	0	0	0	+	+	0	0	0

TABLE 7.6

Improvement Analysis Chart for Future State Solution 3

Measurement	**Improvement**						**Original Value**	**New Value**
	1	2	3	4	5	6		
Cycle time (days)								
	−9.0	−0.5	−3.0	−10.0	+5.0	0	35.0	17.5
Processing time (hours)								
	−1.0	−4.0	−1.1	0	+3.1	0	10.0	7.5
Error rate (errors/1,000)	0	0	0	0	−5	−2	8.0	
	3.0							

TABLE 7.7

Implementation Cost Analysis Chart for Future State Solution 3 (Cost in $1,000)

	Improvement							
Cost Categories	**1**	**2**	**3**	**4**	**5**	**6**	**OCM**	**Total Cost**
Implementation team	10	20	0.5	60	10	0.5	3.5	9,405
Target group	3	0	4	10	0		4.0	21.0
Consulting	0	0	0	100	0	0	50.0	150.0
Equipment	0	0	0	100	0	0	0	100.0
Software	10	0	0	50	0	0	0	60.0
Total	23.0	20.0	54.5	320	10.0	0.5	57.5	523.5

TABLE 7.8

Risk Analysis Chart for Future State Solution 3

Improvement No.	Risk Percentage	Reason
1	20%	Too little data to verify the effectiveness of the improvement
2	5%	High workload in area
3	0%	—
4	20%	New computer and software that we have not used
5	10%	No sustaining sponsor
6	20%	Could be a cost overrun
Total	12%	Corrective action plan did not eliminate all the risks

TABLE 7.9

Benefits-Cost-Risk Analysis Chart for Three Alternative Future State Solutions

		Future State Solutions		
Measurements	**Original Process**	**1**	**2**	**3**
Cycle time (days)	35.0	16.2	19.5	17.5
Processing time (years)	10.0	6.5	8.3	7.5
Errors/1,000	25.1	12.3	9.2	3.0
Cost/cycle	$950	$631	$789	$712
Service response time (hours)	120	65	80	75
Implementation				
Cost in $1,000		$1,000	$100	$423.5
Cycle time (months)		29	6	16
Risk		35%	10%	20%
Workforce restructuring index		65%	5%	45%

TABLE 7.10

Total Improvement Index

Measurement	Improvement Value	Weighted Value	Improvement Index
Cycle time	123%	15%	0.185
Cost	138%	40%	0.552
Error rate	185%	10%	0.185
Service response time	165%	35%	0.578
Total improvement index	1.500		

BUREAUCRACY ELIMINATION METHODS

Our processes are designed to protect the organization from the 0.1% of our employees that are dishonest. So we penalize the 99.9% of our people that are honest and want to do a good job.

Definition

Bureaucracy elimination is an approach to identify and eliminate bureaucracy from business processes.

Just the Facts

The word *streamlining* suggests the ultimate search for efficiency and effectiveness, an absence of fat and excess baggage, the smooth flow and unrestricted directness of effort and motion. Streamlining implies symmetry, harmony of elements, and beauty of design. *Bureaucracy*, on the other hand, means the opposite. It is a major stumbling block to the organized, systemic, organization-wide implementation of business process improvement concepts and methods. Bureaucracy is everywhere, even when we don't recognize it. We must learn to actively search for and recognize it. Then, we need to eliminate it.

The Big B (bureaucracy) stands for bad, boring, burdensome, and brutal. We often think of bureaucracy as departments with layers of officials striving to advance themselves and their departments by creating useless tasks and rigid, incomprehensible rules. We think of long delays in processing as documents go through multiple channels and levels of review, requiring multiple signatures by people who are never available when needed. Their existence seems to add resistance to progress, adding cost but little real value.

Bureaucracy often creates excessive paperwork in the office. Managers typically spend 40 to 50% of their time writing and reading job-related materials; 60% of all clerical work is spent on checking, filing, and retrieving information, while only 40% is spent on process value-added tasks. This bureaucracy results from organizational or individual personalities caused by such psychological factors as:

- Paranoia about being blamed for errors
- Poor training
- Distrust of anyone
- Lack of work
- Inability to delegate
- Lack of self-worth
- Thrill of checking for and finding minuscule mistakes
- Need to overcontrol
- Unwillingness to share information

The sinister effects of bureaucracy are innumerable and profoundly damaging to every organization and to the business process improvement effort. Therefore, you should evaluate and minimize all delays, red tape, documentation, reviews, and approvals. If they are not truly necessary, you should eliminate them. A word of caution: Sometimes an activity may not have an obvious purpose but is, in fact, valuable to some other process in the organization, so don't eliminate a bureaucratic operation without first understanding why it is in existence and what impact eliminating it would have on the organization.

Management must lead an attack against the bureaucracy that has crept into the systems controlling the business. Bureaucracy in government and business continues to worsen. These huge paperwork empires must be destroyed if industry is to flourish. Our copiers are used far too much, and we have too many file cabinets. More than 90% of the documents we retain are never used again.

The attack on bureaucracy should start with a directive informing management and employees that the organization will not put up with unnecessary bureaucracy, that each approval signature and each review activity will have to be financially justified, that reducing total cycle time is a key business objective, and that any bureaucracy activity that delays the process will be targeted for elimination.

There are two approaches to bureaucracy elimination:

- Process focused
- Incident focused

Process-Focused Approach

The process focus starts by flowcharting a process and then analyzing each step (activity) in the process to see if it is a bureaucracy-type step. You can identify bureaucracy by asking such key questions as:

- Are there unnecessary checks and balances?
- Does the activity inspect or approve someone else's work?
- Does it require more than one signature?
- Are multiple copies required?
- Are copies stored for no apparent reason?
- Are copies sent to people who do not need the information?
- Are there people or agencies involved that impede the effectiveness and efficiency of the process?
- Is there unnecessary written correspondence?
- Do existing organizational procedures regularly impede the efficient, effective, and timely performance of duties?
- Is someone approving something he or she has already approved? (for example, approving capital equipment that was already approved during the budget cycle)

The team needs to ask questions about each process step and then carefully consider the responses in order to gain insights that will help streamline the process. Many activities do not contribute to the content of the process output. They exist primarily for protection or informational purposes, and every effort should be made to minimize these activities.

The team is apt to run into resistance related to the bureaucracy elimination activities because of varying opinions and organizational politics. Overcoming resistance to eliminating bureaucracy takes skill, tact, and considerable planning. Bureaucracy's impact on cost and cycle time should be calculated; its impact on the internal and external customer should be understood. Once the full impact of bureaucracy is understood by all concerned, it is often difficult to justify retaining the activity. The entire organization should continually eliminate every example of bureaucracy.

After completing the flow diagram, the team should review it, using a light blue highlighter to designate all activities related to review, approval, second signature, or inspection. The team will soon learn to associate blue on the flowchart with bureaucracy. These blue activities become key targets for elimination.

The managers responsible for each of the Big B activities should justify activity-related costs and delays. Often, a manager will try to push the matter aside, saying, "It only takes me 2 or 3 seconds to sign the document. That doesn't cost the organization anything." The answer to a remark like this is, "Well, if you don't read the document, there is no reason to sign it."

In one organization that started a Big B elimination campaign, a group of 10 capital equipment requests were processed through five levels of management approval signatures. Two requests contained only a legitimate cover sheet with blank pages attached instead of the written justification required. And these two requests made it through all five levels of approval. This experiment shocked management into fully backing the Big B elimination campaign.

The justification for retaining a Big B activity requires some data. How many items are rejected? How much does the organization save when an item is rejected? Rejecting an item does not necessarily mean savings for the organization—quite the contrary. A rejected document may cause more bureaucracy, more delay, and increased costs. Time cycle delay costs are based on the advantage to the customer if the output from the process is delivered early. The justification for each Big B activity should be based on the potential loss or gain to the organization. If it is a break-even point, the activity always should be eliminated. Many organizations require a 3:1 return on any investment (ROI). This rule should apply to bureaucracy activities also. A bureaucracy step should be left in only if there is a sizable, documented savings from the activity.

Even then, the team should look at why the bureaucracy activity is saving money and see if there is any other, less expensive way of accomplishing the same result.

Incident-Focused Approach

With the individual incident-focused approach, the organization sets up a Big B Elimination Team that spearheads the bureaucracy elimination project. All employees in the organization are encouraged to request the Big B Elimination Team to investigate activities that are perceived to be bureaucratic. The Big B Elimination Team investigates each suggestion to determine if the activity should be eliminated or decreased. In conjunction with the person performing the activity under evaluation, a cost-benefit analysis is performed. Bureaucratic activities that do not have a 3:1 ROI are eliminated. This approach has the advantage of getting the total organization involved in the bureaucracy elimination process.

Examples

Example 1

The cost of Big B activities is more than we realize. Reading and approving a purchase order may cost the controller only 1 minute of time, but the process design to obtain that signature costs much more. Let's look a little closer. (See Figure 7.6.)

In this example, the controller needs to justify delaying the purchase order an additional 2 days and increasing its cost by $26.00. The 2-day delay costs the organization 2 days of production savings that the new equipment would have realized if it came on-line 2 days earlier. This can amount to hundreds of dollars of added cost. The real losses caused by Big B activities are always much more than we originally estimate.

Example 2

IBM, Brazil, launched a Big B campaign that eliminated 50 unnecessary procedures, 450 forms, and 2.5 million documents a year.

		Time (minutes)	Cost (dollars)
1.	Manager goes to the supply cabinet to get envelope.	5	6.00
2.	Manager looks up controller's office address and addresses envelope.	5	6.00
3.	Manager takes envelope to mail drop and returns to office.	7	9.80
4.	Mail is picked up and taken to mailroom, sorted, and delivered to controller's secretary.	1800	2.00
5.	Secretary logs in mail and puts into controller's incoming mail folder.	4	.60
6.	Waiting for controller to read document	800	0
7.	Reading and signing documents	1	1.20
8.	Waiting for secretary to process.	800	0
9.	Secretary logs out the document	2	.30
Total:		3424	25.90

FIGURE 7.6
Example of cost of Big B activities.

Example 3

To purchase a ballpoint pen (or anything else, for that matter) at Intel took 95 administrative steps and 12 pieces of paper. When the company eliminated the bureaucracy, the purchase took eight administrative steps and one form. Intel estimated its attack on Big B improved productivity by 30% and saved $60 million a year. It would take the equivalent of $277 million in increased sales for Intel to generate $60 million in profits.

IBM, San Jose, California, set up a bureaucracy elimination committee. Anyone that felt that a specific bureaucracy activity was unjustified was asked to bring it to this committee's attention. The committee would then investigate it to define if it could be eliminated.

Additional Reading

Harrington, H. James. *Business Process Improvement* (New York: McGraw-Hill, 1991).
Harrington, H. James, and James S. Harrington. *Total Improvement Management* (New York: McGraw-Hill, 1995).

CONFLICT RESOLUTION

If we spent as much time loving as we do fighting, the world would double its population.

Definition

Conflict resolution is an approach to obtaining common agreement between two or more parties that have a different position or opinion. It involves techniques like:

- Negotiation
- Mediation
- Diplomacy

It often ends up with all parties finding an acceptable compromise. It is sometimes called dispute resolution. Typical tools include:

- Arbitration
- Litigation
- Ombudsman processes
- Mediation

Just the Facts

Conflict resolution in the Western culture involves communication between the parties and solving the problems by seeking a win-win position.

Example

The tools used for conflict resolution depend on the nature of the dispute, the individuals involved, and the individuals or companies affected by the dispute. Conflict resolution starts with identifying the problem from both sides (points of view). From here it is important to understand what constitutes resolution by all parties. Then it involves communicating with the individuals in order to find common ground.

The appropriate tool to use will depend on all of these factors. A simple model was proposed by Stephen Covey in his book *The Seven Habits of Highly Effective People*. They are:

1. Be proactive: Do something.
2. Begin with the end in mind: What constitutes resolution?
3. Put first things first: Identify points of agreement and conflict.
4. Think win-win: Seek resolutions that will allow a win-win outcome.
5. Seek first to understand, then to be understood: Don't jump at the first resolution.
6. Synergize: Seek ground where all parties are working together.
7. Sharpen the saw. Exercise for physical renewal.

Additional Reading

Covey, Stephen R. *The Seven Habits of Highly Effective People* (New York: Simon and Schuster, 1989).

Johnson, Richard Arvid. *Management, Systems, and Society: An Introduction* (Pacific Palisades, CA: Goodyear, 1976), pp. 142–148.

Knowles, Henry P., and Börje O. Saxberg. *Personality and Leadership Behavior* (Reading, MA: Addison-Wesley, 1971), chap. 8.

Picard, Daniel (ed.). *The Black Belt Memory Jogger* (GOAL/QPC and Six Sigma Academy, 2002).

Wortman, Bill. *The Certified Six Sigma Black Belt Primer* (Quality Council of Indiana, 2001).

CRITICAL TO QUALITY

Measure your performance as your customer would measure it.

Definition

Critical to quality (CTQ) is the analysis of the characteristics of a product or service that are critical to the customer or to the performance of the product or service to guarantee that they meet requirements.

Just the Facts

Often a radar chart is used to identify critical to quality parameters/characteristics related to the item being evaluated. The importance to satisfying the customers' requirements is rated on a scale of 1 to 10 for each of the legs of the radar chart. By connecting the points, you can readily see which parameters are critical to satisfying the customer. In the example (Figure 7.7), you can see that two items would be classified as critical to quality: timely and defect/error-free. The other items, although important to your customer, are not critical.

Critical to Quality (CTQ) Characteristics

A tree diagram is another approach used to define customer CTQs to determine process performance in relationship to voice of the customer (VOC)/voice of the business (VOB). (See Figure 7.8.)

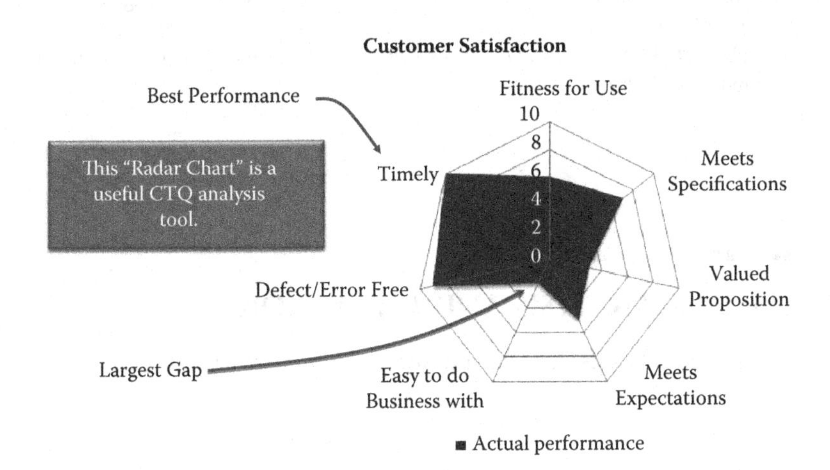

FIGURE 7.7
Customer satisfaction drivers versus actual performance.

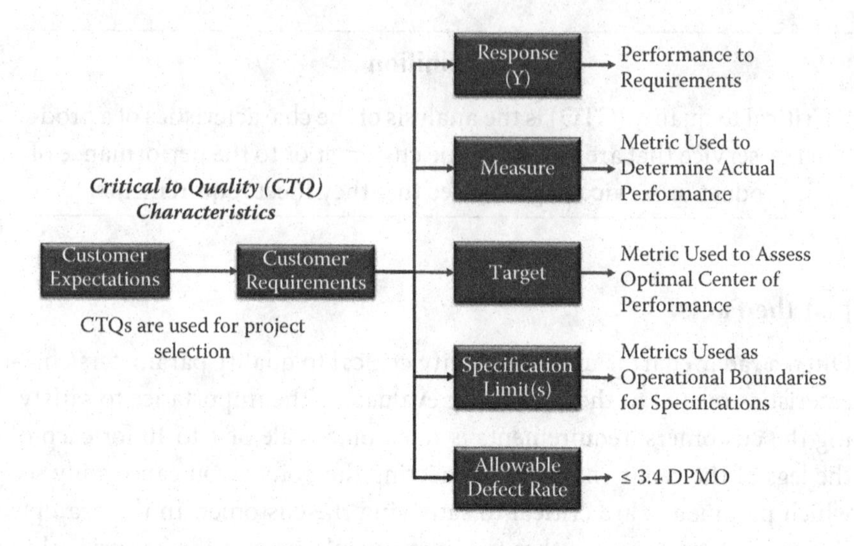

FIGURE 7.8
Customer requirements drive critical to quality metrics.

Example

After identifying the CTQ parameters, evaluate the product or service against the CTQs to ensure that they are met.

Additional Reading

Picard, Daniel (ed.). *The Black Belt Memory Jogger* (GOAL/QPC and Six Sigma Academy, Salem, NH, 2002).

Wortman, Bill. *The Certified Six Sigma Black Belt Primer* (Quality Council of Indiana, West Terre Haute, 2001).

CYCLE TIME ANALYSIS AND REDUCTION

Like any race, the one that gets it to the customer first wins the market.

Definition

Cycle time analysis and reduction is an approach to reduce the time that it takes to move an item through a process.

Just the Facts

Based upon the International Quality Study conducted by Ernst & Young LLP and the American Quality Foundation, there are only six worldwide best practices:

- Top management involvement
- Strategic planning
- Supplier certification
- Process simplification
- Process value analysis
- Cycle time analysis

Time is money. Every time anything is not moving forward, it costs you money. To date, most of our focus has been on reducing processing time because we see it as added labor cost. But cycle time is also very important. Reducing time-to-market, time to respond to a customer request, and time to collect an outstanding bill can mean the difference between success and failure.

Critical business processes should follow the rule of thumb that time is money. Undoubtedly, process time uses valuable resources. Long cycle times prevent product delivery to our customers and increase storage costs. A big advantage Japanese auto companies have over American companies is their ability to bring a new design to the market in half the time and cost. Every product has a market window. Missing the early part of the product window has a major impact on the business. Not only does the company lose a lot of sales opportunities, but it is facing an uphill battle against an already established competitor. With the importance of meeting product windows, you would think that development schedules would always be met. Actually, few development projects adhere to their original schedule.

Applications of Cycle Time Analysis and Reduction

The object of this activity is to reduce cycle time. This is accomplished by focusing the Process Improvement Team's attention on activities with long time cycles and those activities that slow down the process. The timeline flowchart provides valuable assistance in identifying the problem activities. The team should look at the present process to determine why schedules and commitments are missed, then reestablish priorities to eliminate these slippages, and then look for ways to reduce the total cycle time. Typical ways to reduce cycle time are:

- *Do activities in parallel rather than serially.* Often, activities that were done serially can be done in parallel, reducing the cycle time by as much as 80%. Engineering change review is a good example. In the old process, the change folder went to manufacturing engineering, then to manufacturing, then to field services, then to purchasing, and finally, to quality assurance for review and sign-off. It took an average of 2 days to do the review in each area and 1 additional day to transport the document to the next reviewer. The engineering change cycle took 15 working days, or 3 weeks, to complete. If any of the reviewers had a question that resulted in a change to the document, the process was repeated. By using computer-aided design (CAD), all parties can review the document simultaneously and eliminate the transportation time. This parallel review reduces the cycle time to 2 days.

  Another less equipment-intensive approach would be to hold weekly change meetings. This would reduce the average time cycle to 3.5 days and eliminate most of the recycling, because the questions would be resolved during the meeting.
- *Change activity sequence.* The geographic flowchart is a big help to this activity. Often, output moves to one building and then returns to the original building. Documents move back and forth between departments within the same building. In this stage, the sequence of activities is examined to determine whether a change would reduce cycle time. Is it possible to get all the signatures from the same building before the document is moved to another location? When a document is put on hold waiting for additional data, is there anyone else who could be using the document, thereby saving cycle time later on?
- *Reduce interruption.* The critical business process activities should get priority. Often less important interruptions delay them. People working on critical business processes should not be located in high traffic areas, such as near the coffee machine. Someone else should answer their phones. The office layout should allow them to leave their work out during breaks, lunch, or at day's end. The employee

and the manager should agree on a time when the employee will work uninterrupted, and the manager should help keep these hours sacred.

- *Improve timing.* Analyze how the output is used to see how cycle time can be reduced. If the mail pickup is at 10:00 a.m., all outgoing mail should be processed before 9:45 a.m. If the computer processes a weekly report at 10:00 p.m. on Thursday, be sure that all Thursday first-shift data are input by 8:00 p.m. If you miss the report analysis window, you may have to wait 7 more days before you receive an accurate report. If a manager reads mail after work, be sure that all of that day's mail is in his or her incoming box by 4:30 p.m. It will save 24 hours in the total cycle time. Proper timing can save many days in total cycle time.
- *Reduce output movement.* Are the files close to the accountants? Does the secretary have to get up to put a letter in the mailbox? Are employees who work together located together? For example, are the quality, development, and manufacturing engineers located side by side when they are working on the same project, or are they located close to other people in the same discipline?
- *Analyze locations.* Is the process being performed in the right building, city, state, or even the right country? Where the activity is performed physically can have a major impact on many factors. Among them are:
  - Cycle time
  - Labor cost
  - Customer relations
  - Government controls and regulations
  - Transportation cost
  - Employee skill levels

  Performing the activity in less than the optimum location can cause problems, from a minor inconvenience all the way to losing customers and valuable employees. The approach and consideration for selecting the optimum location vary greatly from process to process. As a general rule, the closer the process is located to the customer, the better. The restraints to having the process and its customer in close proximity are economy of scale, stocking costs, equipment costs, and utilization considerations. With today's advances in communication and computer systems, the trend is to go to many smaller locations located either close to the supplier or close to the customers. Even the large manufacturing specialty departments (machine shop, welding department, tool room, etc.) are being separated into small work cells that are organized to fit a process in which a lot size of one is the production plan. Often, the advantages of quick response to customer requests, increased turns per year, and decreased inventories far offset the decreased utilization costs.

Questions like "Should we have a centralized service department or many remote ones?" require very careful analysis. A graphic flowchart helps make these decisions, but the final decision must be based on a detailed understanding of customer expectations, customer impact, and cost comparisons between the options.

- *Set priorities.* Management must set proper priorities, communicate them to employees, and then follow up to ensure that these priorities are lived up to. It is often a big temptation to first complete the simple little jobs—the ones that a friend wants worked on, the ones someone called about—and let the important ones slip. It's the old "squeaky wheel" message. As a result, projects slip, money is lost, and other activities are delayed. Set priorities and live by them.

Cycle Time Analysis and Reduction Process

The cycle time reduction process consists of 16 activities:

1. Flowchart the process that is being studied.
2. Conduct a process walk-through to understand the process and verify the flowchart.
3. Collect cycle time data related to each activity and task. It is often advisable to collect minimum and maximum cycle times in addition to average. This is necessary because, typically, an organization loses customers not over averages but over worst-case conditions.
4. Collect data that define the quantity flow through each leg of the flow diagram.
5. Construct a simulation model that includes all of the data that have been collected.
6. Perform a Monte-Carlo analysis using the simulation model to define the cycle time frequency distribution.
7. Classify each activity or task as real-value-added, business-value-added, or no-value-added. Eliminate as many of the business-value-added and no-value-added activities as possible.

8. Define the average cycle time's critical path through the process using the simulation model.
9. Using the cycle time reduction principles, eliminate the critical path.
10. Repeat activities 8 and 9 until the minimum cycle time is obtained.
11. Define the worst-case critical path through the process using the simulation model.
12. Using the cycle time reduction principles, eliminate the critical path.
13. Repeat steps 11 and 12 until the minimum worst-case cycle time is obtained.
14. Develop a plan to change the process to be in line with the modified simulation model.
15. Pilot the modifications as appropriate.
16. Implement the new process.

Examples

IBM's RPQ or special bit process provides modifications to computers so that they can meet an individual customer's unique needs. Typically, the cycle time to take a customer special requirement and design and price out the modification was taking an average of 90 days. The business was very profitable, as 20% of the bids were closed. IBM decided that 90 days was too long and a team was put together to reduce the cycle time. As a result, 24 months later, any place in the world, it took an average of 15 days to complete a special bid process.

In addition, the cost related to preparing the bid was decreased by 30%, but that was not the big payoff. Along with the quick response to customer requests, the bit closure rate jumped from 20% to 65%—a 325% improvement. Customers love companies who respond quickly to their special requests.

Additional Reading

Harrington, H. James. *Business Process Improvement* (New York: McGraw-Hill, 1991).

Harrington, H. James, and James S. Harrington. *Total Improvement Management* (New York: McGraw-Hill, 1995).

Harrington, H. James, Glen Hoffherr, and Robert Reid. *Area Activity Analysis* (New York: McGraw-Hill, 1998).

Northey, Patrick, and Nigel Southway. *Cycle Time Management* (Portland, OR: Productivity Press, 1993).

FAST-ACTION SOLUTION TECHNIQUE (FAST)

Start your improvement process with some quick wins to get top management's ongoing support.

Definition

FAST is a breakthrough approach that focuses a group's attention on a single process for a 1- or 2-day meeting to define how the group can improve the process over the next 90 days. Before the end of the meeting, management approves or rejects the proposed improvements.

Just the Facts

Fast-action solution technique (FAST) is based on an improvement tool first used by International Business Machines (IBM) Corporation in the mid-1980s. General Electric refined this approach in the 1990s and called it "workout." Ford Motor Company further developed it under the title "RAPID." Today, Cap Gemini Ernst & Young extensively uses this approach (which they call "EXPRESS") with many clients around the world (Figure 7.9). It is also often used by other organizations throughout the Americas.

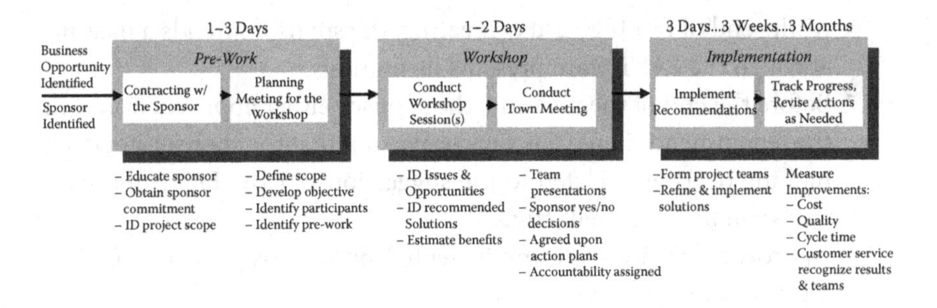

FIGURE 7.9
View of Cap Gemini Ernst & Young's EXPRESS process.

FAST can be applied to any process level, from a major process down to and including the activity level. The FAST approach to business process improvement (BPI) centers around a single 1- or 2-day meeting that identifies root causes of problems and/or no-value-added activities designed into a present process. Typical improvement results from the FAST approach are reduced cost, cycle time, and error rates between 5 and 15% in a 3-month time period. Within 1 or 2 days, the potential improvements are identified and approved for implementation; hence the term *FAST* was given to this approach.

The FAST approach evolves through the following eight activities:

1. A problem or process is identified as a candidate for FAST.
2. A high-level sponsor agrees to support the FAST initiative related to the process that will be improved. (The process must be under the sponsor's span of control.)
3. The FAST team is assigned, and a set of objectives is prepared and approved by the sponsor.
4. The FAST team meets for 1 or 2 days to develop a high-level process flowchart and to define what actions could be taken to improve the process's performance. All recommendations must be within the span of control of the team members and able to be completely implemented within a 3-month time period. All other recommendations are submitted to the sponsor for further consideration at a later date.
5. A FAST team member must agree to be responsible for implementing each recommendation that will be submitted to the sponsor.

6. At the end of the 1- or 2-day meeting, the sponsor attends a meeting at which the FAST team presents its findings.
7. Before the end of the meeting, the sponsor either approves or rejects the recommendations. It is very important that the sponsor not delay making decisions related to the suggestions, or the FAST approach will soon become ineffective.
8. Approved solutions are implemented by the assigned FAST team member over the next 3 months.

Examples

The following are typical documentation results that Cap Gemini Ernst & Young (CGE&Y) recorded as a result of helping clients use FAST (CGE&Y's trademark is EXPRESS).

- Problem 1: Different entities negotiating for identical software.
  - Results:
    - Cost savings of $240,000 the first year
    - Reduced cycle time by 4 weeks

- Problem 2: Long delays and high costs related to suppliers getting approval to change their processes.
  - Results:
    - Cycle time reduced by 2 weeks
    - Reduced mailing and storage cost by $385,000

- Problem 3: The quarterly financial closing cycle was too long.
  - Results:
    - Reduced cycle time by a day
    - Accelerated accrual process by 3 days

Additional Reading

Harrington, H. James. *Business Process Improvement* (New York: McGraw-Hill, 1991).
Harrington, H. James. *FAST—Fast Action Solution Technique* (Paton Professional Press, 2009).
Harrington, H. James, and James S. Harrington. *Total Improvement Management* (New York: McGraw-Hill, 1995).

FOUNDATION OF SIX SIGMA (MINIMIZING VARIATION)

All knowledge is based on analysis of measurement data.

Definition

The fundamental concept that Six Sigma is based upon is by reducing variation, the organization has a higher probability of meeting or exceeding customer requirements.

Just the Facts

The Six Sigma focus is on minimizing variation, and it is represented by the formula where *y* is the dependent variable and *f* is the independent variable. (See Table 7.11.)

TABLE 7.11

Input versus Output Relationship

What Do We Watch?	What Do We Work On?
***Y* variable**	**X_1 ... X_N variable**
Dependent	Independent
Outcome	Process output
Effect	Cause
Symptom	Problem
Monitor	Control
Requirements	Performance

What Does "Good Enough" Mean?

Is 99% good enough? Good enough isn't any of the following:

- 10 million lost articles of mail each day
- 5,000 surgeries each day performed incorrectly
- 200,000 incorrect drug prescriptions dispensed each year
- At least two missed landings at most major airports each day

A good process is one that will repeatedly and reliably produce excellent service and products. Does it have to be perfect? Can any of us perform error-free for the rest of our lives? Would 3.4 errors for million opportunities be good enough? That is the objective of the Six Sigma methodology. (See Figures 7.10 to 7.12.)

Six Sigma is about variation reduction. It is concerned about the process variation that, when depicted on a histogram, forms its width called "common cause variation," and the external variation called "special cause variation."

Example

A process is yielding 50% good parts. An analysis of the data shows that the process is producing product that is out of specification for both the high and low specification limits. A further evaluation shows that 30% of

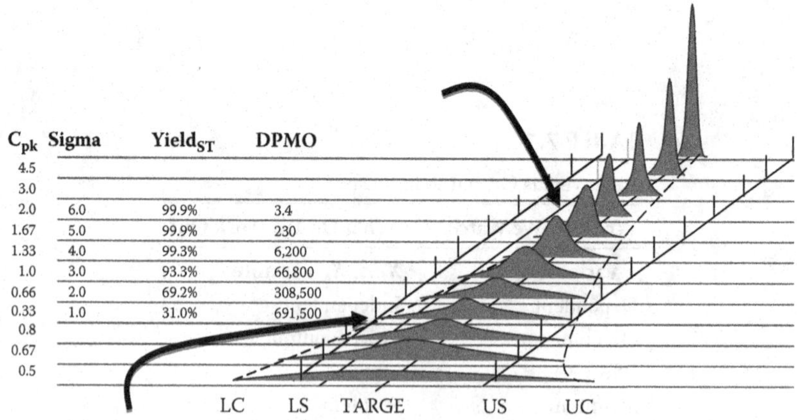

FIGURE 7.10
Sigma level versus defects per million opportunities.

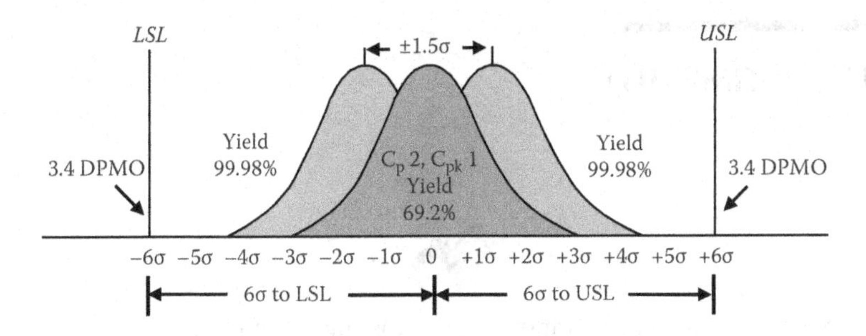

FIGURE 7.11
Output drift over time.

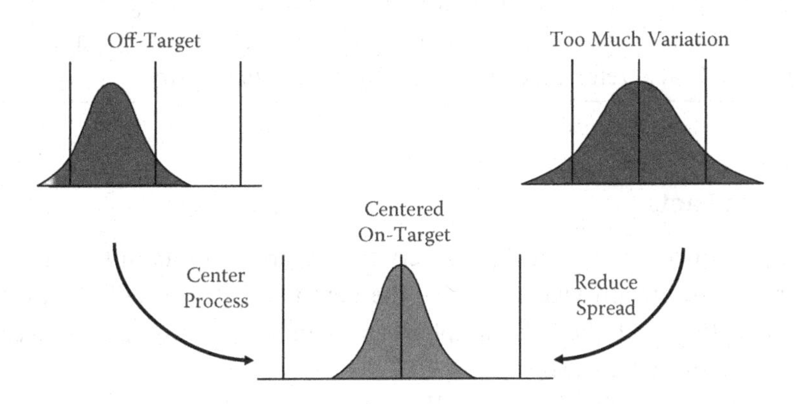

FIGURE 7.12
Results of reducing process variation and centering the output.

the bad product is above the upper specification limits. You will reduce the failure rate from 50% to 30% by simply moving the center of the process toward the center of the specification limits. A further reduction of only 10% of the variation of the process will yield an additional 15%, thereby reducing the defect rate from 50% to 15%.

Additional Reading

Picard, Daniel (ed.). *The Black Belt Memory Jogger* (GOAL/QPC and Six Sigma Academy, Salem, NH, 2002).

Wortman, Bill. *The Certified Six Sigma Black Belt Primer* (Quality Council of Indiana, West Terre Haute, 2001).

JUST-IN-TIME (JIT)

When your product is not moving, it is costing you money.

Definition

Just-in-time (JIT) is a strategy to reduce costs by reducing in-process inventory and associated carrying costs. The just-in-time production method is also referred to as the *Toyota Production System*.

Just the Facts

The JIT process relies on signals between different points in the process that tell production when to make the next part. This usually takes the form of "tickets" but can be simple visual signals, such as the presence or absence of a part on a shelf.

When correctly implemented, JIT focuses on continuous improvement and can improve a manufacturing organization's quality and efficiency.

Key areas include:

1. Flow
2. Employee involvement
3. Quality
4. Environmental concerns
5. Price volatility
6. Demand
7. Supply availability

Example

Instead of storing empty bottles, a drink manufacturer has empty bottles delivered at regular times providing a 4-hour operating window between deliveries. This allows for the following:

- The need to move the bottles from the trailers to storage is reduced.
- Bottles are moved directly to production.
- The supplier knows the production rate required to satisfy the customer.

By producing product to customer demands, it reduces warehouse storage costs.

Having necessary parts available when needed and ordering just enough to keep the process moving with no downtime reduces the inventory of parts that must be maintained. This reduces the within-period costs and storage costs. This also facilitates suppliers in that they can provide parts at a more constant level rather than providing more parts than necessary and then having downtime.

Some of the benefits include:

- Reduced setup time
- Flow of goods from warehouse to shelves improves
- Employees with multiple skills are used more efficiently
- Production scheduling and work hour consistency synchronized with demand
- Increased emphasis on supplier relationships
- Supplies come in at regular intervals throughout the production day

Additional Reading

Hirano, Hiroyuki, and Furuya Makota. *JIT Is Flow: Practice and Principles of Lean Manufacturing* (Vancouver, WA: PCS Press, 2006).

Liker, Jeffrey. *The Toyota Way: 14 Management Principles from the World's Greatest Manufacturer*, 1st ed. (New York: McGraw-Hill, 2003).

Management Coaching and Training Services. The Just-in-Time (JIT) Approach. 2006. Retrieved June 19, 2006, from [www.mcts.com/Just-In-Time.html].

Ohno, Taiichi. *Just-in-Time for Today and Tomorrow* (Productivity Press, 1988).

Wadell, William, and Norman Bodek. *The Rebirth of American Industry* (PCS Press, 2005).

Womack, James P., Daniel T. Jones, and Daniel Roos. *The Machine That Changed the World: The Story of Lean Production* (Harper Business, 1991).

MATRIX DIAGRAM/DECISION MATRIX

Sometimes you need a way to select the right alternative.

Definition

Matrix diagram/decision matrix is a systematic way of selecting from larger lists of alternatives. They can be used to select a problem from a list of potential problems, select primary root causes from a larger list, or select a solution from a list of alternatives.

Just the Facts

The matrix diagram is an approach that assists in the investigation of relationships. While there are many variations of matrix diagrams, the most commonly used is the decision, or prioritization, matrix. These come in two basic formats: the L-Shaped Matrix and the T-Shaped Matrix.

L-Shaped Matrix

We will start by showing a relatively simple L-Shaped Matrix that compares two sets of information. (See Figure 7.13.)

As you can see from the example, we are comparing several automobile dealers (our choices) with a predetermined set of decision criteria. Now all that remains is to determine which type of ranking method we will use. There are four basic types of ranking methods:

1. Forced choice: Each alternative is ranked in relation to the others. The alternative best meeting the criteria gets a score equal to the number of alternatives. Since we have five dealers in our example, the worst would get a 1 and the best a 5.

Criteria / Choice	Recommended By Friends	Good Selection of Cars	Good Service Department	Free Loaner Cars	Free Drop Off and Pick Up
Dealer 1					
Dealer 2					
Dealer 3					
Dealer 4					
Dealer 5					

FIGURE 7.13
Example of an L-Shaped Matrix.

2. Rating scale: Each alternative is rated independently against an objective standard. (For example, a 1 to 10 scale would have 1 = very low (does not meet the standard at all) and 10 = perfect (absolutely meets the standard).
3. Objective data: Here we enter actual data, rather than the opinions of the individual(s) doing the ranking.
4. Yes/no: If the criteria are expressed in absolute terms, so an alternative either meets the criteria or not, a Y for yes or an N for no may be entered to indicate conformance or nonconformance.

Figure 7.14 shows our automobile dealership example using the simple yes/no ranking criteria. As you can see, the easiest ranking method—the yes/no approach—often leaves the user with little information in which to make a decision. In Figure 7.14 all five dealers got three yeses.

Let's try ranking our choices using the forced-choice method. (See Figure 7.15.) Remember, this method ranks each alternative in relation to the others. In this case the dealer meeting our criteria the best will get a 5 and the worst will get a 1 (since we have five choices).

Now we have information that might allow us to make a decision. As you can see, Dealer 4 scored the highest with 18. Does this mean you should automatically buy your car from him? Not necessarily. Although Dealer 4 did score the highest overall, he scored the lowest on "free dropoff and

Criteria / Choice	Recommended By Friends	Good Selection of Cars	Good Service Department	Free Loaner Cars	Free Drop Off and Pick Up
Dealer 1	(Y)	(Y)	N	N	(Y)
Dealer 2	(Y)	N	(Y)	(Y)	N
Dealer 3	N	(Y)	(Y)	N	(Y)
Dealer 4	(Y)	(Y)	(Y)	N	N
Dealer 5	(Y)	N	(Y)	(Y)	N

FIGURE 7.14
Example of an L-Shaped Matrix using yes/no ranking method.

Choice \ Criteria	Recommended By Friends	Good Selection of Cars	Good Service Department	Free Loaner Cars	Free Drop Off and Pick Up	Totals
Dealer 1	5	4	1	2	5	17
Dealer 2	3	1	2	5	3	14
Dealer 3	1	3	3	1	4	12
Dealer 4	4	5	5	3	1	(18)
Dealer 5	2	2	4	4	2	14

FIGURE 7.15
Example of an L-Shaped Matrix using the forced-choice ranking method.

pickup." If this were a critical element to the potential buyer, he or she might want to consider the second choice, Dealer 1.

This is where using the objective data method might be of assistance. The person or group doing the ranking might even consider using a combination of ranking methods. This is certainly an option, but it makes the final selection a bit more complex.

T-Shaped Matrix

The second format we mentioned was the T-Shaped Matrix. While the L-Shaped Matrix compares two sets of information, the T-Shaped Matrix compares two sets of information to a third. An example of this could be a corporation's training program. We could compare the type of training available with departments that need the training and training providers. Figure 7.16 shows an example of the T-Shaped Matrix format.

There are many approaches to designing and developing a matrix diagram. Listed below are five steps you may find useful in developing a matrix diagram that's just right for your purpose.

Step 1: Determine the task. Are we looking at two elements or three? What should the desired outcome look like? Is the matrix to be used a problem-solving tool or a planning graph? Is it a stand-alone tool that leads us to action, or will we use it in conjunction with other tools, such as a tree diagram or relation diagram?

<table>
<tr><td rowspan="4">Training Providers</td><td>OD Dept.</td><td></td><td></td><td>X</td><td></td><td></td><td>X</td></tr>
<tr><td>Quality Group</td><td>X</td><td></td><td></td><td>X</td><td>X</td><td></td></tr>
<tr><td>Direct Supervisor</td><td></td><td></td><td></td><td></td><td></td><td></td></tr>
<tr><td>Outside Resource</td><td>X</td><td>X</td><td>X</td><td></td><td></td><td></td></tr>
<tr><td>Training Available</td><td></td><td>TQM Tools</td><td>Computer Skills</td><td>Specific Job Skills</td><td>Team Building</td><td>Effective Meetings</td><td>Leadership Skills</td></tr>
<tr><td rowspan="5">Departments Requiring Specialty Training</td><td>Engineering</td><td>X</td><td></td><td></td><td>X</td><td></td><td></td></tr>
<tr><td>Manufacturing</td><td>X</td><td></td><td></td><td>X</td><td></td><td>X</td></tr>
<tr><td>Finance</td><td>X</td><td>X</td><td>X</td><td></td><td></td><td></td></tr>
<tr><td>Sales</td><td>X</td><td></td><td></td><td>X</td><td>X</td><td></td></tr>
<tr><td>Marketing</td><td>X</td><td></td><td></td><td>X</td><td></td><td></td></tr>
</table>

FIGURE 7.16
Example of a T-Shaped Matrix.

Step 2: Select the matrix format. If we are reviewing the relationships of two elements, you may want to use the L-Shaped Matrix. If you add a third element, you will want to use the T-Shaped Matrix.

Step 3: Determine the criteria for evaluating alternatives. A typical list of criteria is presented below:

- Customer impact
- Number of customers affected
- Within control of the team
- Within influence of the team
- Cost of quality
- Rework
- Frequency of occurrence
- Cycle time impact
- Revenue impact
- Return on investment
- Complexity of analysis

- Time to develop a solution
- Durability of solution
- Cost to implement solution
- Availability of measurements

The criteria should be worded in terms of the ideal result, not worded neutrally. For example, a criterion could be "easy to implement," but not "ease of implementation."

Step 4: Determine the weights for the individual criterion or use equal weighting.

Step 5: Determine how the individual alternatives will be ranked.

- **Forced choice:** Each alternative is ranked in relation to the others. The alternative best meeting the criterion gets a score equal to the number of alternatives and the worst would get a 1.
- **Rating scale:** Each alternative is rated independently against an objective standard. For example, a 1 to 10 scale would have 1 = very low (does not meet the standard at all) and 10 = perfect (absolutely meets the standard).
- **Objective data:** Enter actual data, rather than the opinions of the individual(s) doing the ranking.
- **Yes/no:** If the criteria are expressed in absolute terms, so an alternative either meets the criterion or not, simply enter Y or N to indicate conformance or nonconformance.

Step 6: Review the results and take action as required.

Guidelines and Tips

Whenever comparing alternatives (forced-choice method or rating scale), the group must agree on the relative importance of the alternatives/criteria for scoring purposes. Relative importance can be established either through consensus discussion or through voting techniques. You will usually want to reach agreement rather quickly on this. The amount of time you spend should be based on the importance of the problem/solution and on the number of alternatives and criteria. If there are a large number of alternatives/criteria, you can reach agreement more quickly keeping in mind the impact of each individual item on the list is smaller.

Depending on the nature and impact of the problem, this process can be simplified for quicker and easier use. For example, the process can be simplified by assuming that the criteria are of equal importance, and therefore the ranking of alternatives can be skipped. You can look for other simplifying assumptions. Just be aware of their impact on results.

There is no one best way to weight criteria or alternatives. In the forced-choice method, you rate each element against the other, based on the number of choices. This is a time-consuming method, though. The rating scale method is quick, but has the drawback that people tend to rank every criterion as very important or high on the scale of 1 through 10.

Again, there is no one best method. Use the method that provides you with the most information. Before using any of the alternative approaches described here, however, think about the implications of the various schemes. If you plan to use prioritization matrices repeatedly, you might set up a simple spreadsheet to assist you with some of the calculations.

Examples

The examples are included in the text.

Additional Reading

Asaka, Tetsuichi, and Kazuo Ozeki (eds.). *Handbook of Quality Tools: The Japanese Approach* (Portland, OR: Productivity Press, 1998).

Eiga, T., R. Futami, H. Miyawama, and Y. Nayatani. *The Seven New QC Tools: Practical Applications for Managers* (New York: Quality Resources, 1994).

King, Bob. *The Seven Management Tools* (Methuen, MA: Goal/QPC, 1989).

Mizuno, Shigeru (ed.). *Management for Quality Improvement: The 7 New QC Tools* (Portland, OR: Productivity Press, 1988).

MEASUREMENTS

One measurement is worth 10 guesses.

Definition

A measurement describes the dimension, quantity, capacity, performance, or characteristic of a process or population.

Just the Facts

There are many tools that provide measurement data that, when analyzed properly, yield "actionable knowledge" (indicate the direction of action to be taken). Some of the key actions and tools include:

- Investigate history data archives
- Check sheets
- Paper files
- Functional experts
- Data collection plans
- Sampling
- Lean metrics
- Sampling plans
- Theory of constraints (TOC)
- Taguchi loss function
- Tree diagrams
- Statistical process control
- Process capability
- Histograms
- Standard run chart graphs
- Scatter diagrams
- Box plots
- Basic descriptive statistics
- Process mapping
- Pace maker identification
- Spaghetti charts
- Basic and advanced charting

Principles of Good Measure

The following is a list of criteria to use in evaluating measurement:

- The measure must be important.
- The measure must be easy to understand.
- The measure is sensitive to the right things and insensitive to other things.
- The measure promotes appropriate analysis and action.
- Data needed must be easy to collect.

Examples

In preparing the measurement system start thinking in scientific terms:

$$Y = f(X1, X2, \ldots, Xn)$$

- *Y* relates to the outputs.
- *X* relates to the process and inputs.
- Identify *X*'s that might explain variation in the output measure.
- Collect data efficiently—both the *Y*'s and the suspected *X*'s.

The focus of the measure activity is to:

- Select key output measures (*Y* metrics)
- Collect baseline data (*X* performance metrics)
- Study the variation in the output measures (*X* process measures)
- Understand the capability of our process (6s)

Quality Measurement

- Effectiveness measures: The degree to which customer needs and requirements are met and exceeded. Some examples include the following:
    - Percent defective
    - Response time

- Efficiency measures: The amount of resources allocated in meeting and exceeding customer requirements. Some examples include the following:
    - Cost per transaction
    - Turnaround time
    - Time per activity
    - Amount of rework

A combination of inputs determines output (Figure 7.17).

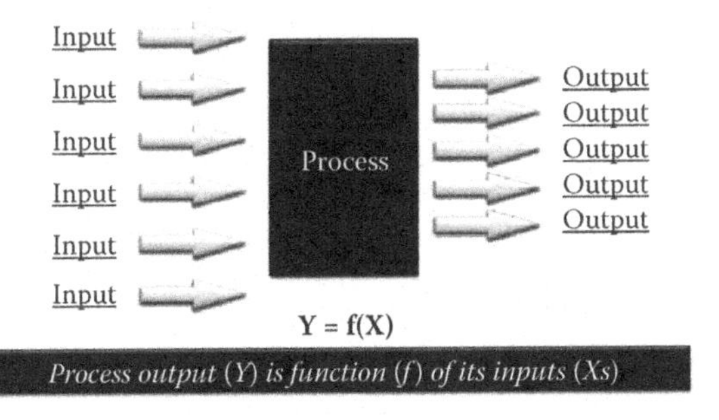

FIGURE 7.17
Processes transform inputs into outputs.

Additional Reading

Picard, Daniel (ed.). *The Black Belt Memory Jogger* (GOAL/QPC and Six Sigma Academy, Salem, NH, 2002).

Wortman, Bill. *The Certified Six Sigma Black Belt Primer* (Quality Council of Indiana, West Terre Haute, 2001).

8

Black Belt Nonstatistical Tools (O Through Q)

In this chapter we discuss the following tools:

- Organizational change management
- Pareto diagrams
- Prioritization matrix
- Project management
- Quality function deployment

ORGANIZATIONAL CHANGE MANAGEMENT (OCM)

> Research confirms that as much as 60% of initiatives and other projects fail as a direct result of a fundamental inability to manager their social implications.
>
> **—Gartner Group**

Definition

Organizational change management (OCM) is a methodology designed to lessen the stress and resistance of employees and management to individual critical changes.

Just the Facts

Organizations today are in a continuous state of change. In the past, individuals were called upon once in a while to adjust and change their work patterns and then had time to master this new process. Today things are changing so fast we seldom become comfortable with one change in the way we do business before it has been changed again. There literally are hundreds of change activities going on in all organizations at the same time, and many individuals are impacted by a number of different processes that are changing simultaneously.

We like to think that the people in our organization are resilient and can handle this rapidly changing environment without additional help. The truth is that everyone is for change as long it does not impact him or her. But when the change becomes personal, everyone is subjected to the four C's:

- **Comfort:** Their comfort level has been destroyed.
- **Confidence:** They lose confidence in their ability to meet the requirements in the future state solution.
- **Competence:** They are not sure if they will have the skills required to perform. They are no longer the individuals that know the most about the job.
- **Control:** They have lost control over their own destiny. Someone else is defining what they will do and how they will do it.

As a result of the disruption of the four C's, the individuals and groups within the organization suffer some severe emotional changes. The stress level within the organization goes way up as people worry about the future. Productivity declines because people spend excessive amounts of time discussing what's going to happen to them and how it is going to impact them. Anxiety levels increase as the uncertainty of the future state remains unclear and the individuals begin to fear loss of security. This results in a great deal of increased conflict between groups and individuals. An attitude develops of "What's wrong with the future state solution," rather than "How can we help to implement the future state solution?" Everyone is most comfortable and happiest in their current state (status quo). Human beings are extremely control-oriented. We feel the most competent, confident, and comfortable when our expectations of control, stability, and predictability are being met.

Definition

Change is a condition in which your expectations based upon past performance are not met. A condition in which you can no longer rely upon your past experience.

We resist change because we don't know if we will be able to prosper or even survive in the new environment. Organizational change management (OCM) is a methodology designed to successfully manage the change process. Today, managers at all levels of the organization must have the ability and willingness to deal with the tough issues associated with implementing major

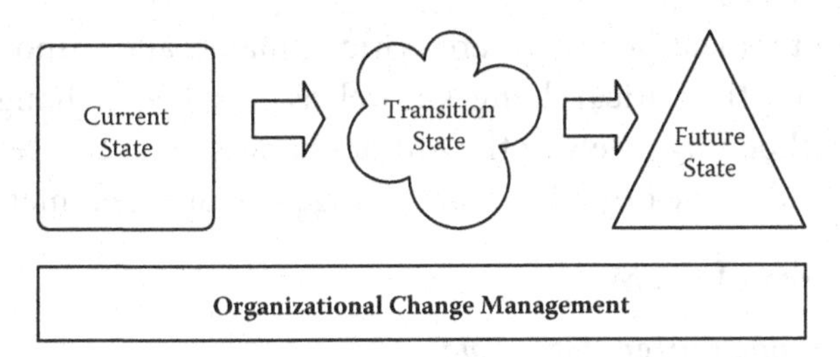

FIGURE 8.1
The change process.

change. They must be capable of guiding their organization safely through the change process. This involves convincing people to leave the comfort of the current state and move through the turbulence of the transition state to arrive at what may be an unclear, distant future state. (See Figure 8.1.)

Specifically, these four states are defined as:

- **Current state:** The status quo, or the established pattern of expectations. The normal routine an organization follows before implementing an improvement opportunity.
- **Present state:** The condition where you know and understand what will happen, when your expectations are being met. (You may not like the current state, but you have learned how to survive in it and you know what to expect.)
- **Transition state:** The point in the change process where people break away from the status quo. They no longer behave as they have in the past, yet they have not thoroughly established a new way of operating.
- **Future state:** The point at which change initiatives are implemented and integrated with the behavioral patterns that are required by the change.

The focus of the OCM implementation methodology is on the transformation between these various states. The journey from the current state to the future state can be long and perilous, and if not properly managed with appropriate strategies and tactics, it can be disastrous.

There are literally hundreds of changes going on within your organization all the time, and the OCM methodology should not be applied to all of them. You need to look very carefully at the changes and select the ones that should be managed. Typical changes that should have the OCM methodology applied to them are:

- When the change is a major change
- When there is a high risk that certain human factors would result in implementation failure

In most organizations there are critical, major, and minor changes. Certainly all of the critical changes as well as most major changes should use this methodology. Few of the minor changes will require the discipline involved in the organizational change management methodology.

Seven Phases of OCM

Phase I: Defining Current State Pain

The individuals involved in the change need to feel enough heat related to their present situation to make them want to leave the security of the status quo and move into the uncertainty in the transformation part (transition phase) of the change management process. Management may reason that the employees have an excellent understanding of the pain related to the status quo, and that is probably true related to the current situation, but they have little understanding of the lost opportunities that will impact them and the organization if the change does not take place. As a result, management needs to surface the pain related to lost opportunities.

The pain related to the current state must be great enough to get the employees to consider leaving the safety of the current state platform and move on to the transformation phase. The highlighting of pain is called "establishing the burning platform." This term comes from an oil rig that was on fire in the North Sea. The platform got so hot that the workers jumped off the rig into the sea because they realized that their chances of survival were better in the cold North Sea than on the burning oil rig platform.

Phase II: Establishing a Clear Vision of the Future State Solution

Employees are going to weigh the pain related to the present situation and compare it to the pain they will undergo during the transformation and the pain related to the future state solution. (Note: Your employees are smart enough to realize that there will always be some pain in using even the very best future state solution.) As a result, management needs to provide the employees with an excellent vision of the future state solution and its impact upon them so they can compare the pain of the current state to the combined pain of the transformation state and future state. If the pain related to the current state is not greater than their view of the pain related to the transition state and future state combined, the change can only be implemented over the objections of the employees, and there is a high likelihood that it will not be successfully implemented. The vision has to crisply define:

- Why is the change necessary?
- What's in it for me?
- Why is it important to the organization?

Phases I and II are often called pain management.

Phase III: Defining Change Roles

The roles of the individuals involved in the change must be clearly defined. These roles are divided into five general categories:

- **Initiating sponsor:** Individual/group that has the power to initiate and legitimize the change for all of the affected individuals.
- **Sustaining sponsor:** Individual/group that has the necessary political, logistical, and economic proximity to the targets to influence their behavior.
- **Change agent:** Individual/group that is responsible for implementing the change management aspects of the project.
- **Change target** (sometimes called a "changee"): Individual/group that must actually change.
- **Advocate:** An individual/group that wants to achieve change but does not have sufficient sponsorship.

Each person needs to be trained in his or her duties related to his or her change management role.

Phase IV: Mapping Change Roles

A change role map should be prepared identifying each individual's role related to the change. (Note: This is a concept developed by Daryl Conner of ODR.) This map generates a visual picture of the individuals, groups, and relationships that must be orchestrated to achieve the change. It is important to note that one individual can serve many roles. For example, every sustaining sponsor is first a target or "change" before he or she can accept the role as a sustaining sponsor. (See Figure 8.2).

Phase V: Defining the Degree of Change Required

The communications system and the change management plan must be designed so that the proper degree of acceptance of the change initiative is

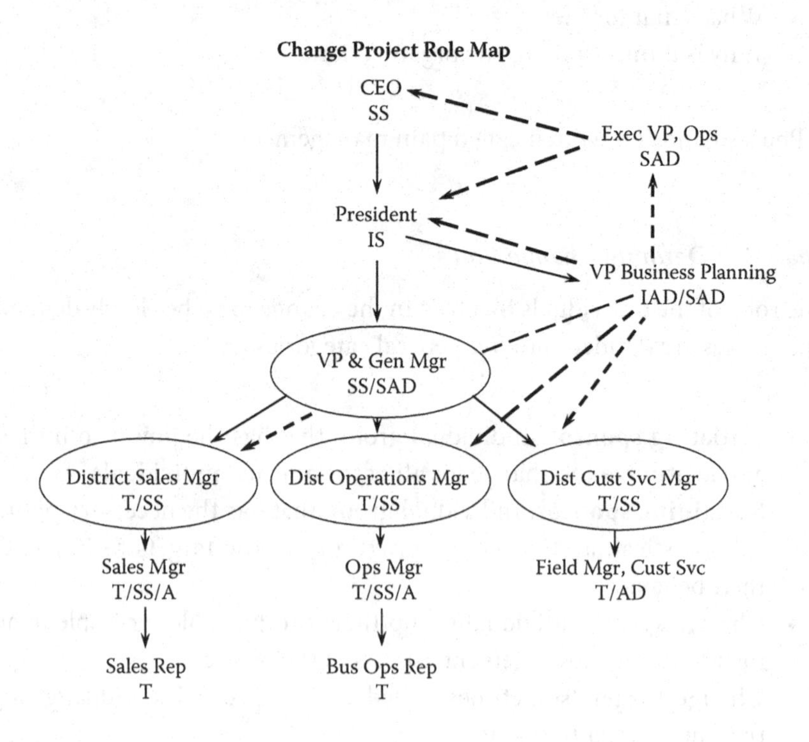

FIGURE 8.2
Typical role map diagram.

accomplished. There is a great deal of difficulty and effort required to have a change internalized in comparison to just having it adopted. For example, there is much less change management required to have a new computer program standardized for use throughout the organization than there is to have the same software package accepted by the individuals as being the very best alternative. Typically, degrees of acceptance can be broken down into:

- **Adoption:** The change has been fully implemented, the new practice is being followed consistently, and its objectives are being met. "We are doing it this way."
- **Institutionalization:** The change has not only been adopted, but is formalized in written policies, practices, and procedures. The organizational infrastructure (hiring, training, performance measures, rewards, etc.) is aligned to support continued conformance to the new practice. "We always do it this way."
- **Internalization:** The new practice is understood to be fully aligned with individual and organizational beliefs, values, and norms of behavior, and as such, commitment to sustain the practice comes from within. "Of course we always do it this way. I believe in doing it this way, and to do otherwise would be inconsistent with the way we like to do things here."

Phase VI: Developing the Organizational Change Management Plan

Any planning activity involves thinking out in advance the sequence of actions required to accomplish a predefined objective. A change management plan is now prepared to meet the degree of acceptance that is required to support the project.

Now the change management plan needs to be integrated into the project management plan. This normally is a timeline chart that reflects the movement throughout the change management process. (See Figure 8.3.) The change management plan should also include any risk related to internal or external events, such as the following:

- Competing initiatives
- Too many initiatives
- Loss of sponsorship

Change Management Activities	Phase I	Phase II	Phase III	Phase IV	Phase V	Phase VI
	Assess	Plan	Redesign	Implement	Audit	Improve
Identify, Document & Communicate Cost of the Status Quo (Business Imperative)						
Create & Communicate the Future-State Vision (People, Process & Technology)						
Clarify the Change and Obtain Initiating Sponsor Understanding and Commitment						
Create Needed Infrastructure and Implementation Architecture						
Conduct a High-Level QMS-Wide Change Risk Assessment (The 8 Risk Factors)						
Create a High-Level QMS-Wide Organizational Change Plan						
Create Role Maps to Identify All Personnel Having Key Change Roles						
Conduct Tier-Level Change Risk Assessments (The 8 Risk Factors)						
Conduct Change Readiness Assessments						
Organizational Alignment Assessment (Structure, Compensation, Rewards etc.)						
Assess Enablers and Barriers						
Develop Tier-Level Transition Management Plans						

FIGURE 8.3
Typical change management plan. (*continued*)

Change Management Activities	Phase I	Phase II	Phase III	Phase IV	Phase V	Phase VI
	Assess	Plan	Redesign	Implement	Audit	Improve
Develop a Communication Plan						
Cascading Sponsorship (Communications, Training, Performance Management)						
Implement the Communication Plan						
Provide Change Management Training for Sponsors, Change Agents and Others						
Form Change Agent, Sponsor and Advocate Teams						
Provide Training for Targets (Those Affected By the Change)						
Implement Organizational Alignment Enablers						
Analyze Effectiveness of Communications and Training Strategies						
Monitor Commitment Levels of Sponsors, Change Agents, Advocates & Targets						
Monitor and Measure Implementation Effectiveness and Schedule Adherence						
Modify Transition Management Plans as Needed to Assure Effectiveness						
Track and Report Planned Versus Actual Activities and Results						
Identify Opportunities for Continuous Improvement to the Change Process						

FIGURE 8.3 ***(continued)***
Typical change management plan.

- Economic trends in the business
- Industry trends

Phase VII: Implementing the Change Management Plan

The OCM plan is now implemented in conjunction with the project plan. Usually a number of surveys are conducted to help refine the OCM plan. Typical surveys or evaluations that could be conducted are:

- Implementation architecture assessment
- Internal/external event assessment
- Vision clarity assessment
- Commitment versus compliance change management
- Strategy analysis
- Predicting the impact of the change
- Organizational effectiveness
- Business imperative analysis
- Sponsor evaluation
- Implementation history assessment
- X-Factor change readiness assessment

Examples

It is evident from Figure 8.3 that a number of change management activities are iterative and not confined to the QMS phase. (See also Figure 8.4.)

FIGURE 8.4
To win the game today, you need to get the most out of every player on the board.

Additional Reading

Beckhard, Richard, and Wendy Pritchard. *Changing the Essence: The Art of Creating and Leading Fundamental Change in Organizations* (San Francisco: Jossey-Bass, 1992).

Bouldin, Barbara M. *Agents of Change: Managing the Introduction of Automated Tools* (New York: Yourdon Press, 1988).

Conner, Darryl R. *Leading at the Edge of Chaos: How to Create the Nimble Organization* (New York: John Wiley & Sons, 1998).

Conner, Darryl R. *Managing at the Speed of Change: How Resilient Managers Succeed and Prosper Where Others Fail* (New York: Villard Books, 1993).

Conner, Daryl, H. James Harrington, and Richard Horney. *Project Change Management—Applying Change Management to Improvement Projects* (New York: McGraw-Hill, 1999).

Harrington, H. James, and James S. Harrington. *Total Improvement Management: The Next Generation in Performance Improvement* (New York: McGraw-Hill, 1995).

Huse, Edgar F., and Thomas G. Cummings. *Organization Development and Change* (New York: West, 1985).

Hutton, David W. *The Change Agent's Handbook: A Survival Guide for Quality Improvement Champions* (Milwaukee, WI: ASQ Quality Press, 1994).

Kanter, Rosabeth Moss. *Rosabeth Moss Kanter on the Frontiers of Management* (Cambridge, MA: Harvard Business School Press, 1997).

Smith, Douglas K. *Taking Charge of Change: 10 Principles for Managing People and Performance* (Reading, MA: Addison-Wesley, 1996).

PARETO DIAGRAMS

As good as you are, you can't do everything. So do the ones with the big pay-back first.

Definition

A Pareto diagram is a type of chart in which the bars are arranged in descending order from the left to the right. It is a way to highlight the "vital few" in contrast to the "trivial many."

Just the Facts

A Pareto diagram is a specialized type of column graph. Data are presented in a manner that allows comparison between a number of problems or a number of causes. The comparison is necessary to set priorities. The Pareto diagram facilitates the process by graphically distinguishing the few significant problems or causes from the less significant many. This diagram is the graphic representation of the Pareto principle.

The Pareto Principle (80/20 Rule)

The Pareto principle states as a "universal," applicable to many fields, the phenomenon of the vital few and the trivial many. In our context, this means that a few significant problems or causes will be most important to our decision-making process.

This principle derives its name from Vilfredo Pareto, a 19th-century economist who applied the concept to income distribution. His observations led him to state that 80% of wealth is controlled by 20% of the people. (Hence, the principle is often referred to as the 80-20 principle.) The name *Pareto* and the universal applicability of the concept are credited to Dr. Joe M. Juran, who applied the Lorenz curve to graphically depict the universal.

The Pareto diagram is distinguished from other column graphs in that the columns representing each category are always ordered from the highest on the left to the lowest on the right. (See Figure 8.5.) The one exception to this rule is the "Other" column, a collection of less important factors. When it is used, it is always the last column on the right. With this arrangement it is possible to make a cumulative or "cum" line (the Lorenz curve), which rises by steps according to the height of the columns to show the total of the percentages, the cumulative percentage. With the cum line, the Pareto diagram becomes a combined column and line graph.

Uses of Pareto Diagrams

Like other graphic representations, Pareto diagrams visualize data to assist investigation and analysis. They are employed to:

- **Establish priorities:** By a comparison of related data, major categories of problems and causes are identified.
- **Show percentage of impact:** The cum line defines the proportionate importance of combined categories, and thus indicates the likely impact of dealing with all of the categories up to that point in the diagram.

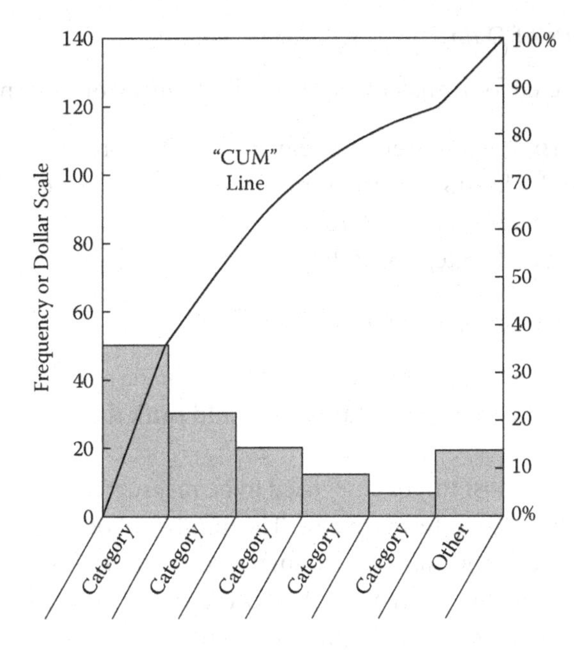

FIGURE 8.5
Pareto diagram form.

- **Show change over time:** Two or more diagrams can be used to demonstrate the result of decisions and actions by showing before and after data. (See Figure 8.6.)
- **Aid communication:** The diagram is an accepted form of communication, readily understood.
- **Demonstrate use of data:** This can be particularly helpful in management presentations to show the activities are solidly rooted in facts, not just opinions.

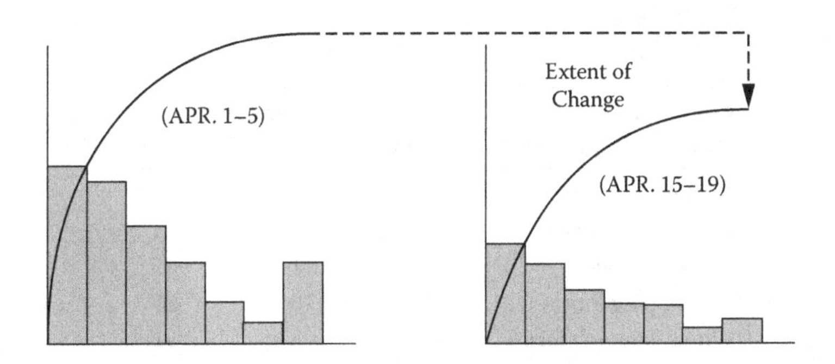

FIGURE 8.6
Performance before (left) and after (right) corrective action.

Classifications of Data

The data collected for transfer to a Pareto diagram are of three major types:

- **Problems:** Including errors, defects, locations, processes, procedures.
- **Causes:** Including materials, machines, equipment, employees, customers, operations, standards.
- **Cost:** Of each category of data.

The purpose of using a Pareto diagram is to:

- Establish the biggest problem, and rank the rest
- Establish the most important cause, and rank the rest

"Biggest" and "most important" need to be measured not solely in terms of frequency but also in terms of *cost.* The number of occurrences may not be as significant as the cost of particular occurrences. Consequently, it is usually important to construct a Pareto diagram using cost data. Mere frequency can be misleading in judging significance.

Constructing a Pareto Diagram

Before a Pareto diagram can be constructed, it is necessary to collect data according to the classifications or categories judged most suitable. With these data the diagram is constructed as follows:

Step 1: Summarize the data on a worksheet by:

a. Arranging the data in order of sequence from largest to smallest and total them. (See Table 8.1, "Number" column.)

TABLE 8.1

Data Table

Problems with Overseas Shipments	Number	% of Total	Cum %
Containers opened	112	40	40
Containers broken	68	24	64
Items missing	30	11	75
No shipping forms	28	10	85
Wrong shipping forms	18	6	91
Other	24	9	100%
Total	280	100%	—

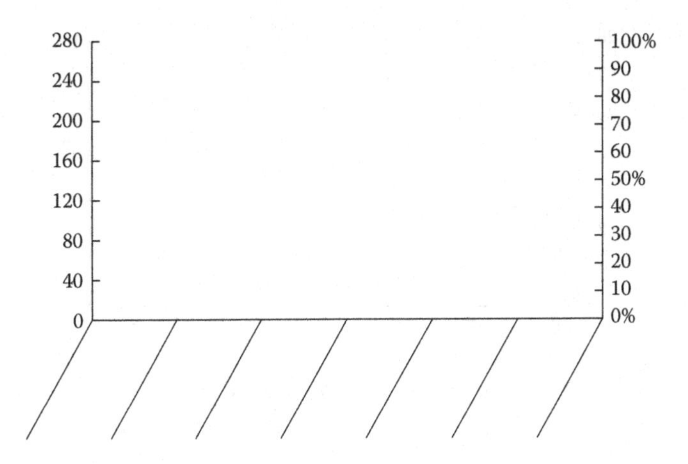

FIGURE 8.7
Vertical scales for a Pareto diagram.

b. Calculating percentages (See Table 8.1, "% of Total" column.)
c. Completing the cumulative percent column. (See Table 8.1, "Cum %" column.)

Step 2: Draw the horizontal and two vertical axes:

a. Divide the horizontal axis into equal segments, one for each category. (See Figure 8.7.)
b. Scale the left-hand vertical axis so that the top figure on the axis is the total of all the occurrences in the categories.
c. Scale the right-hand vertical axis so that 100% is directly opposite the total on the left-hand axis. The percent scale is normally in increments of 10%.

Step 3: Plot the data. Construct a series of columns, putting the tallest column on the extreme left, then the next tallest, and so on. If several minor categories are consolidated into one "Other" column, it is plotted on the extreme right, regardless of its height. (See Figure 8.8.)

Step 4: Plot the cumulative line:

a. Place a dot in line with the right side of each column, at a height corresponding to the number in the cumulative percentage column on the worksheet. In our example, the points would be at 40, 64, 75, 85, and 91%.

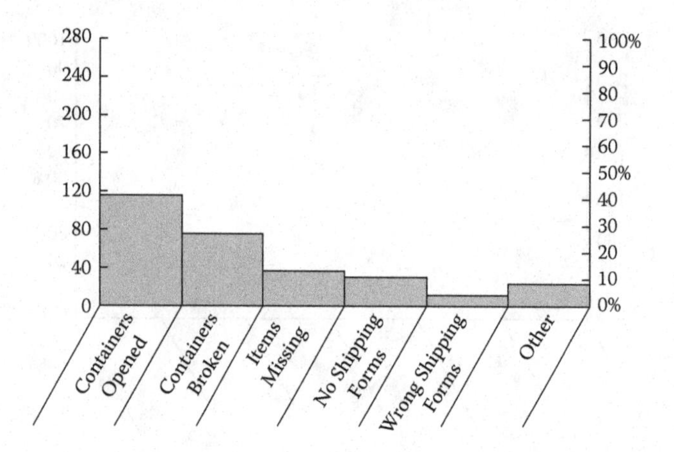

FIGURE 8.8
Plotted Pareto diagram without cumulative line.

b. Beginning with the lower left corner of the diagram (the zero point of origin), connect the dots to the 100% point on the right vertical axis. (See Figure 8.9.)

Step 5: Add labels and the legend:
a. Label each axis.
b. Add the legend. It should include the source of the data, date prepared, where collected, who collected them, period covered, and any other pertinent information. (See Figure 8.10.)

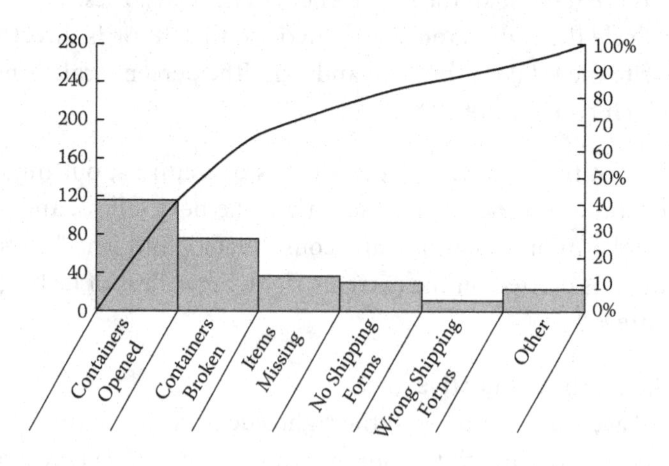

FIGURE 8.9
Pareto diagram with cumulative line plotted.

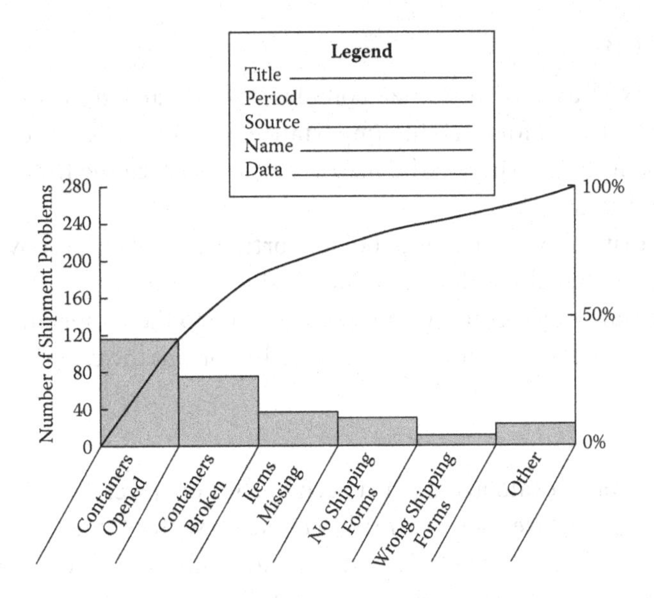

FIGURE 8.10
Completed Pareto diagram.

Example

See the examples in "Just the Facts."

PRIORITIZATION MATRIX

People who don't prioritize work hard but accomplish little.

Definition

The prioritization matrix, also known as the criteria matrix, compares alternatives related to different criteria, such as ease of use, price, service, and any other criterion that is desired to be included in the matrix. It sorts criteria into an order of importance.

Just the Facts

The matrix is used to prioritize complex or unclear issues, where there are several criteria for deciding importance. It requires actual data to aid in establishing the criteria. The matrix helps to reduce the list of possible causes and solutions.

The prioritization matrix is a tool for sorting a set of criteria by order of importance. It is also valuable in identifying each in terms of their relative importance by assigning a numerical value to the importance of each criteria—the larger the number, the more important the criteria.

Example

Each criterion is evaluated against each of a set of key criteria. The scores for each criterion are then summed to come up with a total score.

A good criterion reflects key goals and enables objective measurements to be made. Thus, material cost is measurable and reflects a business profit goal, while simplicity may not reflect any goals and be difficult to score.

When there are multiple criteria, it may also be important to take into account the fact that some criteria are more important than others. This can be implemented by allocating weighting values to each criterion. Figure 8.11 displays a simple matrix.

Additional Reading

Cox, Jeff, and Eliyahu M. Goldratt. *The Goal: A Process of Ongoing Improvement* (Croton-on-Hudson, NY: North River Press, 1986).

Picard, Daniel (ed.). *The Black Belt Memory Jogger* (GOAL/QPC and Six Sigma Academy, Salem, NH, 2002).

Wortman, Bill. *The Certified Six Sigma Black Belt Primer* (Quality Council of Indiana, West Terre Haute, 2001).

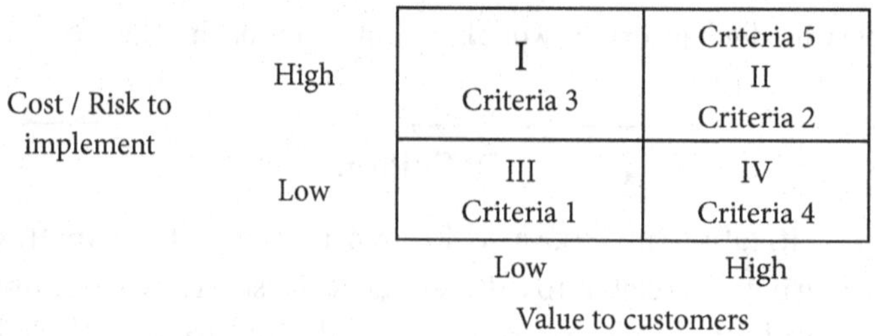

Cost / Risk to implement		Low	High
	High	I Criteria 3	Criteria 5 II Criteria 2
	Low	III Criteria 1	IV Criteria 4

Value to customers

FIGURE 8.11
Simple matrix.

PROJECT MANAGEMENT (PM)

Processes define how we operate. Projects are the way we improve processes.

Definition

Project management is the application of knowledge, skills, tools, and techniques to project activities in order to meet or exceed stakeholders' needs and expectations from a project (Project Management Institute Standards Committee, 1996).

Just the Facts

Performance improvement occurs mainly from a number of large and small projects that are undertaken by the organization. These projects involve all levels within the organization and can take less than 100 hours or millions of hours to complete. They are a critical part of the way an organization's business strategies are implemented. It is extremely important that these multitudes of projects are managed effectively if the stakeholder's needs and expectations are to be met. This is made more complex because conflicting demands are often placed upon the project. For example:

- Scope
- Time
- Cost
- Quality
- Stakeholders with different identified (needs) and unidentified (expectations) requirements

The different stakeholders often place conflicting requirements on a single project. For example, management wants the project to reduce labor cost by 80% and organized labor wants it to create more jobs.

The Project Management Institute in Upper Darby, Pennsylvania, is the leader in defining the body of knowledge for project management. Its Project Management Body of Knowledge (PMBOK) approach to project management has been widely accepted throughout the world. In addition, the International Organization for Standardization's Technical Committee 176 has released an international standard: *ISO/DIS 10006: Guidelines to Quality in Project Management*. These two methodologies complement each other and march hand in hand with each other.

- **Project:** A temporary endeavor undertaken to create a unique product or service.
- **Program:** A group of related projects managed in a coordinated way. Programs usually include an element of ongoing activities.

Large projects are often managed by professional project managers who have no other assignments. However, in most organizations, individuals who serve as project manager only a small percentage of the time manage many projects. In either case, the individual project manager is responsible for defining a process by which a project is initiated, controlled, and brought to a successful conclusion. This requires the following:

- Project completed on time
- Project completed in budget
- Outputs met specification
- Customers are satisfied
- Team members gain satisfaction as a result of the project

A good project manager follows General George S. Patton's advice when he said, "Don't tell soldiers how to do something. Tell them what to do and you will be amazed at their ingenuity." Although a single project life cycle is very difficult to get everyone to agree to, the project life cycle defined in U.S. DOD's document 5000.2 (Revision 2-26-92, entitled "Representative Life Cycle for Defense Acquisition") provides a reasonably good starting point. It is divided into five phases (see Table 8.2).

The following is a life cycle that we like better for an organization that is providing a product and service (see Figure 8.12):

- Phase I: Concept and definition
- Phase II: Design and development

TABLE 8.2

Five Phases of Project Life Cycle

Phase 0: Concept exploration and definition
Phase I: Demonstration and validation
Phase II: Engineering and manufacturing development
Phase III: Production and deployment
Phase IV: Operations and support

Note: This is the representative life cycle for defense acquisition, per U.S. DOD 5000.2.

- Phase III: Creating the product or service
- Phase IV: Installation
- Phase V: Operating and maintenance
- Phase VI: Disposal

Traditionally, projects have followed a pattern of phases from concept to termination. Each phase has particular characteristics that distinguish it from the other phases. Each phase forms part of a logical sequence in which the fundamental and technical specification of the end product or service is progressively defined. Figure 8.13 is the project management life cycle. Figure 8.14 shows the project phase gates for a typical new product development cycle.

The successful project manager understands that there are four key factors that have to be considered when the project plan is developed. All four

Phase					
I	II	III	IV	V	VI
Concept & Design	Design & Development	Manufacturing	Installation	Operation & Maintenance	Disposal
• New product opportunities • Analysis of system concept and options • Product selection • Technology selection • Make/buy decisions • Identify cost drivers • Construction assessment • Manufacturability assessments • Warranty incentives		• Design Trade-offs • Source selection • Configurations and change controls • Test strategies • Repair/throwaway decisions • Performance tailoring • Support strategies • New product introduction		• System integration and verification • Cost avoidance/cost reduction benefits • Operating and maintenance cost monitoring • Product modifications and service enhancements • Maintenance support resource allocation and optimization	• Retirement cost impact • Replacement/renewal schemes • Disposal and salvage value

FIGURE 8.12
Sample applications of PLC.

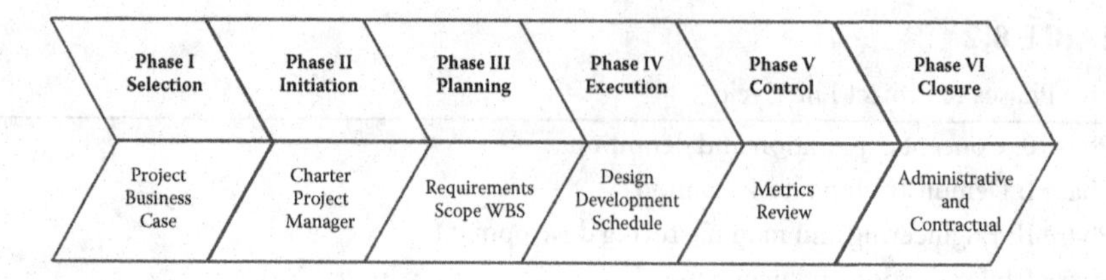

FIGURE 8.13
Project management life cycle.

factors overlap to a degree, but should first be considered independently and then altogether. (See Figure 8.15.)

In order to effectively manage a project, the individual assigned will be required to address the 10 elements defined in Figure 8.16. (The Project Management Institute includes only the first nine elements listed in Figure 8.16.)

Figure 8.17 is the 44-hour processes' flow *PMBOK Guide 2004* as they relate to the five process groups. They do not include OCM as a separate element, but it is placed under the element entitled "project risk management." Of course, the depth and detail that each element needs in order

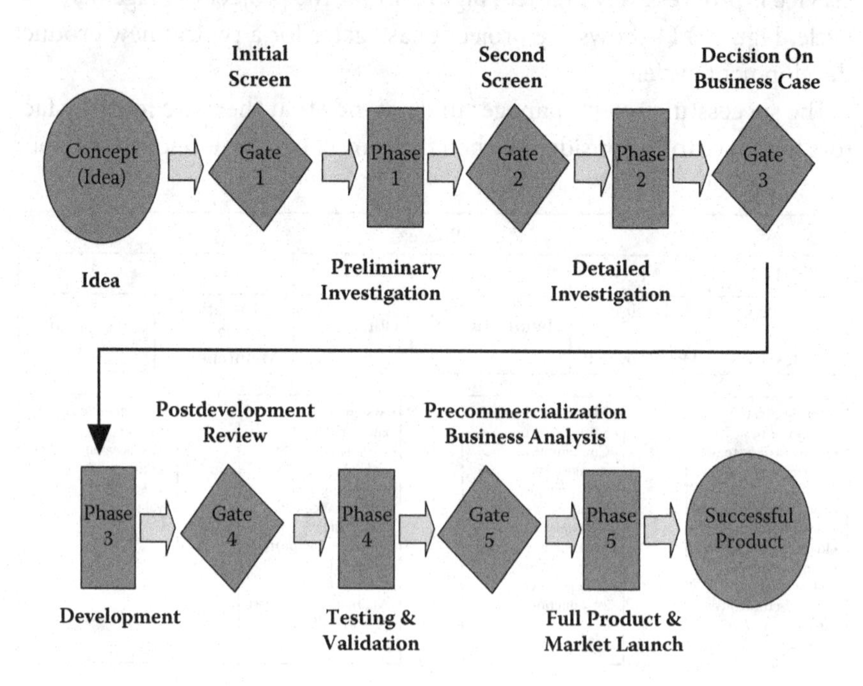

FIGURE 8.14
Project phase gates for typical new product development cycle.

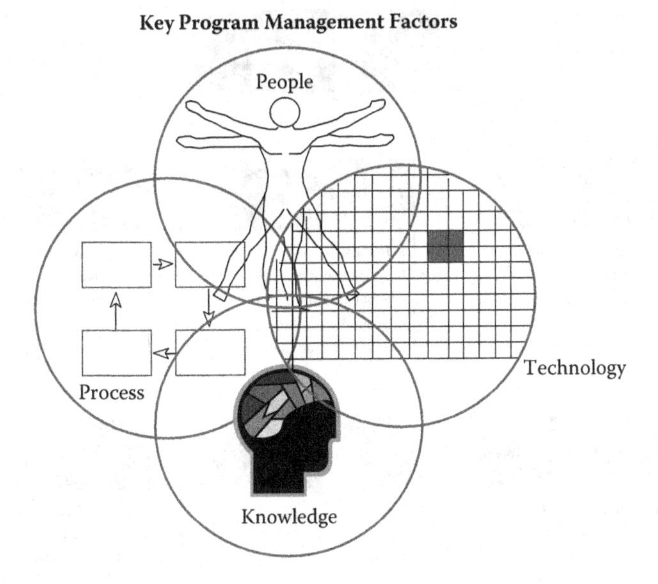

FIGURE 8.15
The key program management factors.

to be evaluated and managed will vary greatly depending upon the scope and complexity of the project.

Project Management Knowledge Areas

The book *A Guide to the Project Management Body of Knowledge*, published by the Project Management Institute, summarizes the project management knowledge areas as follows.

Project Integration Management

This is a subset of project management that includes the processes required to ensure that the various elements of the project are properly coordinated. It consists of:

- Project plan development: Taking the results of other planning processes and putting them into a consistent, coherent document.
- Project plan execution: Carrying out the project plan by performing the activities included therein.
- Overall change control: Coordinating changes across the entire project.

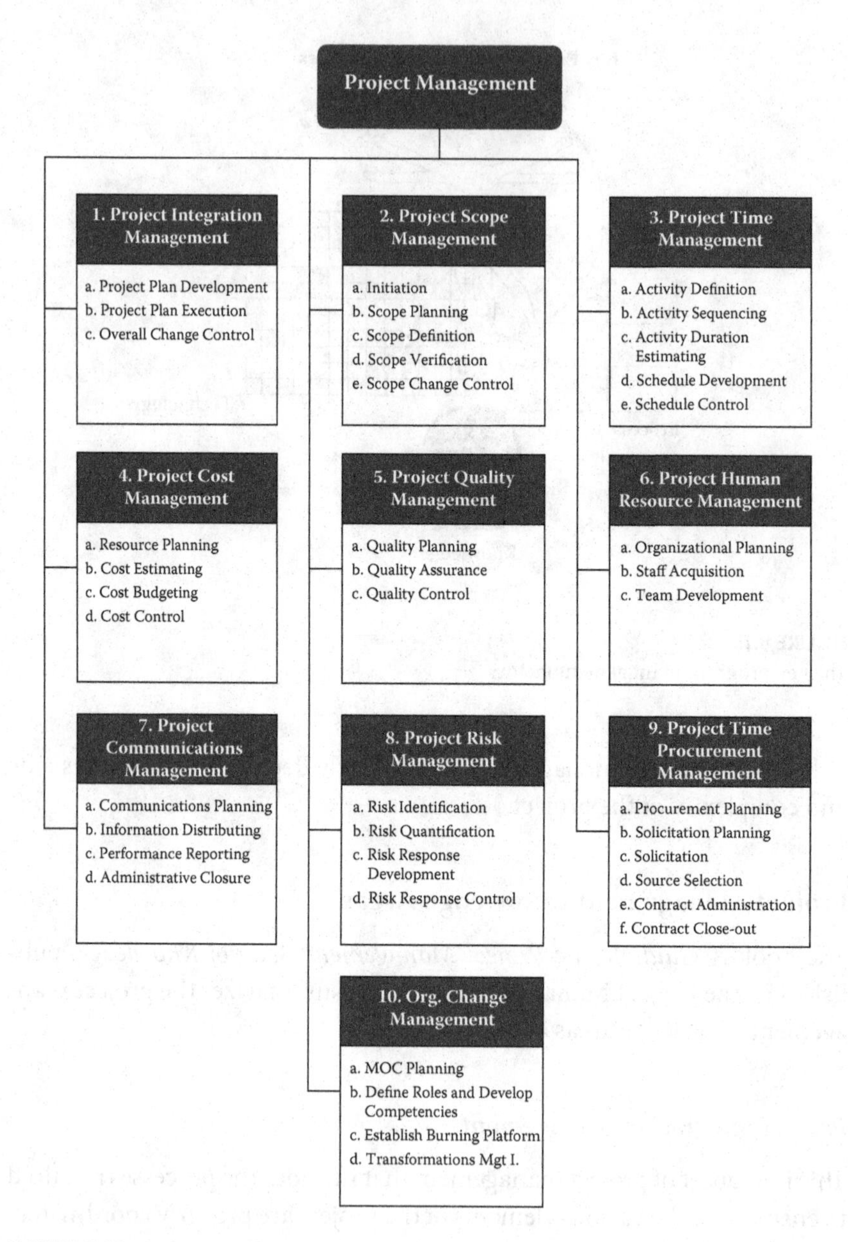

FIGURE 8.16
The 10 elements of project management.

44 Processes in PMI's A Guide to the PMBOK®

Knowledge Area / *Process Group*	**Initiating (2)**	**Planning (21)**	**Executing (7)**	**Monitoring & Controlling (12)**	**Closing (2)**
Ch. 4-Integration	Dev. Project Charter Dev. Prelim. Project Scope Statement	Develop Project Mgt Plan	Direct and Manage Project Execution	Monitor and Control Project Work Integrated Change Control	Close Project
Ch. 5-Scope		Scope Planning Scope Definition Create WBS		Scope Verification Scope Control	
Ch. 6-Time		Activity Definition Activity Sequencing Activity Resource Planning Activity Durat'n Estimating Schedule Development		Schedule Control	
Ch. 7-Cost		Cost Estimating Cost Budgeting		Cost Control	
Ch. 8-Quality		Quality Planning	Perform Quality Assurance	Perform Quality Control	

FIGURE 8.17
Forty-four processes of project management. (*continued*)

Ch. 9-Human Resources		Human Resource Planning	Acquire Project Team Develop Project Team	Manage Project Team	
Ch. 10-Communications		Communications Planning	Information Distribution	Performance Reporting Manage Stakeholders	
Ch. 11-Risk		Risk Mgt Planning Risk Identification Qualitative Risk Analysis Quantitative Risk Analysis Risk Response Planning		Risk Monitoring and Control	
Ch. 12-Procurement		Plan Purchases and Acquisitions Plan Contracting	Request Seller Responses Select Sellers	Contract Administration	Contract Closure

FIGURE 8.17 (*continued*)
Forty-four processes of project management. (*continued*)

Project Scope Management

This is a subset of project management that includes the processes needed to ensure that the project includes all the work required, and only the work required, to complete the project successfully. It consists of:

- Initiation: Committing the organization to begin the next phase of the project.
- Scope planning: Developing a written scope statement as the basis for future project decisions.
- Scope definition: Subdividing the major project deliverables into smaller, more manageable components.
- Scope verification: Formalizing acceptance of the project scope.
- Scope change control: Controlling changes to project scope.

Project Time Management

This is a subset of project management that includes the processes required to ensure timely completion of the project. It consists of:

- Activity definition: Identifying the specific activities that must be performed to produce the various project deliverables.
- Activity sequencing: Identifying and documenting interactivity dependencies.
- Activity duration estimating: Estimating the number of work periods that will be needed to complete individual activities.
- Schedule development: Analyzing activity sequences, activity durations, and resource requirements to create the project schedule.
- Schedule control: Controlling changes to the project schedule.

Project Cost Management

This is a subset of project management that includes the processes required to ensure that the project is completed within the approved budget. It consists of:

- Resource planning: Determining what resources (people, equipment, materials) and what quantities of each should be used to perform project activities.

- Cost estimating: Developing an approximation (estimate) of the costs of the resources needed to complete project activities.
- Cost budgeting: Allocating the overall cost estimate to individual work items.
- Cost control: Controlling changes to the project budget.

Project Quality Management

This is a subset of project management that includes the processes required to ensure that the project will satisfy the needs for which it was undertaken. It consists of:

- Quality planning: Identifying which quality standards are relevant to the project and determining how to satisfy them.
- Quality assurance: Evaluating overall project performance on a regular basis to provide confidence that the project will satisfy the relevant quality standards.
- Quality control: Monitoring specific project results to determine if they comply with relevant quality standards and identifying ways to eliminate causes of unsatisfactory performance.

Project Human Resource Management

This is a subset of project management that includes the processes required to make the most effective use of the people involved with the project. It consists of:

- Organizational planning: Identifying, documenting, and assigning project roles, responsibilities, and reporting relationships.
- Staff acquisition: Getting the human resources needed assigned to and working on the project.
- Team development: Developing individual and group skills to enhance project performance.

Project Communications Management

This is a subset of project management that includes the processes required to ensure timely and appropriate generation, collection, dissemination, storage, and ultimate disposition of project information. It consists of:

- Communications planning: Determining the information and communications needs of the stakeholders: who needs what information, when will they need it, and how it will be given to them.
- Information distribution: Making needed information available to project stakeholders in a timely manner.
- Performance reporting: Collecting and disseminating performance information. This includes status reporting, progress measurement, and forecasting.
- Administrative closure: Generating, gathering, and disseminating information to formalize phase or project completion.

Project Risk Management

This is a subset of project management that includes the processes concerned with identifying, analyzing, and responding to project risk. It consists of:

- Risk identification: Determining which risks are likely to affect the project and documenting the characteristics of each.
- Risk quantification: Evaluating risks and risk interactions to assess the range of possible project outcomes.
- Risk response development: Defining enhancement steps for opportunities and responses to threats.
- Risk response control: Responding to changes in risk over the course of the project.

Project Procurement Management

This is a subset of project management that includes the processes required to acquire goods and services from outside the performing organization. It consists of:

- Procurement planning: Determining what to procure and when.
- Solicitation planning: Documenting product requirements and identifying potential sources.
- Solicitation: Obtaining quotations, bids, offers, or proposals as appropriate.
- Source selection: Choosing from among potential sellers.
- Contract administration: Managing the relationship with the seller.

- Contract closeout: Completion and settlement of the contract, including resolution of any open items.

How OCM Can Help

This is a part of project management that is directed at the people who will be impacted by the project. OCM helps prepare the people not to resist the change—both those who live in the process that is being changed and those who have their work lives changed as a result of the project. In fact, OCM has often prepared the employees so well that they look forward to the change. (Note: This is not part of the PMBOK project management concept.)

- OCM planning: Define the level of resistance to change and prepare a plan to offset the resistance.
- Define roles and develop competencies: Identify who will serve as sponsors, change agents, change targets, and change advocates. Then, train each individual on how to perform the specific role.
- Establish burning platform: Define why the as-is process needs to be changed and prepare a vision that defines how the as-is pain will be lessened by the future state solution.
- Transformations management: Implement the OCM plan. Test for black holes and lack of acceptance. Train affected personnel in new skills required by the change.

Estimate Task Effort and Duration

Generally, the more experience someone has in performing a task, the more accurate the estimate he or she gives the project manager will be. Many variables are used to estimate how much effort (or duration) is needed to complete a particular task and how much duration (or elapsed time) will be required. These variables include known constraints, assumptions, risks, historical information, and resource capabilities. The quality of estimates is dependent on the quality of information and the project life cycle phase the project is in.

For example, if a senior programmer leaves the organization and a less experienced programmer takes his place, the actual amount of time for a programming task might be 25 to 50% higher than originally estimated. In that case, unless a contingency was placed aside, the project would run over the estimate by the amount indicated. When a person assigned to perform a task is known, it's best to let that person estimate the task with the team leader and project manager who review the estimate. There are two well-known reasons for this procedure:

- The person performing the task should have the expertise and past experience with that type of estimated task and should be able to give an educated guess.
- If the person performing the task provides the estimate, he or she will have a sense of owning it and will work hard to achieve it.

If you don't know who will perform the task but people with more expertise are working close to the task that needs estimating, it's wise to let these experts prepare the estimate and then add a 25% contingency to cover a less experienced person carrying out the task.

Certain tasks are effort driven with no constraints around the duration. Others are only duration driven. For example, the walls of a house not being able to be painted until the sheetrock is taped and textured is an effort-driven task. A duration-driven task would be more like painting the walls twice: The second coat can't be completed until the first is dried, no matter how many painters the project manager has assigned to the job.

An accepted practice with task duration is the "80-hour rule," which establishes that each individual should have some task due approximately every 2 weeks, representing a low-level milestone for him or her to reach. The project manager must take all estimates provided by the team members and leaders, roll them up, and review the project plan as a whole. The intention here is to spot an issue with the overall project schedule by examining the proportion of phases within stage(s) or iteration(s).

Develop the Schedule

Once you've estimated the time it takes to perform each activity, defined each activity's inputs and outputs, and where each of these inputs come from and where each output goes, you can start to develop a project schedule. This defines the start and finish dates for each activity. Many things must be considered when developing a schedule. Some of them are:

- Activity duration estimates
- Resource requirements

- Resource availability
- Constraints (imposed dates or milestones)
- Assumptions
- Input from other projects
- Transportation delays
- Risk management plans
- Lead and lag times

Although a number of software tools are available that can help you develop and optimize a schedule, we'll discuss a very simple approach we've found to be effective. We like to start with the output delivery date and work backwards. From that point, we place the activity that will deliver that output to the customer on the right-hand side of a sheet of paper.

As an example, suppose your consulting firm has just conducted an employee-opinion survey for a customer. You've returned to your office with the survey sheets in hand. The next key delivery to your customer is on May 30, when you'll present the customer-survey results. (See Figure 8.18.) Note that May 30 has a boxed-in point representing the deadline for presenting the report. Before that can happen, however, the consultant must travel to the client's office 6 hours away, so we include May 29 to complete that step.

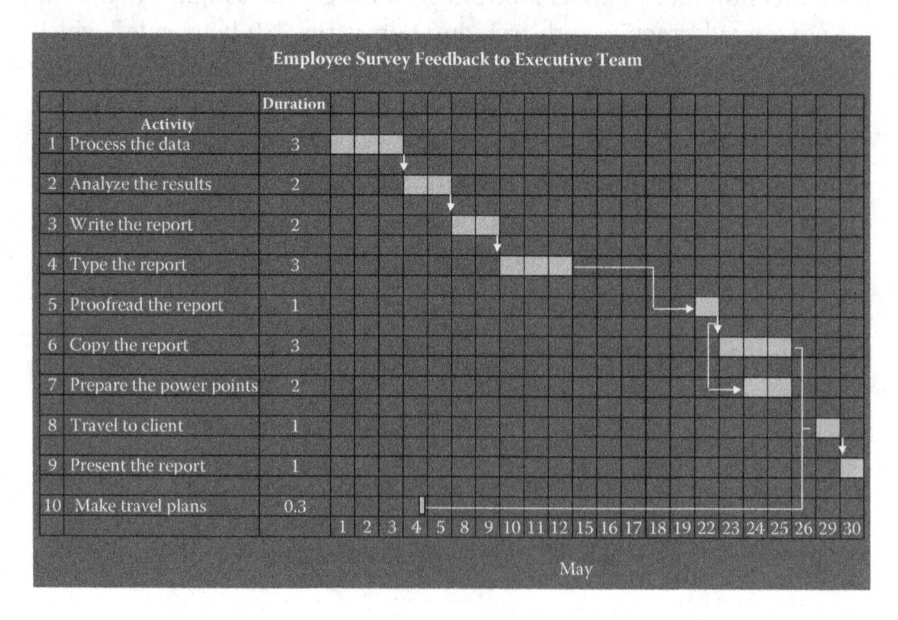

FIGURE 8.18
Employee survey feedback to the executive team.

Note also that that Saturday and Sunday aren't listed on the timeline axis of the graph in Figure 8.18 because we don't want to schedule work on weekends. This builds in a little safety margin in case things go wrong.

The inputs the consultant needs to present the report are the PowerPoint slides he'll be using and copies of the report and travel arrangements. These three inputs are required for the consultant to travel to the client. Note that scheduling travel plans occurs on May 4. This is necessary because company policy requires consultants to make travel plans at least 2 weeks early to ensure minimum fares and best possible connections.

Input necessary to preparing the PowerPoint slides and duplicating the report includes having the final report proofread and approved. The latest date this can happen is Monday, May 22, because the report masters must be taken to the printer on May 23.

The input needed to proofread the report is a completed, typed report. Note that the report's schedule requires that it be completed 5 days before it is proofread. The reason for this delay is because the consultant, who has written the report, will be on assignment and won't be available for this project during the week of May 15.

Basically, when you create a schedule, you determine when something must be delivered and then consider the activities required to deliver that output as well as the inputs required to provide the deliverable. Then you schedule the inputs and define what's required to generate them. This process is repeated until the schedule is complete.

Put another way, to develop a schedule, you define a deliverable and then back up to define what actions are required to produce it, including the actions' duration, the inputs required to provide the output, and any additional transportation time that's involved.

The problem with this approach is that often the total cycle time is unacceptably long and the start points have already passed. As an alternative, one of the following two techniques is used:

- Crushing: This is an approach in which cost and schedule are analyzed to determine how to obtain a greater amount of compression for the least increase in cost.
- Fast tracking: This is a means of determining if activities that would normally be done sequentially can be done in parallel. It also evaluates the possibility of assigning two individuals to an activity to reduce its duration. Fast tracking often results in rework and increased risk.

Five mathematical analysis techniques can help define duration limitations and reduce the total cycle time. They are:

- Critical path method (CPM): This creates a path through the schedule that defines the total project cycle. Understanding the path allows the project team to focus on a smaller group of activities to determine how to reduce the project cycle time. Each time a critical path is eliminated, a new one is created for the project team to work on.
- Critical chain method (CCM): This looks at the schedule to determine what resource restrictions have been imposed on the total cycle time. This allows you to assign additional or more skillful resources to reduce cycle time.
- Graphical evaluation and review technique (GERT): This allows for a probable treatment of both work networks and activity duration estimates.
- Program evaluation and review technique (PERT): This uses a weighted-average duration estimate to calculate activity durations. It ends up with a PERT chart view of the project.
- Simulation modeling: Some people use this method to help reduce cycle time and cost, but we haven't found it particularly useful.

The most frequently used method to document a schedule is with a bar chart called a Gantt chart, which lists each activity that must be performed and visually indicates an activity's duration as well as the interrelationships between activities. (See Figure 8.19.)

Project Management Software

A key part of project management is a very good work breakdown structure (WBS). There are many different software packages that can help with the task of developing the WBS. *Microsoft Project Office* software is the most used product.

IBM has created an *IBM Project Management Center of Excellence* that has developed a Worldwide Project Management Method (WWPMM). The WWPMM contains 13 knowledge areas that group 150 different processes. The IBM Project Management Center of Excellence defines successful operations by four factors:

- All project management professionals are knowledgeable, accountable, and part of a respected learning community.
- Employees in all professions understand and practice their roles and responsibilities.

Combined 3-Year Improvement Plan

Activity #	Activity	Person Responsible
P	**3-Year 90-Day Plan** 4/19	H.I. -EIT
0.2	Develop Plans for Individual Divisions	EIT
BP	**BusinessProcess**	EIT/Bob C.
1.0	BPI	EIT/Tom A.
ML	**Management Support/Leadership**	
1.0	Team Training	EIT/Task Team
2.0	DIT	Dept. Mgrs.
5.1	MBWA	Division President
5.2	Employee Opinion Survey	H.I.
3.0	Strategic Direction	Sam K.
4.0	Performance Planning and Appraisal	Joe B.
6.0	Suggestion System	Task Team
SP	**Supplier Partnerships**	
1.0	Partnership	H.I. –Dave F.
2.0	Supplier Standards	H.I. –Doug J.
3.0	Skill Upgrade	Bob S.
4.0	Cost vs. Price	Jack J.
6.0	Proprietary Specifications	Division President

Timeline columns: 2002 (A M J J A S O N D), 2003 (J F M 2 3 4), 2004 (1 2 3 4). Cycle 1, Cycle 2, Cycle 3, Cycle 4.

= Action
= Ongoing activity

FIGURE 8.19
Combined 3-year improvement plan.

- All management systems, infrastructure, and staff organizations support project management with portfolio, program, and project views.
- Mechanisms are in place to sustain learning and continuous improvement of project management disciplines.

The IBM Project Management Center of Excellence defines a five-level maturity model as follows:

- Level I: Pilot phase. (Ad hoc-type operation.)
- Level II: In deployment. (The project management process is disciplined. Roles and responsibilities are specified in the project plan.)
- Level III: Functional. (The project management process is standardized, consistent, and has defined practices.)
- Level IV: Integrated. (The project management process produces predictable results. Monitoring and management tools are in place.)
- Level V: World class. (The project management process is continuously improving. A knowledge base provides continuous feedback and prompts upgrades that reflect best practices. Lessons learned are continuously included in the process design.)

The IBM Project Management Center of Excellence developed an education curriculum to provide the best project management education available in the industry. This curriculum translates education into effective business advantage through knowledge prerequisites, periodic testing, applied work experience, and a progressive course structure that builds upon previous classes. In 1992 IBM began certifying its project managers. The certification process provided the company with managers who could apply the latest technologies and methods. In addition, certified project managers were required to recertify every 3 years, thereby ensuring constant exposure to the most recent and advanced project management initiatives and experiences.

IBM conducted an extensive benchmarking study of project management software packages available around the world and selected two products:

- Microsoft Project Office
- Systemcorp PMOffice

The advantages and disadvantages of each of these are listed in Table 8.3.

As a result of its increased focus on project management, IBM greatly reduced cost, wasted effort, schedule overruns, and project failures throughout the organization.

TABLE 8.3

Advantages and Disadvantages

Project Office Software	PMOffice Software
• Scope management	• Scope management
• Communications management	• Communications management
• Exception management	• Exception management
• Knowledge-enabled workflow	• Knowledge-enabled workflow
• Time management	• Cost management
	• Quality management
	• Human resource management
	• Integration management
	• Customized reporting
Advantages	**Advantages**
• Lower cost	• More functionality
• Less training	
• Minor organization change	
Disadvantages	**Disadvantages**
• Less enterprise reporting capability	• 3–6 months to implement
• Less resource management capability	• Significant organization change

Project Management Software Selection

Essex Electrical tried many combinations to make its IT project successful. As the company president, Harold Karp, put it, "I've had teams that were headed up by just business executives, and I've had them headed up by just an IT executive. In both cases the teams achieved only limited success." When they switched over to a project management approach, IT operating expenses dropped by 35%, and sales gained 1%, which translated into margins of 2.5 to 3%.

Well-managed projects do make a difference, according to a December 2003 article in *CRM Magazine*: As a result of effectively implementing a customer relationship management project, Waters Corp. was able to:

- Achieve an overall ROI of 35%
- Nearly triple sales
- Increase e-commerce revenue by 300%
- Generate additional revenue of $2 million in the service department alone

Projects represent a significant investment of strategic importance to any organization. The investment is comprised of resources, capital expenditure, time commitment, and individual dedication to make and embrace a change that will enable the organization to improve business performance and strengthen its competitive position.

Because projects are so critical to an organization's continued growth, it's important to generate a project postmortem. This document consolidates and outlines essential project best practices and identifies any negative lessons learned so that other project managers don't have to make the same mistakes.

A postmortem verifies the following:

- Did the project realize business benefits?
- Did the project deliverables meet requirement objectives?
- What best practices could be passed on to other project managers?
- What lessons learned should other project managers look out for?
- Did the project result in customer satisfaction?
- Which resources excelled during this project?

Answering these questions for senior management will help them understand the project outcome. The information will also enable the organization to continue using its best resources on other projects.

The Project Management Body of Knowledge (PMBOK) and other standards define 75 different tools that a project manager should master. Few project managers I've met during the past 50 years have mastered all these tools. Today very few project managers are certified as such by their peers. The Project Management Institute has an excellent certification program that we highly recommend. As an executive, you should select certified project managers to run your projects.

> Project management is one of the most important improvement tools available but also one of the most misused. It's as if we're trying to use a hammer to remove a screw.

PMBOK Tools and Techniques

Some of the more commonly used project management tools and techniques (as recommended by the Project Management Institute and others) are listed in Table 8.4. Evaluate yourself to determine your project management maturity level. For each one, check off your present level:

- Do not know it
- Know it but have not used it
- Used it
- Mastered it

If you would like more information on these or other improvement tools, contact the Harrington Institute: www.harrington-institute.com.

Using the sum of the four individual point scores, the following is your project manager maturity level:

- Excellent project manager: 175–225
- Acceptable project manager: 125–174
- Acceptable project team member: 100–124
- Unacceptable project manager: 50–100
- Unacceptable project team member: 0–50

TABLE 8.4

PMBOK Tools

	Don't Know It	Know It but Haven't Used It	Used It	Mastered It
1. Arrow diagramming method (ADM)				
2. Benchmarking				
3. Benefit-cost analysis				
4. Bidders conferences				
5. Bottom-up estimating				
6. Change control system				
7. Configuration management				
8. Checklists				
9. Communications skills				
10. Computerized tools				
11. Conditional diagramming methods				
12. Contingency planning				
13. Contract change control system				
14. Contract type selection				
15. Control charts				
16. Control negotiation				
17. Cost change control system				
18. Cost estimating tools and techniques				
19. Decision trees				
20. Decomposition				
21. Design of experiments				
22. Duration compression				
23. Earned value analysis				
24. Expected monetary value				
25. Expert judgment				
26. Flowcharting				
27. Human resource practices				
28. Information distribution tolls and techniques				
29. Independent estimates				
30. Information distribution systems				

(*continued*)

TABLE 8.4 (CONTINUED)

PMBOK Tools

	Don't Know It	Know It but Haven't Used It	Used It	Mastered It
31. Information retrieval systems				
32. Interviewing techniques				
33. Make-or-buy analysis				
34. Mathematical analysis				
35. Negotiating techniques				
36. Network templates				
37. Organizational procedures development				
38. Organizational theory				
39. Parametric modeling				
40. Pareto diagrams				
41. Payment system analysis				
42. Performance measurement analysis				
43. Performance reporting tools and techniques				
44. Performance reviews				
45. Pre-assignment technique				
46. Precedence diagramming method (PDM)				
47. Procurement audits				
48. Product analysis				
49. Product skills and knowledge				
50. Project management information system (PMIS)				
51. Project management information system organizational procedures				
52. Project management software				
53. Project management training				
54. Project planning methodology				
55. Project selection methods				

TABLE 8.4 (CONTINUED)

PMBOK Tools

	Don't Know It	Know It but Haven't Used It	Used It	Mastered It
56. Quality audits				
57. Quality planning tools and techniques				
58. Resource leveling heuristics				
59. Reward and recognition systems				
60. Schedule change control system				
61. Scope change control system				
62. Screening system				
63. Simulation modeling				
64. Stakeholder analysis				
65. Stakeholder skills and knowledge				
66. Statistical sampling				
67. Statistical sums				
68. Status review meetings				
69. Team-building activities				
70. Trend analysis				
71. Variance analysis				
72. Weighting system				
73. Work authorization system				
74. Work breakdown structure templates				
75. Workaround approaches				
Total				
Times weight	0	1	2	3
Point score				

Sum of the four individual total: —

Any project manager who has a point score below 125 needs project manager training.

A project is managed through a series of interaction of iterative process groups through its life cycle phases. (See Figure 8.20.) Figure 8.21 provides a complete view of how the project management process model is linked to developing a new product, service, or system.

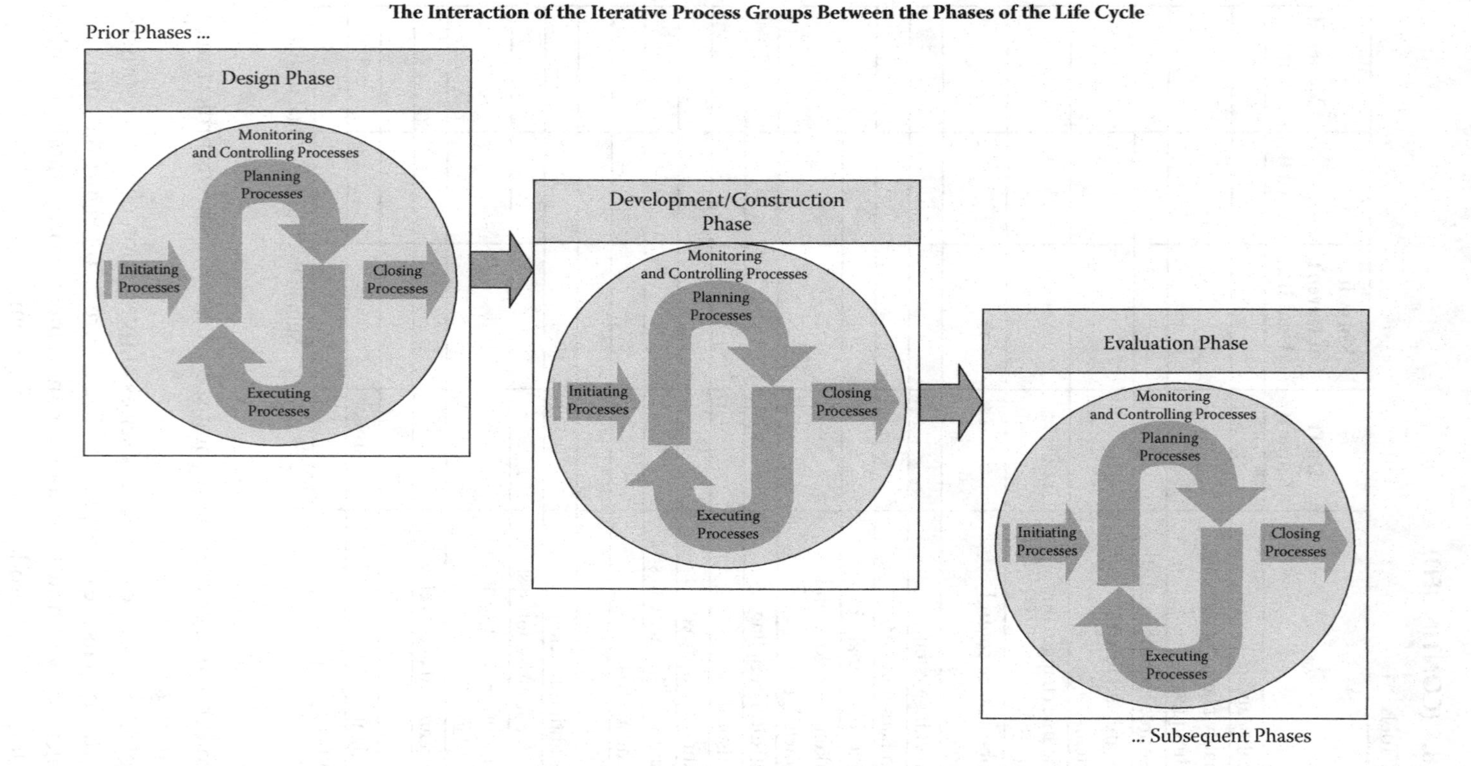

FIGURE 8.20
Interaction of iterative process groups among a project life cycle's phases.

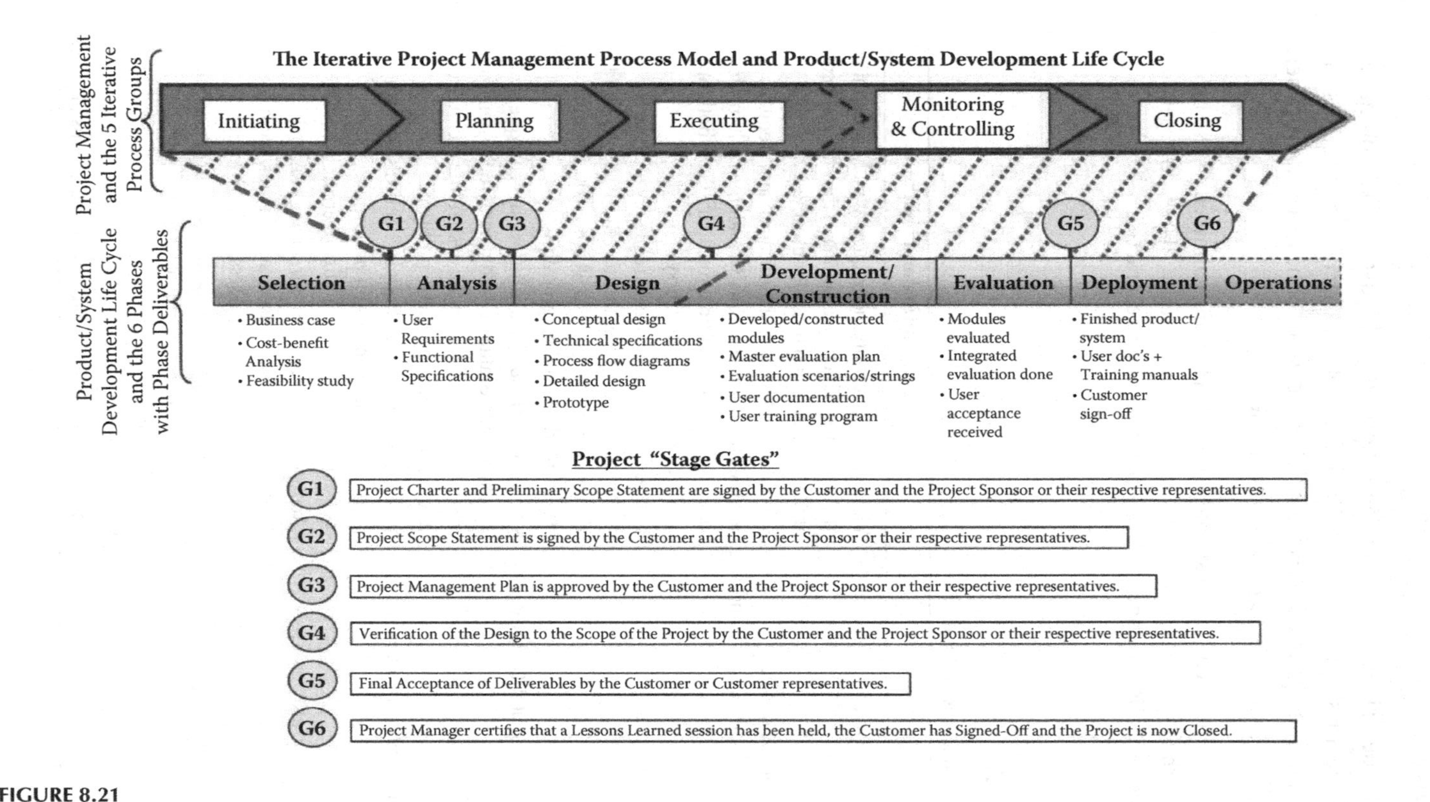

FIGURE 8.21
The iterative project management process model and product/system development.

Examples

Table 8.5 describes the approaches required to do just one part of managing a project risk management. With a list of approaches so long, it is easy to see that managing a project is not for the weak of heart or the inexperienced. (See Figures 8.22 to 8.24 and Table 8.6.)

> When I am asked, "How should I do TQM?" I am reminded of the joke about the New Yorker who was asked by a passerby, "How do I get to Carnegie Hall?" His answer was—practice, practice, practice.

TABLE 8.5

Methods Used in Risk Analysis

Method	Description and Usage
Event tree analysis	A hazard identification and frequency analysis technique that employs inductive reasoning to translate different initiating events into possible outcomes.
Failure mode and effects analysis and failure mode and effects criticality analysis	A fundamental hazard identification and frequency analysis technique that analyzes all the fault modes of a given equipment item for their effects both on other components and the system.
Fault tree analysis	A hazard identification and frequency analysis technique that starts with the undesired event and determines all the ways in which it could occur. These are displayed graphically.
Hazard and operability study	A fundamental hazard identification technique that systematically evaluates each part of the system to see how deviations from the design intent can occur and whether they can cause problems.
Human reliability analysis	A frequency analysis technique that deals with the impact of people on system performance and evaluates the influence of human errors on reliability.
Preliminary hazard analysis	A hazard identification and frequency analysis technique that can be used early in the design stage to identify hazards and assess their criticality.
Reliability block diagram	A frequency analysis technique that creates a model of the system and its redundancies to evaluate the overall system reliability.
Category rating	A means of rating risks by the categories in which they fall in order to create prioritized groups of risks.

TABLE 8.5 (CONTINUED)

Methods Used in Risk Analysis

Method	Description and Usage
Checklists	A hazard identification technique that provides a listing of typical hazardous substances and/or potential accident sources that need to be considered. Can evaluate conformance with codes and standards.
Common mode failure analysis	A method for assessing whether the coincidental failure of a number of different parts or components within a system is possible and its likely overall effect.
Consequence models	The estimation of the impact of an event on people, property, or the environment. Both simplified analytical approaches and complex computer models are available.
Delphi technique	A means of combining expert opinions that may support frequency analysis, consequence modeling, and/or risk estimation.
Hazard indices	A hazard identification/evaluation technique that can be used to rank different system options and identify the less hazardous options.
Monte-Carlo simulation and other simulation techniques	A frequency analysis technique that uses a model of the system to evaluate variations in input conditions and assumptions.
Paired comparisons	A means of estimation and ranking a set of risks by looking at pairs of risks and evaluating just one pair at a time.
Review of historical data	A hazard identification technique that can be used to identify potential problem areas and also provide an input into frequency analysis based on accident and reliability data, etc.
Sneak analysis	A method of identifying latent paths that could cause the occurrence of unforeseen events.

Typical Index for a Project Plan	
Title Page	
1.	Foreword
2.	Contents, distribution and amendment record
3.	Introduction
3.1	General description
3.2	Scope
3.3	Project requirement
3.4	Project security and privacy
4.	Project aims and objectives
5.	Project policy
6.	Project approvals required and authorization limits
7.	Project organization
8.	Project harmonization
9.	Project Implementation strategy
9.1	Project management philosophy
9.2	Implementation plans
9.3	System integration
9.4	Completed project work
10.	**Acceptance procedure**
11.	Program management
12.	Procurement strategy
13.	Contract management
14.	Communications management
15.	Configuration management
15.1	Configuration control requirements
15.2	Configuration management system
16.	**Financial management**
17.	Risk management
18.	Project resource management
19.	Technical management
20.	Test and evaluation
21.	Reliability management
21.1	Availability, Reliability and Maintainability (ARM)
21.2	Quality management
22.	**Health and safety management**
23	Environmental issues
24.	Integrated Logistics Support (ILS) Management
25.	Project team organization
25.1	Project staff directory
25.2	Organization chart
25.3	Terms of Reference (TOR)
	a) for staff
	b) for the project manager
	c) for committees and working groups
26.	Management reporting system
27.	Project diary
28.	Project history

FIGURE 8.22
Model project plan activity per four major phases.

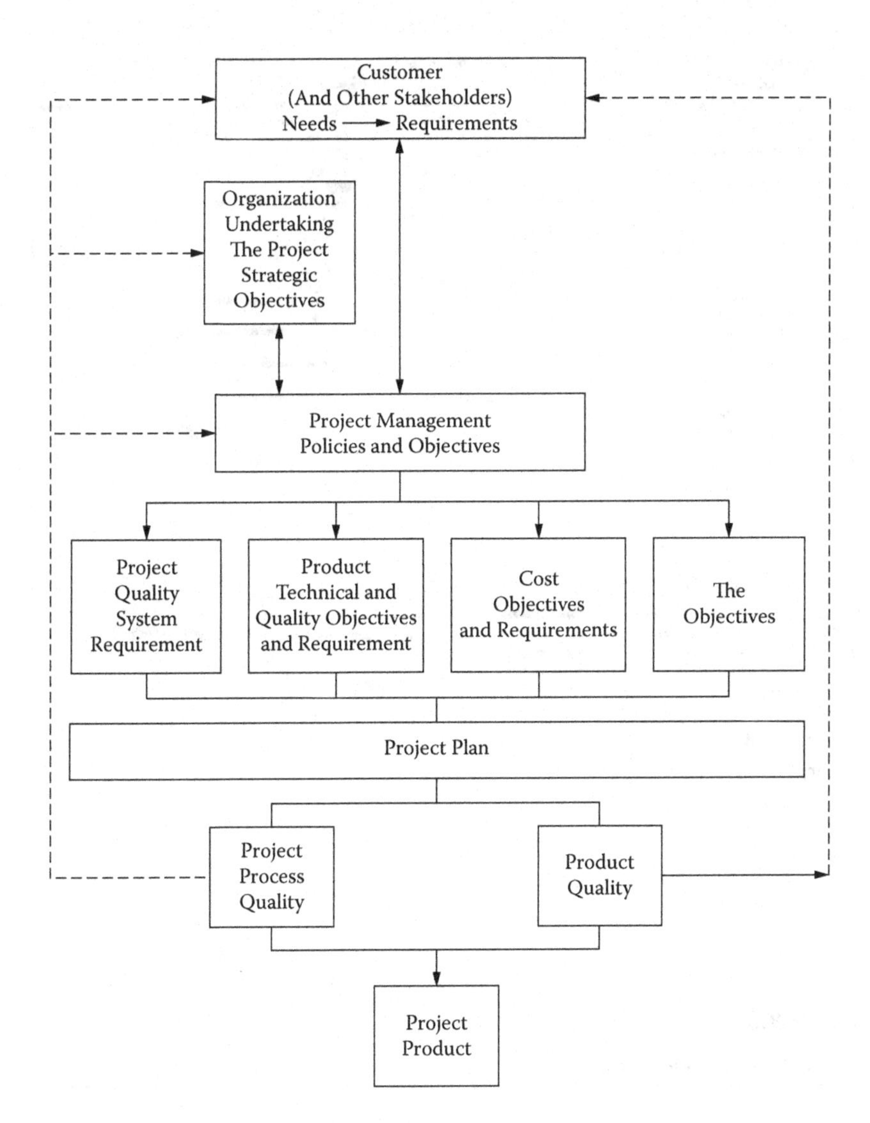

FIGURE 8.23
Block diagram of the project management process.

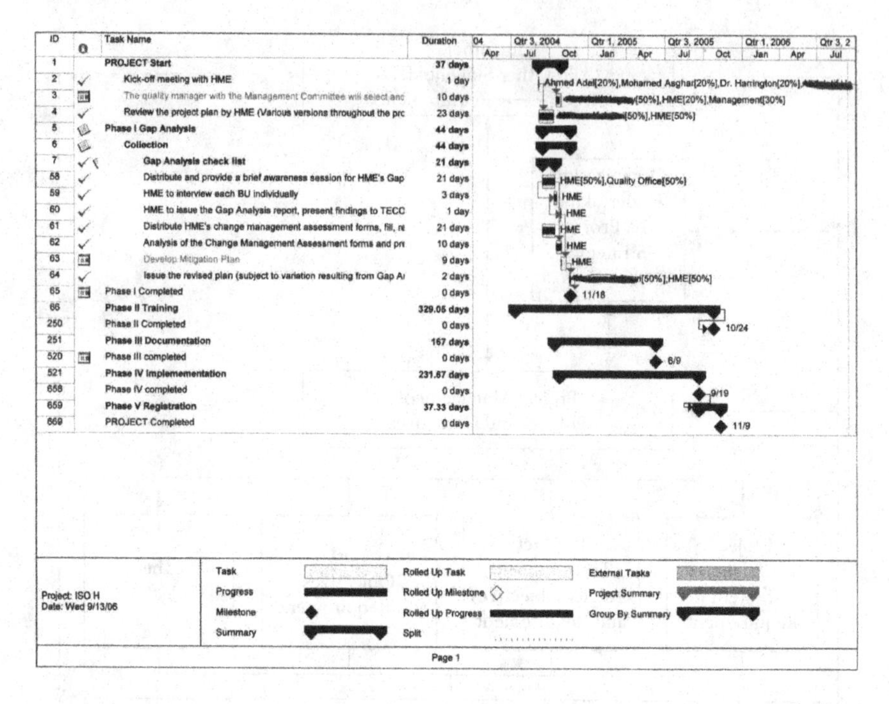

FIGURE 8.24
Excerpt from a typical work breakdown structure.

TABLE 8.6
Project Processes and Phases

	Phases			
Processes	**Conception**	**Development**	**Realization**	**Termination**
Strategic Project Process				
Strategic project process	Ä	x	x	x
Operational Process Groups and Processes within Groups				
Scope-Related Operational Processes				
Concept development	Ä			
Scope definition	Ä	x		

TABLE 8.6 (CONTINUED)

Project Processes and Phases

	Phases			
Processes	**Conception**	**Development**	**Realization**	**Termination**
Task definition	x	Ä	x	
Task realization		Ä	Ä	x
Change management		Ä	Ä	
Time-Related Operational Processes				
Key event schedule planning	x	Ä	x	
Activity dependency planning	x	Ä		
Duration estimation	x	Ä		
Schedule development		Ä	x	
Schedule control		x	Ä	x
Cost-Related Operational Processes				
Cost estimation	Ä	x		
Budgeting		Ä	x	
Cost control		x	Ä	x
Resource-Related Operational Processes (Except Personnel)				
Resource planning	x	Ä		x
Resource control		x	Ä	x
Personnel-Related Operational Processes				
Organizational structure definition	x	Ä	Ä	
Responsibility identification and assignment	x	Ä	x	
Staff planning and control		x	Ä	x
Team building	x	Ä	Ä	x
Communication-Related Operational Processes				
Communication planning	x	Ä		

(*continued*)

TABLE 8.6 (CONTINUED)

Project Processes and Phases

Processes	Phases			
	Conception	Development	Realization	Termination
Meeting management	x	Ä	Ä	x
Information distribution		Ä	Ä	x
Communication closure			x	Ä
Risk-Related Operational Processes				
Risk identification	Ä	Ä	x	
Risk assessment	Ä	Ä	x	
Solution development		Ä	x	
Risk control		x	Ä	
Procurement-Related Operational Processes				
Procurement planning	x	Ä		
Requirements documentation	x	Ä		
Supplier evaluation	x	Ä		
Contracting		Ä	x	
Contract administration		x	Ä	x
Product-Related Operational Processes				
Design	x	Ä		
Procurement	x	Ä		
Realization			Ä	
Commissioning				Ä
Integration-Related Operational Processes				
Project plan development	Ä			
Project plan execution		Ä	Ä	Ä
Change control		Ä	Ä	
Supporting Processes				

Legend: Ä = key process in the phase, x = applicable process in the phase.

Additional Reading

Badiru, A.B., and G.E. Whitehouse. *Computer Tools, Models, and Techniques for Project Management* (Blue Ridge Summit, PA: TAB Books, 1989).
Block, Robert. *The Politics of Projects* (New York: Yourdon Press, 1983).
Cleland, David, and Roland Gareis (eds.). *Global Project Management Handbook* (New York: McGraw-Hill, 1994).
Dinsmore, Paul C. *The AMA Handbook of Project Management* (New York: AMACOM Books, 1993).
Dinsmore, Paul C. *Human Factors in Project Management* (New York: AMACOM Publications, 1990).
Focardi, S., and C. Jonas. *Risk Management: Framework, Methods and Practice* (New York: McGraw-Hill, 1998).
Harrington, H. James, and Thomas McNellis. *Project Management Excellence* (Chico, CA: Paton Press, 2006).
ISO/DIS 10006: Guidelines to Quality in Project Management (Geneva: International Organization for Standardization, 1998).
Project Management Institute. *A Guide to the Project Management Body of Knowledge* (Sylva, NC: Project Management Institute, 1996).
Project Management Institute Standards Committee. *A Guide to the Project Management Body of Knowledge* (Upper Darby, PA: Project Management Institute, 1996).

It is always the last song that an audience applauds the most.

—Homer

QUALITY FUNCTION DEPLOYMENT (QFD)

Prepared by Dave Farrell

Definition

Quality function deployment (QFD) is a structured process for taking the voice of the customer and translating it into measurable customer requirements and measurable counterpart characteristics, and "deploying" those requirements into every level of the product and manufacturing process design and all customer service processes.

Just the Facts

The words *quality function* do not refer to the quality department, but rather to any activity needed to assure that quality is achieved, no matter what department performs the activity.

In QFD, all operations of the company are linked to, and driven by, the voice of the customer. This linkage brings quality assurance and quality control to all relevant functions in the organization and shifts the focus of quality efforts from the manufacturing process to the entire development, production, marketing, and delivery process. QFD provides the methodology for the entire organization to focus on what the customers like or dislike, and puts the emphasis on designing in quality at the product development stage, rather than problem solving at later stages.

Typical users of QFD in the United States include AT&T Bell Labs, Black & Decker, Chrysler, DEC, DuPont, Eaton, Ford, General Electric, General Motors, Hewlett-Packard, Johnson Controls, Kraft Foods, Proctor & Gamble, Rockwell, Scott Paper, Sheller-Globe, Texas Instruments, and Xerox. Organizations such as these consistently report benefits, including:

- Measurable improvement in customer satisfaction/market share
- Reduction in time-to-market (product development cycles)
- Significant reduction in the number of engineering changes
- Start-up costs reduced
- Identification of competitive advantage and marketing strategy
- Reduction in warranty claims and customer complaints
- Increase in cross-functional teamwork

The House of Quality, a matrix format used to organize various data elements, so named for its shape, is the principal tool of QFD. (See Figure 8.25.) Although "house" designs may vary, all contain the same basic elements:

- Whats: The qualities or attributes the product or service must contain, as required by the customer. Whats are compared to competitors' qualities and ranked by importance.
- Hows: The technical means for satisfying the whats, including the exact specifications (how much) that must be met to achieve them.

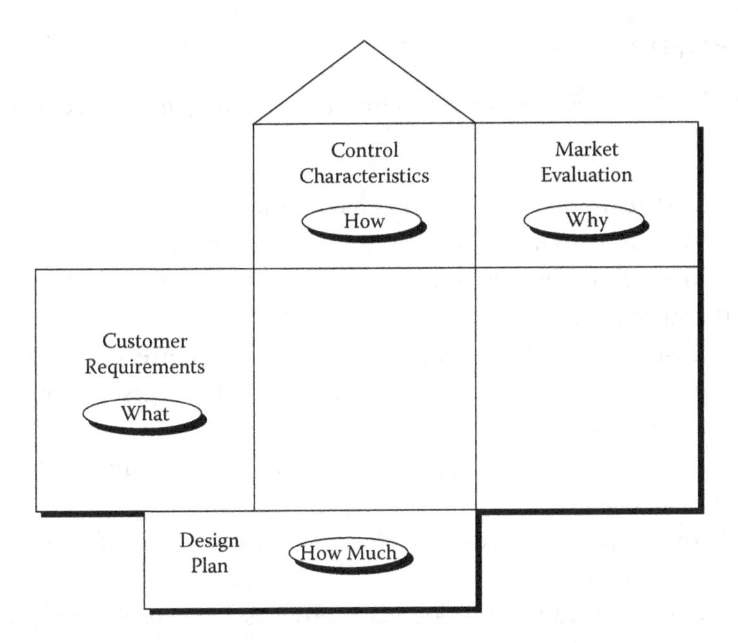

FIGURE 8.25
Basic QFD matrix.

- Correlation matrix: An evaluation of the positive and negative relationships between the hows, sometimes called the "roof."
- Relationship matrix: An evaluation of the relationships between the whats and hows. It identifies the best ways to satisfy the customer and generates a numerical ranking used as a guide throughout the development process.

These Houses of Quality are used in each of the four primary phases of QFD:

1. Product planning phase: Where customer requirements are identified and translated into design specifications or product control characteristics in the form of a planning matrix.
2. Parts development phase: Translates the outputs of the product planning phase (design specifications) into individual part details that define part characteristics.
3. Process planning phase: Defines the process for making each component and establishes critical component parameters.
4. Operating instructions (production planning): Define the production requirements for each component/operation.

Using QFD

Richard Tabor Green points out in his book *Global Quality* that QFD was designed to:

- Translate observed customer behavior into customer wants
- Translate customer articulations into customer wants
- Translate competitor success at meeting customer wants into a threat assessment
- Translate competitor process capabilities into a threat assessment
- Translate customer wants and competitor threat assessment into a market opportunity map
- Translate a market opportunity map into product lines and release schedules
- Translate product lines and release schedules into product development projects
- Translate product development projects into part and process specifications
- Translate part and process specifications into operation specifications
- Translate operation specifications into delivery and service specifications

We agree that the House of Quality looks very complex and confusing, but it does contain a great deal of information. To help you understand this complex picture, we will discuss the major parts of the House of Quality. By understanding the parts, you will be able to see how they blend together to provide a very effective weapon.

Effective application of QFD is likely to require significant change in the way an organization operates. Substantial training and planning are involved. There are new tools to be learned. Cross-functional teamwork capability will be put to the test. Purchasing, sales, and engineering organizations will be involved in product or service design and development to a degree never previously experienced. All of these factors suggest that success, even of a limited project, is dependent on a high level of management commitment. As with any significant change effort, effective sponsorship by key management personnel of each affected organization is essential. Attention should be given at the outset to the principles and methodologies of organizational change management, which are beyond the scope of this chapter.

To maintain sponsor commitment, senior management personnel should be involved in the selection of initial projects. The selection team should identify a wide variety of potential projects and formally apply an agreed upon set of evaluation criteria (for example, market share, product life cycle, cost reduction potential, revenue enhancement potential, market share implications, need for time-to-market reduction, competitive position, organizational commitment, potential for success, alignment with business plan/strategy, etc.).

With the project selected, it is now time to select the QFD team leader and members. Care should be taken to include participants from all key organizations involved from the product (or service) concept/design through to delivery and after-sale service. It is likely that the core team will be augmented by additional people at different stages of the QFD process, as additional detail is required.

The core team's initial activity should include the development of a team charter, including mission, objectives, guidelines for team conduct, work plan, and a timetable. The team charter should be shared with the executive group to assure alignment on mission, objectives, and expectations. Consensus around a team charter also clarifies the resources that will be required, and avoids midproject surprises that can derail the effort. The team should also plan for periodic status reporting (at a minimum, upon the completion of each QFD phase) to the executive group to maintain sponsor awareness and commitment.

Time spent early in the process in team building and member training is an investment well worth the effort. It is important that team members understand the QFD methodology at the outset, have an agreed upon roadmap of the process they are going to follow, and possess basic team skills. In addition, competence in the related methodologies and tools described in this book will substantially enhance the effectiveness of the team and the success of the QFD project.

Voice of the Customer

This is not only where QFD begins, but it is the foundation on which every step that follows is based. In this phase we abandon previous practices of assuming that we know what the customer wants and needs, or worse yet, a marketing philosophy predicated on the notion that we can "sell" the customer on what we already have to deliver. We begin by really listening to the customer's expression of his or her requirements, in his or her own terms. And we probe for a level of detail heretofore unidentified. A variety of sources and market research tools should be considered, including follow-up letters to customers, observations

TABLE 8.7

Requirements for Typical Toaster

Dimension	Possible Customer Requirement Statement
Performance	• Speed of making toast • Always produces requested darkness
Features	• Handles a wide variety of sizes • Up to four slices at a time • Handles frozen foods (e.g., waffles)
Reliability	• No breakdowns • Works well with fluctuating currents
Durability	• Long service life • Hard to damage
Serviceability	• Easy to clean • Easy to repair—can do it yourself • Availability of spare parts
Aesthetics	• Fit with a variety of kitchen decors • Variety of color choices • Clean lines, smooth design
Safety	• Child can operate • No shock hazard

of the customers using the product, interviews, focus groups, "buff" magazines, independent product reviews, trade shows, sales contacts, printed and/or telephonic surveys, etc. Also, the past history of unsolicited customer feedback should be reviewed, such as customer returns, complaint letters, and accolades. Product or service dimensions around which customer input is sought can include any or all of the following: performance, features, reliability, conformance, durability, serviceability, safety, environmental, aesthetics, perceived quality, and cost. Examples of the application of these dimensions in the case of a toaster are shown in Table 8.7.

Customer requirements can also be expressed at multiple levels of detail, commonly called primary, secondary, and tertiary. The customer requirements, so expressed, become the vertical axis of the top portion of the planning matrix, as shown in Figure 8.26.

While collecting customer requirements information, the relative importance of each characteristic should be identified (using a numerical rating scale) to enable prioritization of those characteristics. This

PERFORMANCE	**Speed of Making Toast**	Fast
		Not slowed down by multiple slices
		Not slowed down by thick slices
		First slice as fast as subsequent slices
	Always produces requested darkness	Should have a "warm" setting
		Darkness exactly like dial setting
		Darkness not affected by thickness
		Darkness not affected by temperature
		Can darken one side only
FEATURES	**Handles a wide variety of sizes**	Will warm rolls and muffins
		Toast regular bread
		Will handle oversized breads
	Handles frozen foods (e.g. waffles)	Handle any temperature of material
		Frozen items done as well as bread

FIGURE 8.26
Breakdown of a voice of the customer for a toaster.

information will be used in conjunction with the competitive evaluation to assure favorable performance measures on those characteristics most important to the customer. The prioritization data will be entered on the right side of the top portion of the matrix in the column labeled "Customer Importance Rating."

Next, the final product control characteristics (design requirements) are identified and arranged across the top horizontal row of the matrix. (See Figure 8.27.) These are measurable characteristics that must be deployed throughout the design, procurement, manufacturing, assembly, delivery, and service processes to manifest themselves in the final product performance and customer acceptance.

Since not all of the relationships between customer requirements and the corresponding final product control characteristics will be equally strong, the next step involves building a relationship matrix to display all of those relationships (using either symbols or numerical values) to further enable focus on those which are most highly related to high-priority requirements. It is a way to validate that design features cover all characteristics

	Final Product Control Characteristics										
Customer Requirements	Thermostat Accuracy	Heating Wire Thickness	Heating Wire Coverage	Variable Power Delivery	Airtight Case	Interior Dimensions	Horizontal Layout	Variable Heating Area	Shut-off	Humidity Sensor	Exterior Surface
Fast											
Not slowed down by multiple slices											
Not slowed down by thick slices											
First slice as fast as subsequent slices											
Should have a "warm" setting											
Darkness exactly like dial setting											
Darkness exactly like dial thickness											
Darkness exactly like dial temperature											
Can darken one side only											
Will warm rolls and muffins											
Toast regular bread											
Will handle oversize breads											
Handle any temperature of material											
Frozen items done as well as bread											

FIGURE 8.27
Typical final product characteristics for a toaster.

needed to meet customer requirements. An example of the relationship matrix is shown in Figure 8.28.

At a very minimum, every important customer requirement should have at least one product characteristic that has a medium or strong relationship to it. If this is not the case, additional product characteristics should be added to correct this void. The relationship matrix also serves the important function of identifying conflicting requirements, for example, high strength versus low weight. When such conflicts are identified, the application of Taguchi methods as mentioned above is indicated to optimize design characteristics.

While analyzing the relationship matrix and its implications, it is important to keep open to the possibility of modifying or adding to the list of product control characteristics.

The next step is completion of the market evaluation, including both the customer importance rating and competitive evaluations, on the right side of the top portion of the matrix. (See Figure 8.29.) This section displays

Customer Requirements	Thermostat Accuracy	Heating Wire Thickness	Heating Wire Coverage	Variable Power Delivery	Airtight Case	Interior Dimensions	Horizontal Layout	Variable Heating Area	Shut-off	Humidity Sensor	Exterior Surface
	Final Product Control Characteristics										
Fast	○	●	●	●	○	○	△	○			△
Not slowed down by multiple slices		●	●	●	○			●			
Not slowed down by thick slices		●	●	●	○			●			
First slice as fast as subsequent slices		●	●	●	△						
Should have a "warm" setting	●			●							
Darkness exactly like dial setting										5	
Darkness exactly like dial thickness								5			
Darkness exactly like dial temperatu…		4	4	5	2			3			
Can darken one side on…	2		1					2	5		
…ons and muffins	3	3	3	4		4	5	5			
Toast regular bread	4	5	5	5	2			1			
Will handle oversize breads	4	5	5	5	2	4	5	5		5	
Handle any temperature of material	5	3	3	5	2					5	
Frozen items done as well as bread	5			5	2					5	

● – Strong relationship ○ – Medium relationship △ – Weak relationship

5 – Strong relationship 3 – Medium relationship 1 – Weak relationship

FIGURE 8.28
Toaster voice of the customer/product control characteristics relationship matrix.

the comparative strengths and weaknesses of the product in the marketplace around those requirements most important to the customer. The data for the competitive evaluation can come from the same sources as the customer requirements data, from independent or internal product testing, and from independent benchmarking of product characteristics as described briefly above. As such, the comparisons can be based on both objective and subjective data.

The company's and competitor's present performance data for the final product control characteristics are entered in the lower section of the matrix in the columns for each relevant characteristic, as shown in Figure 8.30.

The present performance data are next used to establish performance targets for relevant characteristics that, when achieved, will position the product

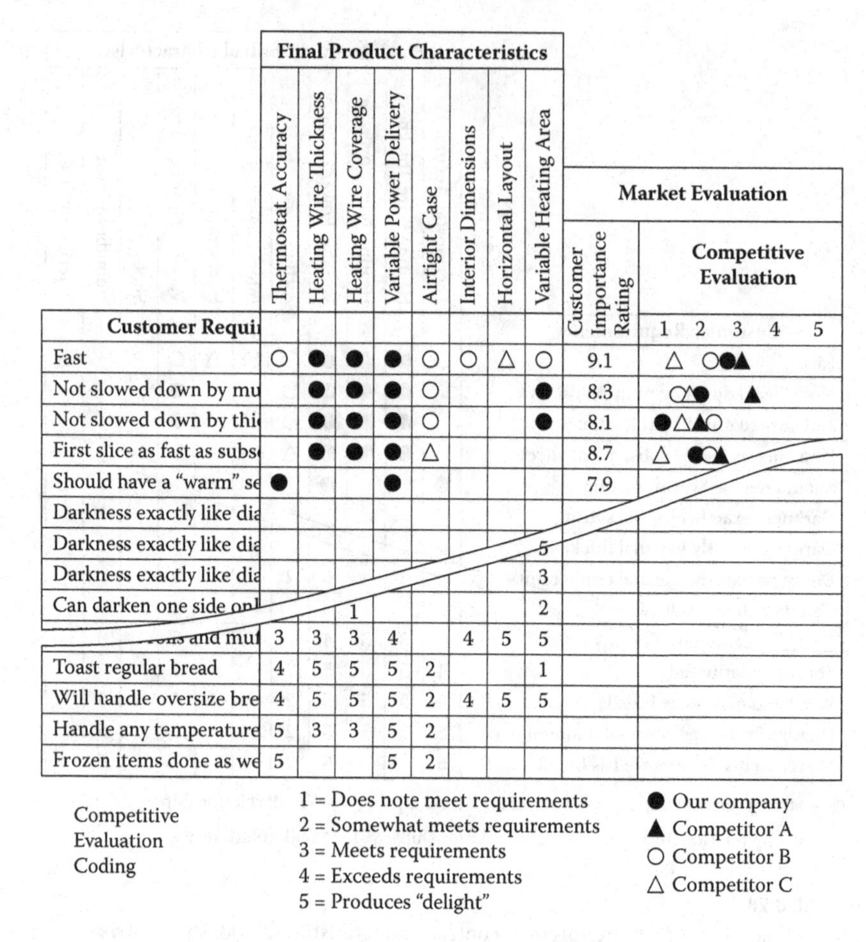

Final Product Characteristics
Thermostat Accuracy
Heating Wire Thickness
Heating Wire Coverage
Variable Power Delivery
Airtight Case
Interior Dimensions
Horizontal Layout
Variable Heating Area
Market Evaluation
Customer Importance Rating
Competitive Evaluation
1 2 3 4 5
Customer Requi
Fast
Not slowed down by mu
Not slowed down by thi
First slice as fast as subs
Should have a "warm" se
Darkness exactly like dia
Darkness exactly like dia
Darkness exactly like dia
Can darken one side on
and mu
Toast regular bread
Will handle oversize bre
Handle any temperature
Frozen items done as we
9.1
8.3
8.1
8.7
7.9
Competitive Evaluation Coding
1 = Does note meet requirements
2 = Somewhat meets requirements
3 = Meets requirements
4 = Exceeds requirements
5 = Produces "delight"
Our company
Competitor A
Competitor B
Competitor C

FIGURE 8.29
Comparison to the competition.

as highly competitive or best in class. Target data are displayed immediately below the corresponding competitive data, as shown in Figure 8.31.

The final element of the planning matrix is the selection of those product quality characteristics that are to be deployed through the remainder of the QFD process. Those that are a high priority to the customer, have poor competitive performance, or require significant improvement to achieve established target levels should be taken to the next level of QFD analysis.

Part Deployment Phase

The matrix begins with the outputs of the planning matrix, in particular the overall product quality characteristics, and defines their deployment down to

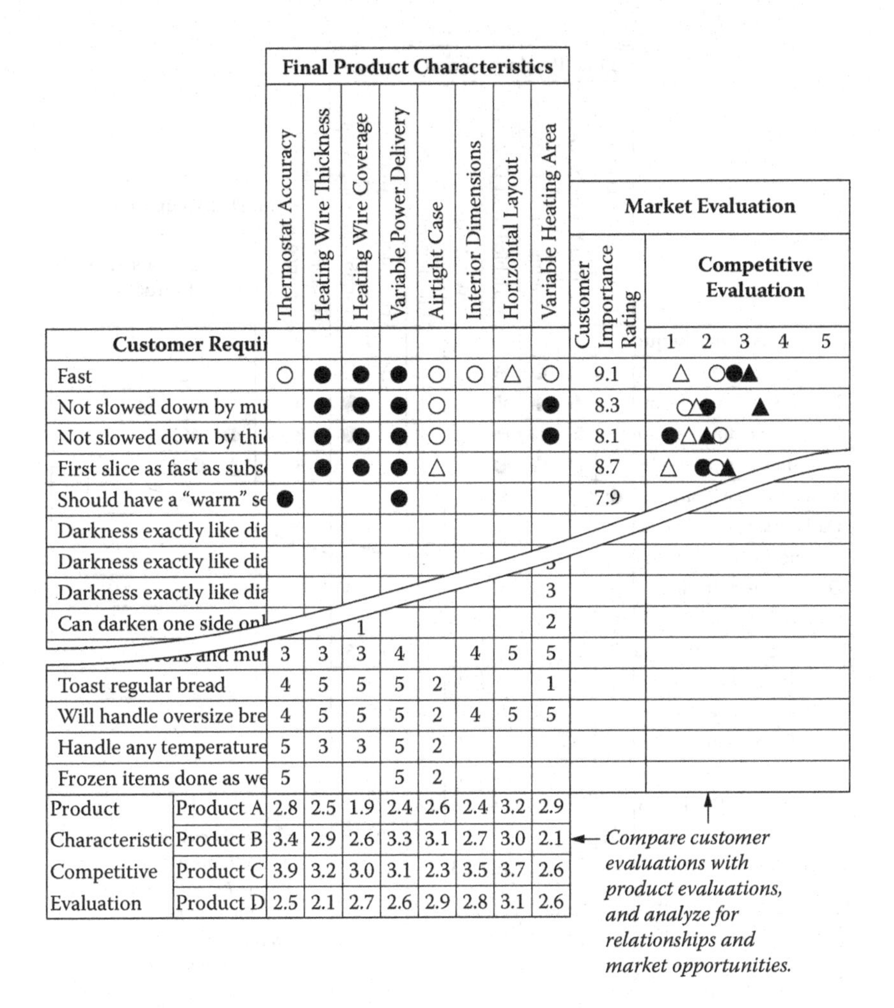

FIGURE 8.30
Product characteristic competitive evaluation.

the subsystem and component levels. In the process, the component part characteristics that must be met in order to achieve the final product characteristics are identified, and the matrix indicates the extent of the relationship between the two, as shown in Figure 8.32. It is the critical component characteristics that will be deployed further and monitored in the later stages of QFD.

It should be noted that a focus on high-priority characteristics is crucial from this point forward. Experience has shown that a number of QFD teams have lost focus or energy at this point due to the apparent complexity of the process or the sheer number of charts or possible correlations. To start, work to identify the three to five most critical finished component

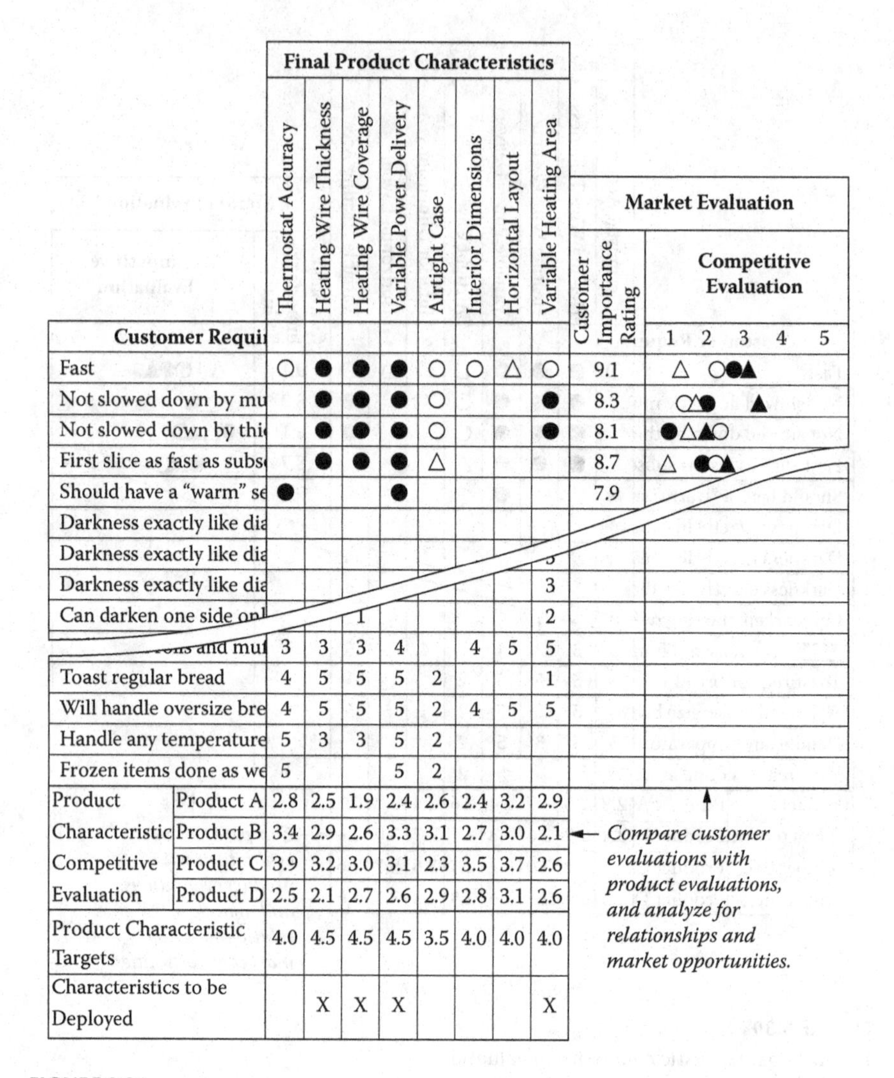

FIGURE 8.31
Performance targets displayed.

characteristics and concentrate on these. When the results have been deployed through the remaining steps with success, you can go back and address the next-higher-priority part characteristics.

The Process Plan and Quality Plan Matrices

In the previous step, critical component part characteristics were identified. We are now ready to identify the process used to produce those parts, the steps in the process that are critical to those characteristics,

	Component Part Characteristics										
	Heating Assembly			Power Transformer				Heating Cavity			
	Wire Routing	Wire Dimensions	Wire Material	Winding	Insulation	Case	Connectors	Walls	Backs	Hinges	Height Control
Final Product Control Characteristics											
Heating Wire Thickness		●	●		○			○			
Heating Wire Coverage	●	●	○	○							
Variable Power Delivery		●	●	●	●	○					
Variable Heating Area			△	●			●		●	●	●

● – Strong relationship ○ – Medium relationship △ – Weak relationship
or
5 – Strong relationship 3 – Medium relationship 1 – Weak relationship

FIGURE 8.32
Breakdown of product control characteristics to component parts characteristics.

the appropriate control points in the process to assure conformance to requirements, and the process monitoring plan. These data are displayed in the process planning matrix. (See Figure 8.33.)

The Roof of the House of Quality

To complete the House of Quality, we need to put a roof on it. The roof of the House of Quality is used to analyze the interactions between characteristics. (See Figure 8.34.) Basically two characteristics can have one of three types of relationships:

1. As one changes, it has no impact upon the other.
2. As one changes for the better, it has a negative impact on the other.
3. As one gets better, it has a positive impact upon the other.

		Heating Assembly							
		Product Control Characteristics					Process Monitoring Plan		
No.	**Process Steps**	Wire Length	Wire Thickness	Ductility	Radius	Control Points	Check Points	Monitor Method	Freq.
1	Mount Wire Stock					Alignmt	Specing	Visual	100%
2	Set Counter	●				"0" Meter	Meter	Visual	100%
3	Set Die		●		○	I.D.	Temp.	Histogrm	1/hr
4	Draw Wire		●	○	○	O.D.	Temp.	$\bar{X}$ & R	5/hr
5	Cut wire	●				Length			
6	Shape to Specifications				●	Radius			
7	Mount to Heating Frame								

Note: *The above data drives the Operating Instructions.*

● – Strong relationship ○ – Medium relationship △ – Weak relationship
or
5 – Strong relationship 3 – Medium relationship 1 – Weak relationship

FIGURE 8.33
Process planning matrix.

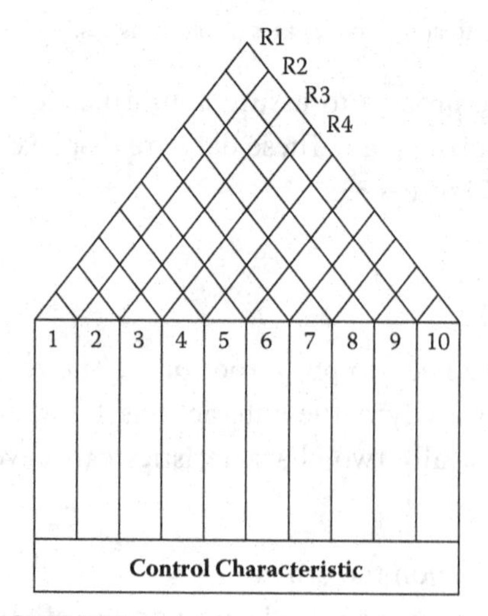

R1	**R2**	**R3**	**R4**
1 *to Y*	2 *to Y*	3 *to Y*	4 *to Y*
1 – 2			
1 – 3	2 – 3		
1 – 4	2 – 4	3 – 4	
1 – 5	2 – 5	3 – 5	4 – 5
1 – 6	2 – 6	3 – 6	4 – 6
1 – 7	2 – 7	3 – 7	4 – 7
1 – 8	2 – 8	3 – 8	4 – 8
1 – 9	2 – 9	3 – 9	4 – 9
1 – 10	2 – 10	3 – 10	4 – 10

FIGURE 8.34
The roof of the House of Quality.

In Figure 8.34 row 1 is made up of ten diamonds. The diamond at the lower left-hand end of the row analyzes the interactions between characteristics 1 and 2. The second diamond analyzes the interactions between characteristics 1 and 3, etc. The diamond on the lower left-hand end of the row labeled R2 analyzes the interaction between characteristics 2 and 3, etc. By analyzing the interactions between each diamond that makes up the House of Quality roof, all the interactions between the different characteristics are analyzed. To pictorially show the interactions, the following codes are recorded in the related diamonds:

- No interactions
- Negative interaction
- Positive interaction

Some people choose to put nothing in the diamond that has no interaction, but we feel this is poor practice because there is no visual evidence that the interactions were analyzed. Some individuals carried out the analysis to further classify the interactions in five categories.

1. Strong positive
2. Strong negative
3. No interaction
4. Negative
5. Strong negative

In the more sophisticated models point scores are assigned that indicate the interrelationship's strengths or weaknesses. For example:

- Plus 4 points is given to relationships that have a strong positive impact upon each other.
- Plus 2 points is given to relationships that have a medium positive impact upon each other.
- No points are given when there is no interaction.
- Minus 2 points is given to relationships that have a medium negative impact on each other.
- Minus 4 points is given to relationships that have strong negative impacts on each other.

With the addition of the roof on top of the House of Quality, we have now completed the quality function deployment of one level. In reality, the House of Quality should be broken down from:

- Voice of a customer to design specifications
- Design specifications to parts characteristics
- Parts characteristics to manufacturing operations
- Manufacturing operation to production requirements

At each one of these breakdowns a new House of Quality is often prepared to help understand the product and how it relates to the voice of the customer.

As with previous matrices, symbols or numerical values can be used in the process plan matrix to show the strengths of the relationships. Control points are established at the steps in the process that are critical to meeting component characteristics. They establish the data and the strategy for achieving product characteristics that meet high-priority customer requirements.

In the quality plan matrix, the process steps can be displayed in flowchart format, and the control points and checking methods for each control point are taken to a more specific level of detail. It is the latter information that forms the basis for developing the final QFD document, operating instructions.

Operating Instructions

Unlike previously described matrices, the operating instructions in QFD do not have a single prescribed format. They may be designed to meet the specific characteristics of the process and needs of the process operators.

The essence of this final step, other than implementation, is to deploy the results of the quality plan to the people who will be executing it. The instructions should be in sufficient detail to provide needed information, in a useful format, on what to check, when to check it, how to check it, what to check it with, and what parameters are acceptable. This step makes the final connection between the work of the operator and his or her ultimate objective of satisfaction of the customer's requirements.

Summary

QFD is one of the most rapidly growing methodologies in the quality arsenal. Its successful application requires management's willingness to invest the three T's—time, tools, and training. There are a number of issues to be considered before adopting the process. Training and implementation time and costs can be significant. Do you have sufficient sponsor commitment to sustain the effort over time? How successful has the organization been at other cross-functional team efforts? How are you going to incorporate ongoing process improvements into the QFD documentation system? QFD can be complex to administer. Who is going to be responsible for that task? Product revisions, although fewer in number, require effort to integrate into the entire system. The increased effort at planning stages, rather than implementation, requires a cultural shift in many organizations.

As Ted Kinni put it, in concluding his November 1, 1993, *Industry Week* article, "Simply put, QFD is work. On the other hand, its growing popularity indicates that faster and less expensive development cycles, improved quality and reliability, and greater customer satisfaction are well worth the effort."

Examples

Examples are included in the text. Figure 8.35 is a detailed view of the House of Quality. Figure 8.36 shows how QFD drives down through four levels of documents to keep them all focused on the voice of the customer.

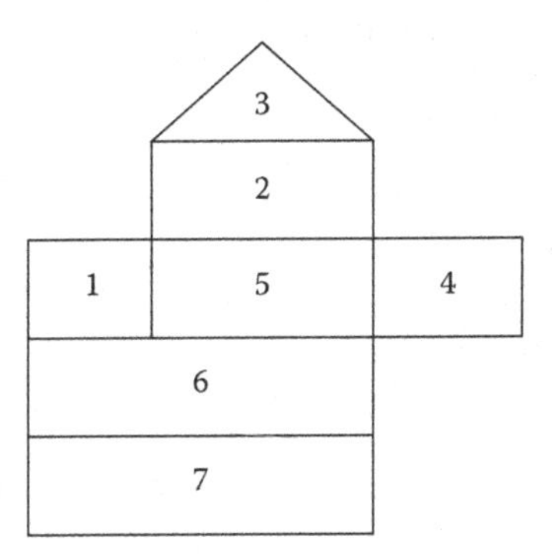

FIGURE 8.35
House of Quality for the product planning level.

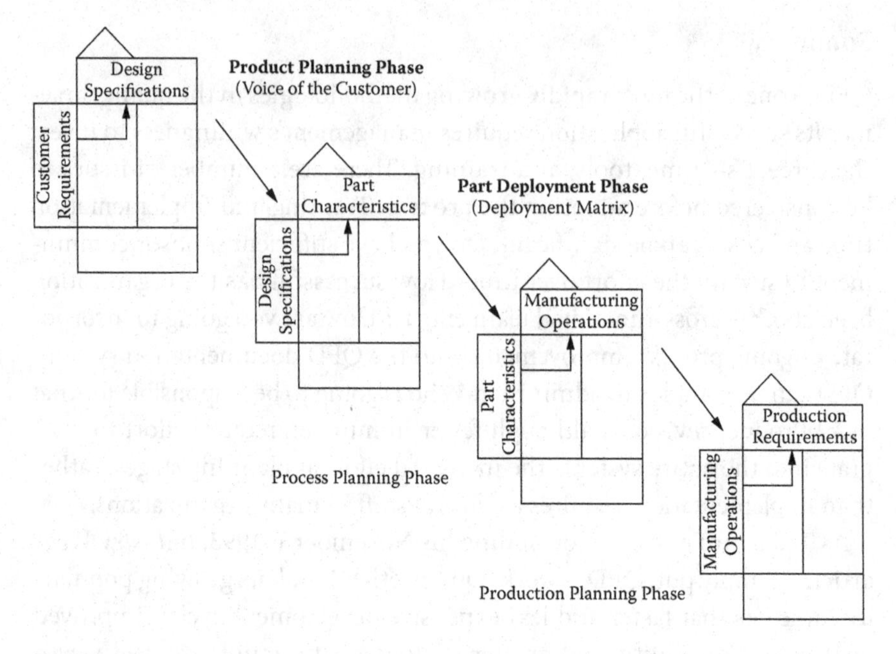

FIGURE 8.36
How QFD drives down through four levels of documents.

1. Voice of the customer/customer requirements
2. Product characteristics
3. Correlation between characteristics
4. Market evaluation
5. Characteristic relationships
6. Design competitive assessment
7. Limiting factors

Additional Reading

Akao, Yoji (ed.). *Quality Function Deployment: Integrating Customer Requirements into Product Design* (Portland, OR: Productivity Press, 1990).

Bossert, James L. *Quality Function Deployment: A Practitioner's Approach* (Milwaukee, WI: ASQ Quality Press, 1990).

Cohen, Lou. *Quality Function Deployment: How to Make QFD Work for You* (Milwaukee, WI: ASQ Quality Press, 1995).

9

Black Belt Nonstatistical Tools (R through Z)

INTRODUCTION

The tools that are presented in this chapter include the following:

- Reliability management systems
- Root cause analysis
- Scatter diagrams
- Selection matrix/decision matrix
- SIPOC
- SWOT
- Takt time
- Theory of constraints
- Tree diagram
- Value stream mapping

RELIABILITY MANAGEMENT SYSTEM

> Customers buy the first time based upon quality. They come back to buy again based upon reliability.

Definition

A reliability management system is a system for designing, analyzing, and controlling the design and manufacturing processes so that there is a high probability of an item performing its function under stated conditions for a specific period of time.

JUST THE FACTS

Although everyone is talking about quality today, it isn't quality that gets your customers to come back—it's reliability. Quality is what makes the first sale, but it is reliability that keeps the customers. Reliability is the most important thing to customers of both products and services. Customers expect that their new car will start when they pick it up at the dealer, but it is even more important that it starts every time they get into it to drive to the airport for a flight that they are already late for (reliability).

We are amazed at the way reliability is being overlooked. In fact, it's hard to find a book on quality management that directly addresses reliability. In 20 books on quality that we pulled at random from our bookshelves, only two had reliability even listed in their index: Joe Juran's old faithful *Quality Control Handbook* (McGraw-Hill, 1988) and Armand Feigenbaum's classic *Total Quality Control.*

Unfortunately, management has been trained to understand quality but has little understanding about the methodologies that result in improved reliability; as a result, they do not ask the right questions of their engineers and supporting managers. This results in misguided focus all the way through the organization.

- **Rule 1:** To get quality, everyone must understand what quality is.
- **Rule 2:** To get reliability, everyone needs to understand what reliability is.

The maximum reliability is determined by the product design. The production process design can only detract from the inherent reliability of the original design.

It is not poor quality that is causing North America to continue to lose market share—it is poor reliability. We analyzed the midsized passenger

TABLE 9.1

Reliability Weighting Auto Table

Excellent	+2 points
Very good	+1 points
Good	0 point
Fair	−1 point
Poor	−2 points

automobiles reliability data reported in the *Consumer Reports Buying Guide 1998*. We calculated an index by weighing the complexity of the unit from 1 to 3. We scored each unit's reliability as shown in Table 9.1.

The very best index a brand of automobile could get is 180. Using this index, we discovered that Japanese brands have an index that is almost 100% higher than North American brands (Japanese brands index = 113.8, North American brands index = 60.9). The two lowest reliability indexes were for two European cars. The Japanese Subaru Legacy had the highest reliability index, which was 133 out of a maximum possible points of 180. Only one Japanese car had a rating lower than 118 points, the Mitsubishi Gallant, which had a rating of 62. All of the North American brands index ratings were below 72.

The difference between Japanese and North American midsized passenger cars is only 5 percentage points. It is so small that the consumer cannot detect the difference. It is the lack of reliability that is turning once-loyal consumers of Ford, GM, or Chrysler autos to Toyota, Nissan, or Honda. The result is that U.S. brands are continuously losing their market share in passenger cars even within the United States. (See Figure 9.1.)

In the 1990s it was well known that the reason U.S. automakers were losing market share was because of their poor reliability record compared to Japanese cars. Now 20 years later the U.S. automobile industry has made some progress, but is it enough? I will let you be the judge. The following is the analysis of today's 3-year-old cars' error-free performance by automakers as defined in *Consumer Reports Magazine* (April 2012, p. 78). They are listed in order from the best to the worst.

1. Toyota
2. Subaru
3. Honda
4. Nissan
5. Volvo

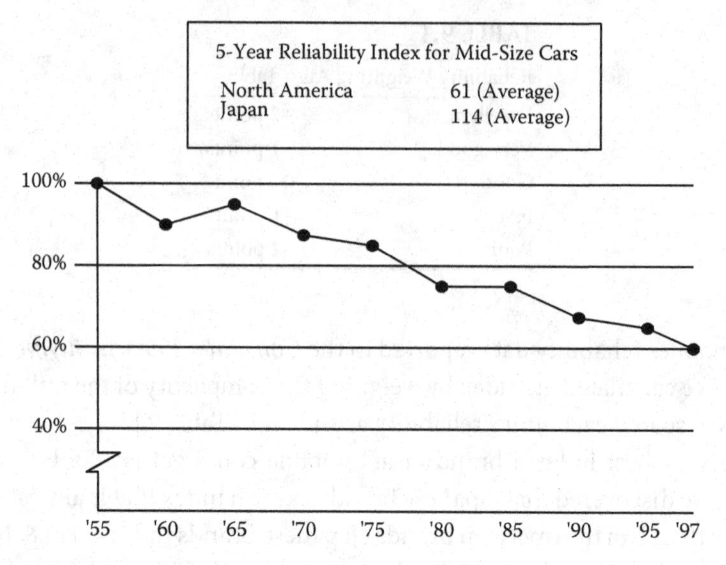

FIGURE 9.1
U.S. domestic brands' share of U.S. car sales. (From *Consumer Reports*, 1998.)

6. Ford
7. Hyundai
8. Mercedes-Benz
9. Mazda
10. Chrysler
11. General Motors
12. Volkswagen
13. BMW

> Meeting reliability requirements has become one of the major demands upon modern product technology. Buyers who once concentrated their purchases upon products that were primarily innovative or attention-getting now concentrate upon such products which also operate reliably.
>
> —**Armand V. Feigenbaum**
> *Total Quality Control*
> (McGraw-Hill, NY, 1991)

To provide a competitive, reliable item or service requires that the organization delivering the output understands and improves the processes that are used to design and build reliability into its output. Figure 9.2 represents a typical reliability system or, as we like to call it, reliability cycle. We have selected the word *cycle* because each generation builds upon the experiences and knowledge generated in the previous cycle.

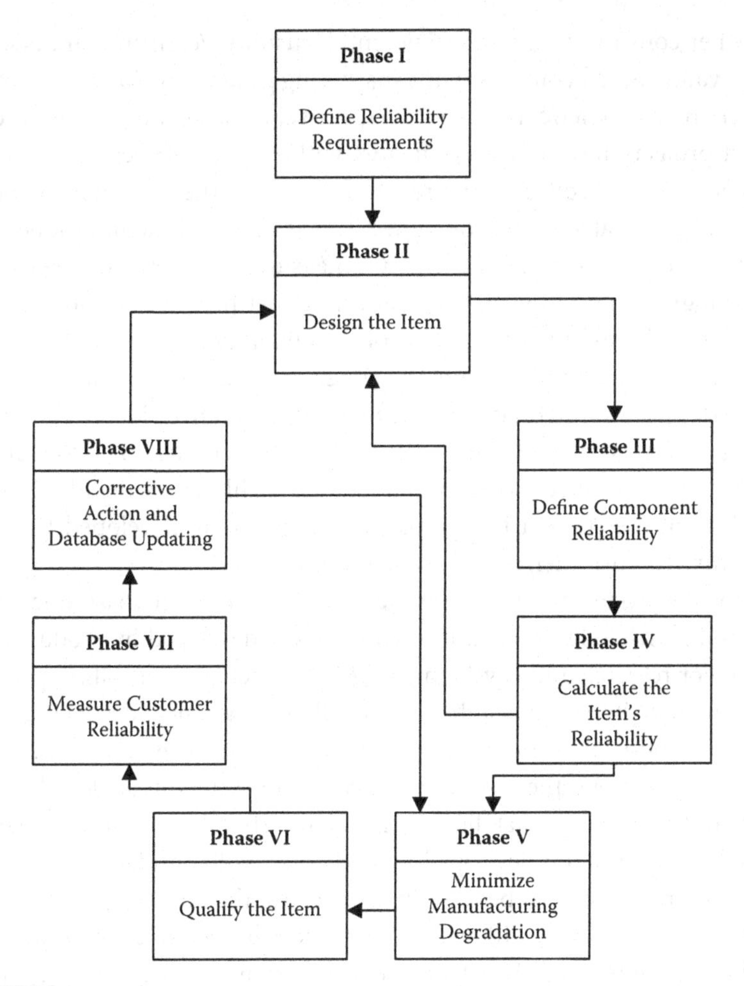

FIGURE 9.2
The reliability cycle.

Phase I: Defining Reliability Requirements

The textbook approach is for the marketing organization to identify a product opportunity and develop a product perspective that defines the cost, performance, and reliability requirements needed to service a specific market. Typical reliability measurements would be:

- Mean time to failure
- Mean time between failures

This perspective should be based upon a detailed knowledge of the external customer/consumer's needs and environment, plus a great deal of customer/consumer contact and potential customer/consumer inputs. The major value-added contents that a marketing group contributes to a product are the customer/consumer's perspective, needs, expectations, and an expert projection of how these factors will change between the time the marketing group collected the research data and the date that the product will be available to the consumer. This is often a difficult task because many product development cycles are measured in years, not weeks, and technology is improving so fast that it is sometimes difficult for marketing's "crystal ball" to accurately project reliability requirements. On the other hand, there should be no one closer to the customer/consumer and in a better position to do it than the marketing organization. A good marketing organization not only will understand the customer/consumer's present and projected expectations, but also will have a detailed understanding of the present technologies that are being developed that will drive further customer/consumer requirements.

Using the product perspective as the product performance basic document, the product reliability specifications are usually developed by product engineering or research and development (R&D). Often, the reliability specifications are more impacted by inherent reliability of the embedded technologies and competition than they are by customer/consumer expectations. In the better organizations, the consumer's reliability expectations are less demanding than the product reliability specifications. That does not mean that the reliability specification developed by R&D does not consider the customer/consumer requirements, because in most cases, they do, but the engineer's thought process is very different. The engineer has a tendency to say: "The product's reliability is *X*. What percentage of the market will consider buying our product?" The marketing-driven approach will state: "If you can provide product that will do this function, at this cost, and at this level of reliability, the market for the product would be *y* customers/consumers, and we should be able to capture *z* percent of the market." It is then left up to R&D to design a product that will meet or exceed the market expectations.

> The initial new-design control activity involves establishing the requirements for MTTF and whatever other reliability targets may be indicated to meet the reliability required for the product. To be meaningful, the reliability targets must be within reach at a planned date.
>
> **—Armand V. Feigenbaum**

Phase II: Designing Reliability into the Item

A major consideration in every design is the reliability requirement for the end item. Major considerations in the design of an item are:

- Cost
- Performance
- Reliability
- System compatibility
- Availability
- Manufacturability

The item's intrinsic reliability is defined by the design. The design will dictate, for example, the materials selection, the item operating temperature, the structure of an item, the way it is maintained, and the way components of the item are used.

Of course, maintainability is part of reliability because downtime is a typical reliability measurement, and of course, easy maintenance has a big input on the customer/consumer's cost of ownership.

Phase I is the most important phase in the reliability cycle because it starts the whole cycle, and defines the requirements for each of the other phases, but Phase II runs a close second. Proper preventive maintenance may mask some of the reliability problems, but other activities, like manufacturing, field repairs, and shipping, only serve to degrade the intrinsic reliability of the design. Even preventive maintenance often has a temporary negative effect on the item's reliability.

A well-designed, preventive maintenance system removes components just before they reach their end-of-life points where the components' failure rates take off.

Note: There is a **Rule 3:** The item's intrinsic reliability is defined by the design.

In order for the R&D engineers to create an acceptable design, a great deal of component reliability data need to be at their fingertips. In addition, they need to have an excellent understanding of reliability considerations (for example, impact on components' reliability when the component is used at derated values, or impact of changing environmental conditions, or where and when to use a redundant circuit).

Phase III: Defining Component Reliability

It is easy to see that the engineering departments need to understand the bathtub curve failure rates of all components if they are going to design a product that meets customer/consumer reliability requirements.

Definition

A bathtub curve is a picture of an item's failure rate versus time. It shows how the failure rate decreases during the item's early life to its intrinsic failure rate level and remains at the level until the item starts to wear out and its end-of-life rate begins to increase.

Establishing the bathtub curve error rates of all components is a critical part of a reliability program. Although early-life and end-of-life failure rates are important design considerations, the most important consideration is the intrinsic failure rate. For a typical electronic component, the early-life phase lasts anywhere between 100 and 200 hours of operation, and the end-of-life phase begins about 8 to 10 years after its first power-on point. Some place between 5 and 7 years of the electronic component's life cycle, it will be performing at its intrinsic failure rates. It is for this reason that most reliability projects use only intrinsic failure rates in calculating the total item's projected reliability. This approach greatly simplifies the calculations. Unfortunately, the first thing that the consumer is subjected to is the high early-life failure rates. As a result, consumer impression of the product is heavily weighted on the early-life failures. "I just got this product and it already isn't working. I'll take it back and get another brand."

A properly designed manufacturing process can do a lot to keep early-life failures from reaching the consumer. There are many ways that intrinsic (inherent) component failure rate data can be obtained:

- Components' supplier databases
- Reliability data collected by clearinghouses
- U.S. government databases
- Performance of the components in the organization's previous products
- Component reliability testing by the organization

Phase IV: Calculating the Item's Reliability

There are two ways to calculate an item's reliability:

- Testing at the completed item level and statistically analyzing the results
- Testing at the lower levels and calculating the item's reliability

Usually, a combination of both these approaches is used to give a maximum level of confidence at minimum cost. Testing the completed item in a simulated customer/consumer environment provides the most accurate data, but it is expensive, very time-consuming, and often impractical. For example, if the product's life expectancy were 10 years, the organization would have to test a very large sample of the item for 10 years before it could publish reliability figures. For this reason, stress testing and accelerated life testing of a small sample of the product is most frequently used to support shipment to customers/consumers.

This approach makes all of your customers/consumers part of the organization's product reliability test sample. To minimize the chance of not meeting the reliability specifications in the customer/consumer's environment, the final item's tests are supplemented and verified by mathematically combining testing at the lower levels of the item (component, assembly, unit, etc.). This is the best way to estimate the item's reliability during the design stage of the item's life cycle.

Once the error rate for the individual component is determined, the reliability of assembly can be calculated. In order to accomplish this calculation, the engineer must understand how the components are used and what environmental conditions they will be subjected to. An assembly's reliability can be calculated by mathematically combining the reliability projections for the components that make up the assembly once they have been adjusted to reflect the way they are used. Assembly reliability can be combined mathematically to define unit-estimated reliability. These data usually do not take into consideration any degradation that would occur as a result of the manufacturing and servicing activities. If the reliability estimates do not indicate that the product will meet the reliability specification, Phase II (design the item), Phase III (define component reliability), and Phase IV (calculate the item's reliability) will be repeated. (See Figure 9.2.)

Phase V: Minimizing Manufacturing Degradation

As we have stated, the item's intrinsic reliability is defined by the design. The manufacturing process can only degrade the item's performance from this intrinsic reliability value. Typical manufacturing things that could cause degradation of the item are:

- Stresses that are applied to the component during manufacturing that exceed the component specification or expected use, but the unit does not stop functioning altogether. (Example: A flow soldering machine that is set too high can cause damage to many flow solder joints within the components and the connections on the circuit board without causing them to completely fail, but it could greatly decrease their mechanical integrity, increasing the item's failure rates after it is delivered to the customer.)
- Tests that do not functionally test all the conditions that the customer will apply to the item. This is referred to as test coverage when applied to electronic systems.
- Assembly errors. (Example: A wire pulled too tight around a sharp edge could result in cold flow, causing the unit to short out.)

New manufacturing processes should be qualified to minimize the degradation they have on the design. Typical things that would be considered during a manufacturing process qualification activity would be:

- Are the documents complete and understandable when used by the manufacturing operators?
- Are the tools and test equipment capable of producing high-quality products that repeatedly meet the engineering specification?

- Are the controls in place to ensure that out-of-specification conditions can be quickly identified, preventing the escape of deviant product from the process?
- Have process control requirements been met? The process should be capable of producing products at a Cpk of 1.4 minimum. World-class Cpk is 2.0 or six sigma.

Phase VI: Qualifying the Item

To qualify an item, two factors have to be evaluated:

- The product design needs to be evaluated to ensure it is capable of meeting its design requirements. This is called verifying *the design.*
- The manufacturing processes and their support processes need to be evaluated to ensure that they do not degrade the basic design to an unacceptable level. This is called *manufacturing process validation.*

Once the design has been released and the item's production process has been certified, the item should be subjected to a qualification test to determine if the theoretical reliability estimates were correct and to provide a minimum degree of assurance that the specified reliability performance has been designed and built into the item. At the components and small assemblies levels, the sample size may be large enough for the organization to have a high degree of confidence that the reliability projections will be met. For large, complex assemblies and items (for example, cars, planes, computers, etc.), it is often too costly and time-consuming to evaluate a large enough sample size to have a meaningful level of confidence that reliability projections can be met. To offset this risk, many organizations conduct a number of evaluations throughout the product development cycle in order to gain a higher level of confidence that the item will meet reliability expectations in the customer's environment.

> The most important reliability tests are those applied to the system or end product as a whole; however, many tests may be conducted at the part, subassembly, and assembly levels prior to receipt of hardware for system tests. It is not feasible, from cost and schedule viewpoints, to conduct statistically significant tests at all equipment levels. Accordingly, it is suggested that

major emphasis be placed upon tests at the component-part level and the complete system level.

—**J.M. Juran**

Phase VII: Measuring Customer/Consumer Reliability

Once the floodgates are open and the product is being delivered to the external customers, the true reliability of the item can be measured. Unfortunately, the customer sees reliability as a black and white consideration. For example, a man/woman gets into the car, and the car starts or doesn't. He/she puts on the brakes, and the car stops or doesn't. The customer doesn't care if it is an early-life failure or end-of-life failure or whether he/she was using the item as it was intended to be used or not. The customer just expects it to work each time he/she wants it to operate. Now, the real test of your reliability system is put on the line. It is very important that systems be put in place to collect these data in a way that will allow the reliability problems related to the item to be corrected and to provide information that will allow future items to benefit from the mistakes made on this item. It is a lot easier to say that the data system should be put in place than it is to make it work.

In today's fast-reacting environment, it is absolutely essential to have pertinent information available to the total organization on an ongoing basis. Today's best practices put the repair action data from yesterday's failures on the president and other key executives' desks today. IBM's field reporting system is an excellent one to benchmark.

Phase VIII: Corrective Action and Database Updating

The database that is developed in Phases III, IV, V, VI, and VII is established for three purposes:

- To measure current product reliability
- To correct current reliability or perceived reliability problems
- To collect component and assembly-error rate information for future use

The key to good corrective action is defining the root cause of the problems. In most cases, this means the suspected component needs to undergo a detailed failure analysis activity where it is characterized and very often dissected. We like to be able to recreate every external failure condition.

When you can take a good component and get it to fail in exactly the same way that the component under study failed in the field operation, you have gained true knowledge of the failure mechanism and can solve the problem. In most cases, a good failure analysis system is a key ingredient of an effective corrective action cycle, but correcting the problem is only the start of the total cycle. Truly, you need to prevent problems from reoccurring on the current process and all future products. In this case, we need to look at what an effective preventive action cycle is. Seven critical ingredients that are required in any long-range plan to permanently eliminate problems are:

- Awareness
- Desire
- Training in problem solving
- Failure analysis
- Follow-up system
- Prevention activities
- Liberal credit

The reliability cycle in Figure 9.2 depicts the process that every item whose reliability is of concern to the customer and/or management should go through. You will note that each cycle is triggered with a new reliability requirement and that the cycle is short-circuited when an item's calculated reliability does not meet reliability requirements or when the product's actual reliability performance fails to meet consumer expectations.

In a pure service environment, the reliability cycle needs to be modified to focus upon the processes that produce the service. In these cases, a fault tree analysis-type activity becomes very important. Too often, management believes that reliability is a hardware measurement and does not apply to the service industry. Nothing could be further from the truth. The five things that are most important to having customers return to a service provider are:

- Accuracy of information provided
- Timeliness of service
- Responsiveness of personnel
- Reliability of organization's process
- Physical appearance

Reliability can be your competitive edge. Is your reliability management system up to it?

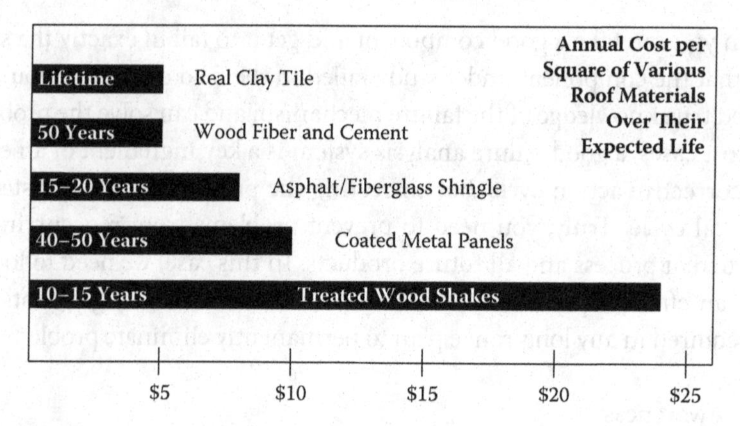

FIGURE 9.3
Yearly cost to own a roof with different roofing materials.

Examples

While customers/consumers treat product reliability as a specified quantity, they fail to consider reliability costs of various producers during purchase negotiations. Quality may be free, but reliability is not. The basic bromide of "you get what you pay for" is substantially untrue. Your house shingles that last for 50 years initially installed cost 60% more than the house shingles that last 20 years. In the long run, the less reliable materials cost 500% more to own. (See Figures 9.3 to 9.8.)

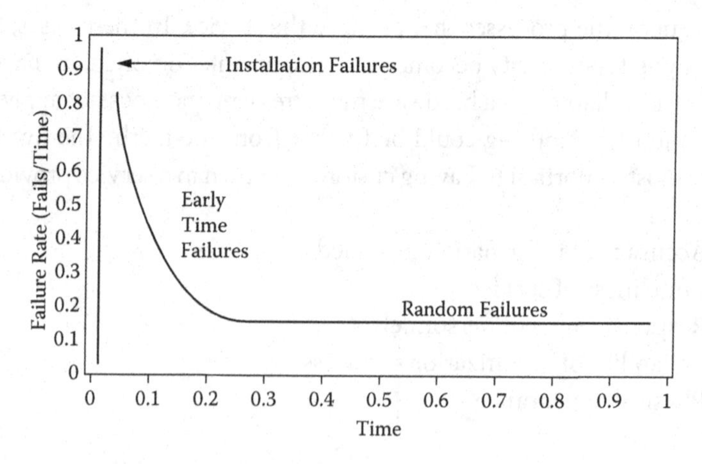

FIGURE 9.4
Product that does not wear out.

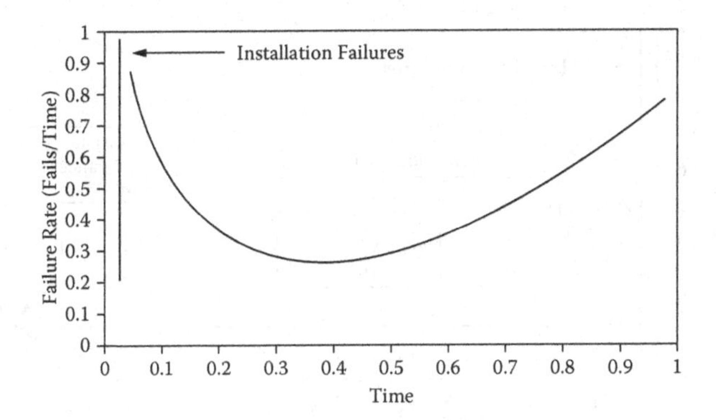

FIGURE 9.5
Product designed for complete wear-out.

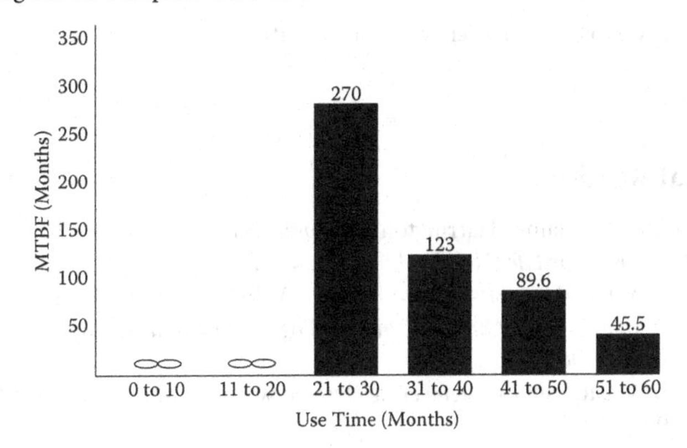

FIGURE 9.6
MTBF of automobile batteries.

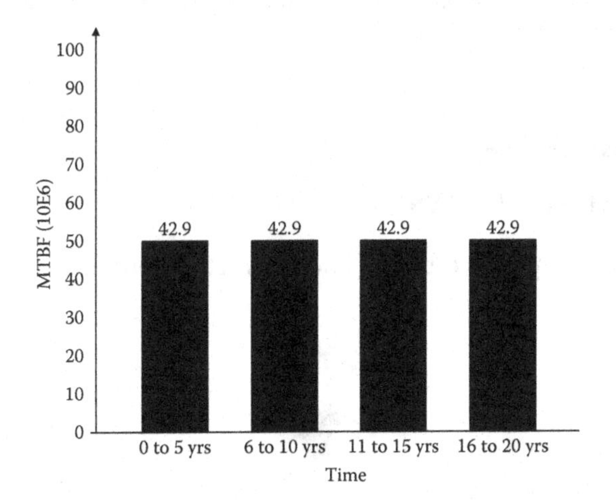

FIGURE 9.7
MTBF of a semiconductor component.

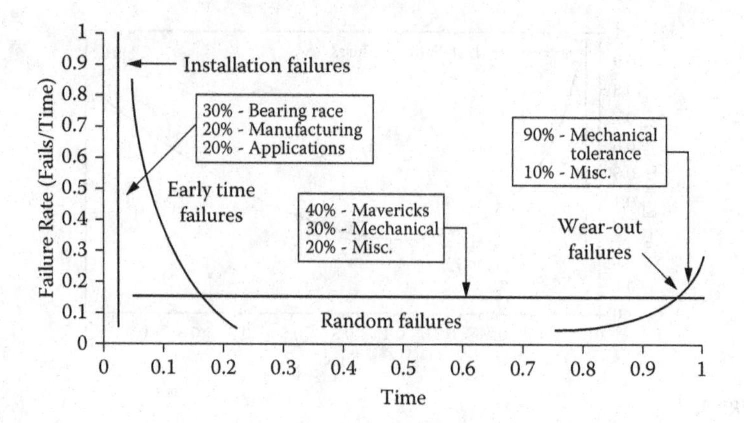

FIGURE 9.8
Input information to a product reliability specification.

Additional Reading

Anderson, Les, and H. James Harrington. *Reliability Simplified—Going Beyond Quality to Keep Customers for Life* (New York: McGraw-Hill, 1998).

Dovich, Robert A. *Reliability Statistic* (Milwaukee, WI: ASQ Quality Press, 1990).

Feigenbaum, Armand. *Total Quality Control—Engineering and Management* (New York: McGraw-Hill, 1961).

Kececioglu, Dimitri. *Reliability Engineering Handbook*, vol. 1 (Milwaukee, WI: ASQ Quality Press, 1991).

Lloyd, David K., and Myron Lipon. *Reliability: Management, Methods and Mathematics* (Milwaukee, WI: ASQC Quality Press, 1984).

Modarres, M. *What Every Engineer Should Know about Reliability and Risk Analysis* (New York: Marcel Dekker, 1992).

ROOT CAUSE ANALYSIS

Treating the symptoms often will not correct the problem.

Definition

Root cause analysis is the process of identifying the various causes affecting a particular problem, process, or issue and determining the real reasons that caused the condition.

Just the Facts

Everyone talks about industrial problem solving, and some books explain how to do it. These books typically provide precise methods for selecting the problem to work on and ways to protect customers from receiving defective products. They all advise you not to treat the symptoms, but to define the root cause so that the real problem can be corrected. It sounds so simple; all you have to do is define the root cause, but most books never tell you how. Why do they avoid giving details about this crucial activity? The reason is simple. Defining the root cause is often very difficult and complex; there is no one right way that works all the time. The practitioner must be skilled at selecting the most effective approach.

There are a number of ways to get to the root of a problem. A good failure analysis laboratory can provide the insight necessary to understand how a failure such as a broken bolt occurred. Duplicating the failure under laboratory conditions also has proved to be an effective way to define the root cause of problems. You know you have found the root cause when you can cause the problem to occur and go away at will. Either of these approaches works well, but they require expensive laboratories and highly trained personnel.

Excessive variation is at the heart of most problems, at least the difficult ones. Variation is part of life. No two items or acts are exactly identical. Even identical twins have very different fingerprints, voice patterns, and personal values. No two screws made on the same machine are exactly the same. Equipment may not be sensitive enough to measure the variation, but it exists. Some variation is good; it keeps our lives from being monotonous. No one would like steak, mashed potatoes, and peas three times a day, every day of the week. They are good once in a while but would get old and boring if eaten at every meal.

Some variation, within specific limits, has little or no effect on output. In other cases, variation can cause an entire plant to come to a halt. The variation we're concerned about here is the variation that *causes* problems resulting in waste. There is no such thing as a random problem, just problems whose occurrence is more or less infrequent, meaning that the combination of specific variables occurs more or less infrequently. The art of defining the root cause is the art of variables analysis and isolation.

The root cause of a problem has been found when the key variables that caused the problem have been isolated. Over the years, there have been many methods developed to isolate key variables. Design of experiments and Taguchi methods are popular today. But the difficulties and effort required to prepare and conduct these studies cause them to be used on only a small fraction of the problems. Engineers, managers, production employees, and sales personnel solve most of their problems by brute force and a lot of luck. Even then, most of the time, the answer that is implemented is not the best solution to the problem.

While this part of our book covers root cause analysis, not statistical process control, we need to understand that by studying different types of variation, the source of the variation can be identified. Then the problem solver can quickly and effectively reduce the many potential sources to a critical few, and often to a single factor, thereby greatly simplifying the problem evaluation cycle and reducing the amount of data for collection. The results can be profound:

- Problems can be solved faster.
- Fewer samples are required.
- Less-skilled people can solve very complex problems.
- Preventive and corrective action plans can be evaluated quickly.
- Nontechnical people can easily understand the results of a technical evaluation.

How to Do a Root Cause Analysis in Six Steps

Step 1: Identify the potential root cause for the problem. The most effective method of root cause analysis is to determine how the root

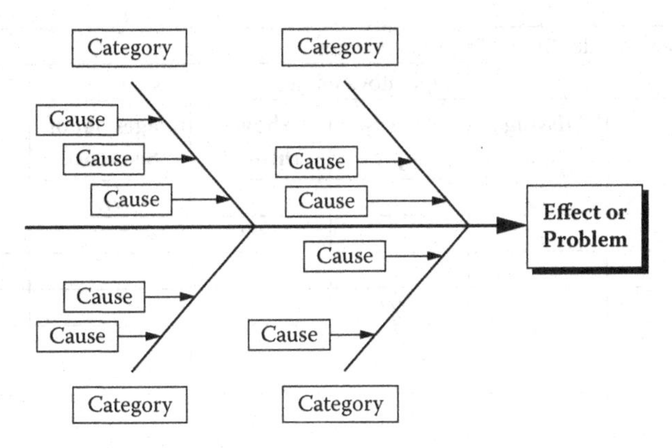

FIGURE 9.9
Example of a cause-and-effect diagram.

cause will be identified or what approach will be used. One of the most frequently used tools for identifying root cause is the cause-and-effect or fishbone diagram. (This tool is part of the Green Belt training.) Its primary function is to provide the user with a picture of all the possible causes of a specific problem or condition. (See Figure 9.9.)

Step 2: Gather data on the frequency of the problem. Use check sheets or other approaches to track occurrences of problems and identify the causes. Figure 9.10 shows how a check sheet can be used for several problems that may result from one or more of several causes.

Step 3: Determine the impact of the problem. Use a scatter diagram or similar tool. (See Figure 9.11.) Scatter diagrams are explained later in this chapter.

Step 4: Summarize all major root causes and relate them back to the major opportunity areas. The purpose of this is to:

- Identify root causes that impact several problems.
- Ensure that the major root causes are identified in all opportunity areas.
- Aid in selection of the key root cause to eliminate it.

Check Sheet for Identifying Defective Copies					
Machine no.:		Operator's name:		Date:	
	Missing Pages	**Muddy Copies**	**Show-Through**	**Pages out of Sequence**	**Total**
Machine jams	///			////	7
Paper weight	//		////		6
Humidity		//		/	3
Toner		////			4
Condition of original	////	////	//	////	14
Other (specify)					
				Total	34
Comments					

FIGURE 9.10
Example of a check sheet.

Step 5: Prioritize the root causes. Use a prioritization matrix. (See Table 9.2.) This procedure consists of the following four steps:

- List the criteria to be used to evaluate the causes.
- Weight each criterion according to its relative importance. Put the weight for each criterion in that column heading.

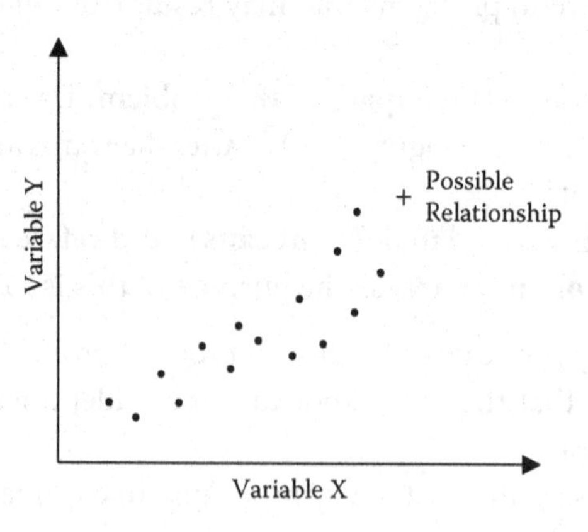

FIGURE 9.11
Example of a scatter diagram.

TABLE 9.2

Example of a Prioritization Matrix

Criteria / Root cause	Criteria 1	Criteria 2	Criteria 3	Criteria 4	Totals
Root cause 1					
Root cause 2					
Root cause 3					
Root cause 4					
Root cause 5					
Root cause 6					
Root cause 7					
Root cause 8					

- Using one criterion at a time, rank the order of all the causes—with 1 being the least important. Enter the ranking in the column under the criterion in question.
- Multiply each rank order figure for each cause by the weight of each of the criteria to arrive at a total for each cause. Enter these totals in the final column of each row.

Step 6: Select the key root cause to eliminate. This decision should be based on the analysis of all available data. If you use a prioritization matrix, you may simply decide according to the totals in the final column.

Examples

The fault tree analysis is another alternative and sometimes is more effective than the approach just described. (See Figure 9.12.) We like to think of it as the "What could cause this?" approach.

The following is an example of the "What could cause this?" approach in use.

Symptom: TV will not turn on.

What could cause this? (Level 1)

- TV is defective
- Electrical power is out
- Remote control is defective

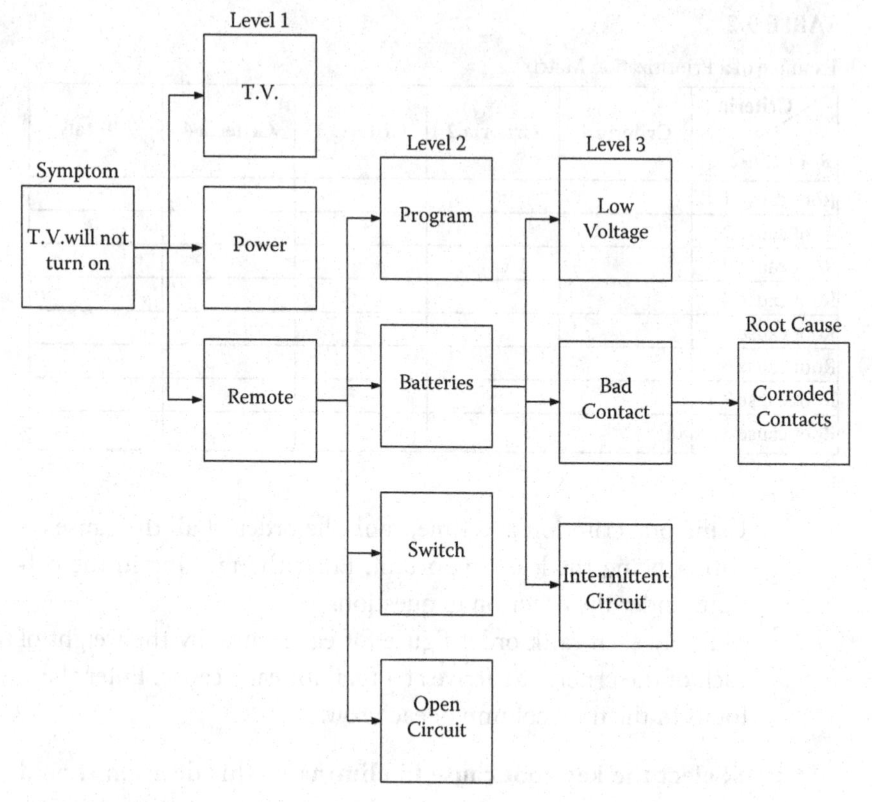

FIGURE 9.12
Fault tree analysis.

Investigation: TV turns on when the on button on the TV is pushed, but not when the on button on the remote control is pushed.

What could cause this? (Level 2)

- Not programmed to the TV
- Discharged batteries
- Defective on switch
- Open circuit

Investigation: Replaced batteries. Remote control now turns on the TV. Put back in the old batteries and the remote turned on the TV.

What could cause this? (Level 3)

- Batteries' voltage is low so the remote control works intermittently.
- Bad contact
- Intermittent circuit

Investigation: Inspected battery terminals and found that they were corroded. Cleaned terminals. Checked the age of the batteries. They were less than 2 months old. Checked the voltage and current of the batteries. They checked out well. Put the old batteries in the remote control and it turned on the TV at a distance of two times normal usage.

Root cause of failure: Corroded terminals.

SCATTER DIAGRAMS

Scatter diagrams could be called *Data-Organized Diagrams.*

Definition

Scatter diagrams are a graphic tool used to study the relationship between two variables.

Just the Facts

The scatter diagram is used to test for possible cause-and-effect relationships. It does not prove that one variable causes the other, but it does show whether a relationship exists and reveals the character of that relationship.

The relationship between the two sets of variables can be evaluated by analyzing the cluster patterns that appear on the graph when the two sets of data are plotted with each axis being used for one of the sets of data. The direction and tightness of the cluster gives an indication of the relationship between the two variables.

Steps to Prepare a Scatter Diagram

1. Collect paired samples of data.
2. Construct the horizontal and vertical axes of the diagram. The vertical axis is usually used for the variable on which we are predicting or measuring the possible effect. The horizontal axis is used for the variable that is being investigated as the possible cause of the other variable.
3. Plot the data on the diagram. Circle data points that are repeated.
4. Analyze the cluster pattern that appears.

Guidelines and Tips

Though a scatter diagram is completed to study the cause-and-effect relationship between two variables, you should be cautious about the statement that "Variable 1 causes Variable 2." There might be other reasons why two variables appear to be related, such as a third variable not represented in the plot, but related to both of the other variables.

Keep in mind that the full range over which Variable 1 varies is sometimes key in detecting a correlation between two variables. For example, experimental studies are often done over a wider range than normal production.

Also keep in mind that correlations do not have to be linear. Notice the last example, showing two variables that are correlated, but not in a linear fashion. Look for patterns that might indicate a relationship between two variables.

Example

See Figure 9.13.

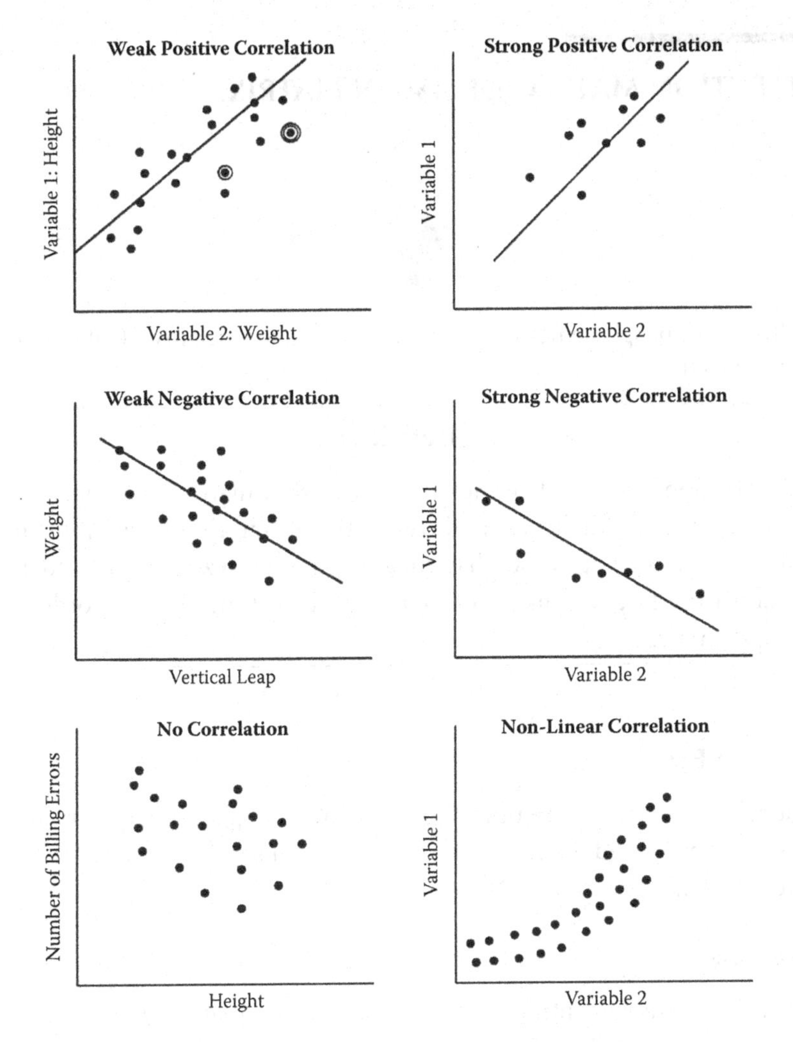

FIGURE 9.13
Sample scatter diagrams.

Additional Reading

Brassard, Michael. *The Memory Jogger Plus* (Milwaukee, WI: ASQ Quality Press, 1989).
Harrington, H. James. *The Improvement Process* (New York: McGraw-Hill, 1987).
Lynch, Robert F., and Thomas J. Werner. *Continuous Improvement Team and Tools: A Guide for Action* (Milwaukee, WI: ASQ Quality Press, 1991).

SELECTION MATRIX (DECISION MATRIX)

Too many people do nothing because they have so much to do that it overwhelms them.

Definition

A selection matrix, also known as a decision matrix, is a chart that systematically identifies and rates the strength of relationships between information so that they can be analyzed. It is used for examining large numbers of factors and ranking them in order of importance.

Just the Facts

The selection matrix is frequently implemented in planning quality activities to set goals and develop process flows. It is useful in selecting alternative solutions. (See Table 9.3)

Example

The steps in implementing a selection or decision matrix are:

- Identify the possible alternatives.
- Agree on the decision/selection criteria.
- Evaluate each criterion and assign a relative weighted value.
- Establish a scoring system.
- Identify and rank the alternatives.
- Total the scores.

Additional Reading

Cox, Jeff, and Eliyahu M. Goldratt. *The Goal: A Process of Ongoing Improvement* (Croton-on-Hudson, NY: North River Press 1986).

Picard, Daniel (ed.). *The Black Belt Memory Jogger* (GOAL/QPC and Six Sigma Academy, 2002).

Wortman, Bill. *The Certified Six Sigma Black Belt Primer* (Quality Council of Indiana, 2001).

TABLE 9.3

Selection Matrix

Criterion	Weight	Alternative 1	Alternative 2	Alternative 3
Company performance	3	3 × 3 = 9	1 × 3 = 3	3 × 3 = 9
Ease of implementation	1	3 × 1 = 3	5 × 1 = 5	5 × 1 = 5
Negative elements	2	3 × 2 = 6	1 × 2 = 2	3 × 2 = 6
Total		18	10	20

Scoring: 5 = high, 3 = medium, 1 = low.

SIPOC DIAGRAM

Understand your suppliers and your customers and you are 75% of the way to success.

Definition

A SIPOC diagram is a tool used by a team to identify all relevant elements of a process improvement project before work begins. It helps define a complex project that may not be well scoped.

Just the Facts

There are two approaches to understanding the present process: One is descriptive and the other is graphic. A good way to understand the process is to describe it. One benefit of describing the process is that it sometimes leads to the discovery of obvious problems and solutions that can be fixed quickly. A flowchart of the process is particularly helpful in obtaining an understanding of how the process works because it provides a visual picture.

There are four types of flowcharts that are particularly useful:

- Top-down flowchart
- Deployment matrix flowchart
- Process map
- SIPOC diagram

Of the four types, the SIPOC diagram method is the one that is most often used. The SIPOC diagram will assist with improvements and simplification by providing:

- A high-level description of the business process addressed by the project.
- An accurate picture of how work is currently done in order to pinpoint the location or source of error as precisely as possible by building a factual understanding of the current process conditions.
- Knowledge that will allow the problem solvers to narrow the range of potential causes to be investigated. The key is to understand how the process presently works. Before the Six Sigma Team (SST) can attempt to improve the process, it must understand how it works now and what it is supposed to do.

The team should ask and answer key questions:

- What does the process do?
- What are the stages of the process?
- What are the starting and finishing points of the process?
- What are the inputs and outputs from the process?
- Who are the suppliers and customers of the process?
- Who uses the product and who pays for it?
- Are there obvious problems with the process?

A SIPOC diagram shows only the essential steps in a process without detail. Because it focuses on the steps that provide real value, it is particularly useful in helping the team to focus on those steps that must be performed in the final "improved" process. The SIPOC diagram provides a picture of the process that the team can use to work on and simplify. It allows people to focus on what should happen instead of what does happen. Usually, most processes have evolved in an ad hoc manner. When problems have occurred, the process has been fixed. The end result is that a simple process has evolved into something complex. A flowchart is a first step to simplify things.

The SIPOC diagram is impressively astute at identifying the part in the process that affects customer satisfaction the most. It illustrates the

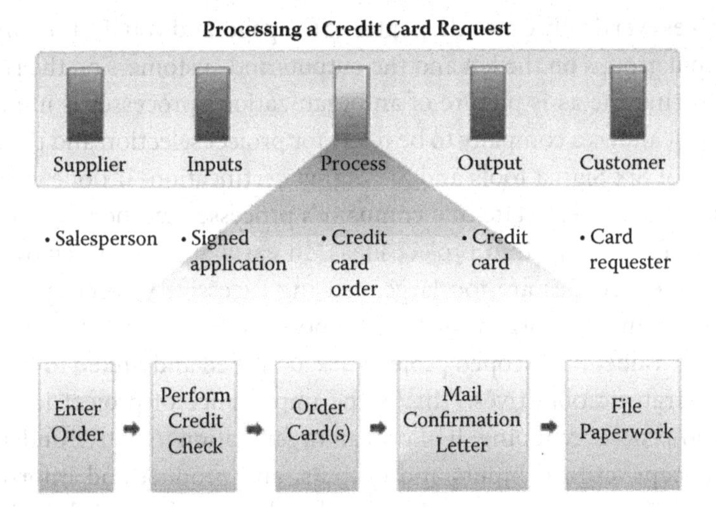

FIGURE 9.14
SIPOC diagram for credit card processing.

upstream inputs to the process as well as the outputs and the customers served. This global view assists in identifying exactly where to make baseline measurements.

Two examples illustrate how a SIPOC diagram can keep the focus on the performance of the inputs and outputs so glaring problems can be identified. (See Figures 9.14 and 9.15.)

This is useful because it shows who is responsible for each activity, how they fit into the flow of work, and how they relate to others in accomplishing the overall job. To construct a SIPOC diagram, you list the major steps in

Suppliers	Inputs	Process	Outputs	Customers
Manufacturer →	Copier	Making a Photocopy	Copies	You
Office Supply Company →	Paper			File
↘	Toner			Others
Yourself →	Original			
Power Company →	Electricity			

Process Steps

Put. original on glass — Close Lid — Adjust Settings — Press **Start** — Remove originals & copies

FIGURE 9.15
SIPOC table for credit card processing.

the process vertically down the center of the page, and you list the suppliers and input groups on the left and the outputs and customers on the right.

Capturing the as-is picture of an organization's processes is important because it allows a company to be ready for project selection and the introduction of Six Sigma tools and Six Sigma certification. If done correctly, defining the current state of a company's processes can help break down strategic focus areas into project ideas. In early Six Sigma deployments, often project scopes are too large, causing excessive project cycle times and loss of internal organizational support. The goal should be ensuring that high-value, well-scoped projects are identified and linked to the company's strategic objectives. This is the importance of project identification and process mapping: It allows an organization to better understand all the steps, critical inputs and outputs, and product and information flows from supplier to customer. Armed with a detailed and shared visual understanding of how work actually occurs, the organization can more easily identify project ideas for improvement.

Once projects are identified, a discussion with key stakeholders can take place to validate initial findings and prioritize projects. This healthy discussion allows individuals to come together and objectively discuss ongoing activities and gaps. Not only will many Lean Six Sigma projects be identified, but other projects that the organization could address will come to light. For the Six Sigma project ideas selected, team charters can be drafted that provide the business case for each project and serve as the guiding framework for improvement efforts. It also is at this point that baseline metrics are established, allowing one to track project and process improvement performance, but it all starts with mapping the current state.

The SIPOC Approach Expanded

As previously discussed, the purpose of mapping an organization's current process is to position the organization to quickly define, document, analyze, prioritize, and recommend solutions and follow-up plans to move the company toward its financial- and customer-focused goals. Any process mapping activity starts with a simple assessment that can be conducted by interviewing the key stakeholders of the processes. A key activity for this assessment is capturing those critical to quality (CTQ) factors of internal clients' processes and services to their customers. This lays the foundation for collecting data, developing metrics to measure success, and ultimately building value stream maps. But before a company can leverage the Six Sigma process to identify and execute critical process improvement initiatives—let alone do detailed mapping techniques such as value stream mapping—it needs to capture the basics from initial interviews and assessment. This is where one would use a SIPOC diagram, and this initial phase of process mapping is the foundation leveraged throughout the initial phase of a Six Sigma deployment. Think of the SIPOC as a simple process mapping tool used to map the entire chain of events from trigger to delivery of the target process.

Basically the SIPOC diagram used is a combination of matrix, flowchart, and summarization and includes:

- **Suppliers:** Significant internal/external suppliers to the process.
- **Inputs:** Significant inputs to the process, like material, forms, information, etc.
- **Process:** One block representing the entire process.
- **Outputs:** Significant outputs to internal/external customers.
- **Customers:** Significant internal/external customers to the process.

A SIPOC diagram helps to identify the process outputs and the customers of those outputs so that the voice of the customer can be captured. When mapping the detailed level of the SIPOC diagram, one can choose the swim lane or other related method. A swim lane flowchart is a type of process flow diagram that depicts where and who is working on a particular activity or subset of the process. (See Figure 9.16.)

The swim lane flowchart differs from other flowcharts in that processes and decisions are grouped visually by placing them in lanes. Parallel lines divide the chart into lanes, with one lane for each person or group. Lanes are arranged either horizontally or vertically, and labeled to show how the chart is organized.

This method requires that the business unit capture all information without directly relating it to a certain process, output, etc., similar to a brainstorming session. This method works best with high-level mapping and is vertical in nature. The swim lane method is best suited for lower-detail-level mapping. Swim lanes allow the business unit to capture all information directly related to a specific process, output, etc. This method requires more space and several mapping sessions due to the amount of time required to map each process; it is horizontal in nature.

A team should initially avoid the swim lane flowchart unless the objective is detailed lower-level mapping, as this method takes many hours and sessions to complete.

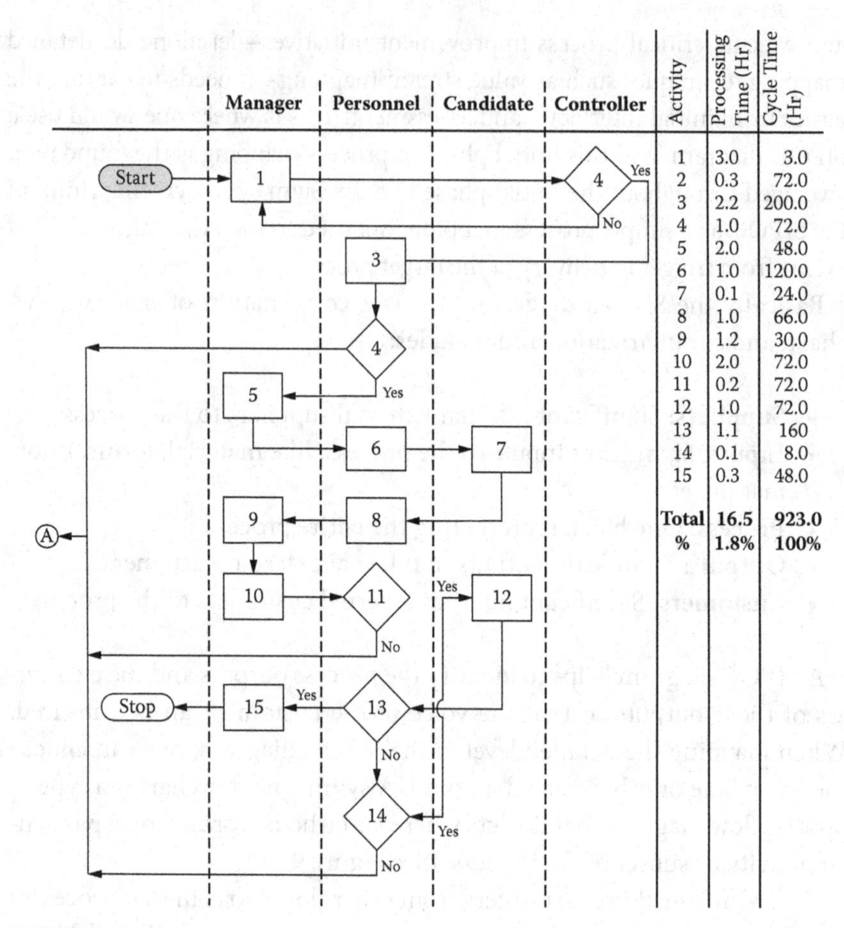

Activity	Processing Time (Hr)	Cycle Time (Hr)
1	3.0	3.0
2	0.3	72.0
3	2.2	200.0
4	1.0	72.0
5	2.0	48.0
6	1.0	120.0
7	0.1	24.0
8	1.0	66.0
9	1.2	30.0
10	2.0	72.0
11	0.2	72.0
12	1.0	72.0
13	1.1	160
14	0.1	8.0
15	0.3	48.0
Total	**16.5**	**923.0**
%	**1.8%**	**100%**

FIGURE 9.16
Swim lane flowchart.

Part of the power of the SIPOC diagram is that it is simple to do, but it is full of information that allows the participants in a process to learn together. This enables them to come to consensus, not only about the makeup of the SIPOC diagram itself, but on the lessons learned and opportunities as well. Places in the processes for potential improvement can then be discussed and prioritized in a nonthreatening fashion. By having the business units participate in the session and rank/prioritize opportunities together, they tend to be clearer and more descriptive in a shorter period of time. So, the SIPOC diagram acts as a dynamic tool to create dialogue and acceptance of a new approach to change, in addition to simply capturing the as-is state.

Building a SIPOC Diagram

When creating a SIPOC diagram, a SST does not necessarily begin at the beginning. The team should more likely start in the middle and ask questions about the process itself. The team may label the process with the summaries of the most critical three to six steps. Next is documenting what is delivered to whom. The team can brainstorm and prioritize the most critical one to three customers and identify, prioritize, and align the outputs most significant to those customers. Later, the team can verify these initial assumptions with voice-of-the-customer tools from the Six Sigma process and/or designate them as critical to quality, speed, or cost. Finally, the team can identify what input or information is needed to perform that process and who provides that input. This brainstorming and prioritization of significant inputs finishes the activities around building a SIPOC diagram.

The following are some further concepts about building a SIPOC diagram:

- Outputs: Outputs are defined as anything the business unit distributes. Frequency/timing is listed along with the output. Examples of outputs would be reports, ratings, products, documents, etc.
- Recipients (customers): A recipient is defined as anyone who receives outputs. It is important to note that the recipient must get the output directly from the business unit and does not necessarily have to be a user of the output. If the output is received from a third party, it is not a recipient. Examples of recipients would be a manager, CEO, board of directors, or another department.
- Triggers: Triggers are anything that starts the business unit's process. A trigger could be the receipt of a report, a certain day of the month, etc.
- Estimated time: The estimated time is how long it takes to complete process steps—this can be continuous, days, weeks, years, etc.
- Fail points: Fail points are ranked/prioritized and then numbered based on the priority.

In the end, the reason a SST frequently begins with building a SIPOC diagram as a first step in the process mapping exercise is threefold in nature:

1. A SIPOC diagram quickly and easily captures the current or as-is state of the organization and processes in question.
2. The SIPOC exercise brings associates together in a nonthreatening way that builds teamwork and momentum to the cause around culture and learning about Six Sigma.

3. The SIPOC exercise allows the team to review all the processes in a manner in which next steps can be identified, and limited resources assigned during the next phase of the rollout to those processes with an objectively identified listing of the most critical project opportunities.

Example: Mama Mia Case Study

One of our first steps was to go out and ask our customers what they liked and didn't like about Mama Mia's. Surveys, focus groups, and follow-up interviews with customers identified their most important requirements as fresh-tasting pizzas on time. Then we interviewed our internal customers, including the food preparation personnel, the VP procurement, the VP food and beverage, and delivery personnel. They indicated that having enough supplies on hand to fill orders and freshness were important requirements. (See Figure 9.17.)

But we had to redefine our terms to clearly understand what a botched delivery meant. To us, it was whenever the right pizza didn't get to the right house. We felt if we made a mistake, but eventually fixed it, that was still a successful mission. By our count, we were getting it right about 97% of the time, which is 3.4 sigma. For Mama Mia's customers, a botched delivery was whenever they received anything less than a perfect pizza the

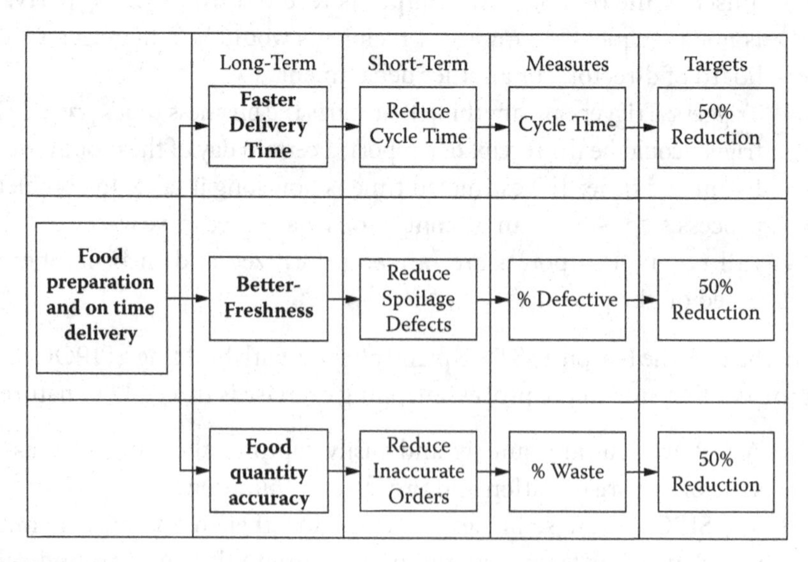

FIGURE 9.17
Customer requirements—operationally defined—delivery example.

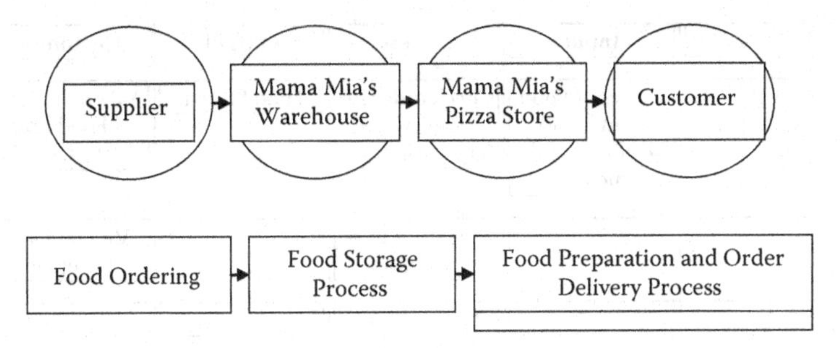

FIGURE 9.18
Mama Mia's SIPOC diagram—Food ordering delivery process.

first time, on time. By their definition, we were only getting it right 87% of the time. We had a problem: How to get the pizza to the customers on time, every time? (See Figure 9.18.)

The VP of procurement creates and places a weekly order from Pizza Supplies R Us. She negotiates for price and quantity of the order. The order is scheduled to be delivered to Mama Mia's warehouse. The order is tracked and received at the warehouse and is unloaded. (See Figure 9.19.)

Supplier	Input	Process	Output	Customer
Pizza Supplies R Us	Food	Order Created	Food Delivered	VP Operations – Food Prep Personnel
		Supplier Contacted		VP Procurement
		Order Negotiated		VP Food and Beverage
		Orders Placed		
		Delivery Scheduled		
		Delivery Tracked		
		Delivery occurs		
		Truck Unloaded		

FIGURE 9.19
Food order SIPOC diagram.

Supplier	Input	Process	Output	Customer
VP Procurement	Crates of Food from Truck Delivery and unloaded	Check order for completeness and freshness	Food Stored	VP Operations – Food Prep Personnel
		Put stock into inventory		VP Food and Beverage
		Inspect stock for freshness – weekly		
		Determine if food within shelf life		
		Dispose of spoiled food		
		Notify VP Procurement over spoiled items if within shelf life		

FIGURE 9.20
Food storage SIPOC diagram.

Mama Mia's SIPOC—Food Storage Process

The VP of food and beverage checks the order for completeness and freshness. The stock is put into inventory. The pizza supplies are inspected weekly for freshness. The food and beverage VP determines if the food is within shelf life, disposes of spoiled food, and notifies the VP procurement of any spoiled items within shelf life. (See Figure 9.20.)

Mama Mia's SIPOC—Food Preparation and Order Delivery

The store manager orders supplies from the warehouse. The supplies are delivered and checked for completeness. The food preparation personnel stock and refrigerate food supplies. The customer places his or her order on the phone or in person and provides address if needed. The food preparation person provides cost and estimated delivery time. The pizza is baked, boxed, and assigned to the delivery person, who scopes out the directions from an old plastic map. The pizza is delivered and money is collected. The delivery person returns to the store. (See Figure 9.21.)

Supplier	Input	Process	Output	Customer
VP Food and Beverage	Food in warehouse	Food supplies ordered from warehouse	Food Delivered	Food preparation personnel
	Current inventory level	Food supplies delivered		External customer
		Store Manager checks for order completeness		
		Stock and refrigerate food supplies		
		Customer places order and provides address on phone or face to face		
		Provide cost and expected delivery time		
		Food is prepared		
		Pizza is boxed and assigned to a delivery person		
		Scope out directions		
		Food is delivered and money is collected		
		Delivery person return to store		

FIGURE 9.21
Food preparation and order delivery SIPOC diagram.

SWOT—STRENGTHS, WEAKNESSES, OPPORTUNITIES, AND THREATS

Winners win because they know their weaknesses and plan so that they are not a factor.

Definition

SWOT analysis is a strategic planning method that evaluates the Strengths, Weaknesses, Opportunities, and Threats that will be encountered in a project or in a business venture. It starts by specifying the objective of the business venture or project and identifying the internal and external factors that are favorable and unfavorable to achieve that objective. The SWOT technique is credited to Albert Humphrey, from Stanford University, who devised the technique by studying data from Fortune 500 companies from the 1960s through the 1970s.

Just the Facts

A SWOT analysis starts by defining a desired objective. Strategic planning should incorporate a SWOT analysis as part of the planning. The elements of SWOT can be defined as:

S—Strengths: The elements of the business that constitute its advantages.
W—Weaknesses: The disadvantages that a business has.
O—Opportunities: The chances of achieving greater profits in the market.
T—Threats: Market elements that could be problematic to achieving the desired results.

TABLE 9.4

SWOT Matrix

	Opportunities	**Threats**
Strengths	O-S strategies	T-S strategies
Weaknesses	O-W strategies	T-W strategies

O-S strategies describe opportunities that build on the company's strengths.
O-W strategies identify weaknesses and means of overcoming them.
T-S strategies identify how to use the company's strengths to reduce external threats.
T-W strategies define plans to prevent the weaknesses from achieving the goal.

The SWOT Matrix

The SWOT analysis can be placed into a matrix of these factors. An example of a SWOT matrix is shown in Table 9.4.

Example

Strengths are resources and capabilities that form a basis for developing a competitive advantage such as:

- Patents
- Brand recognition
- Positive customer base
- Proprietary knowledge
- Access to distribution networks

Weaknesses are an absence of strengths such as:

- Patent protection liabilities
- Negative customer recognition
- High cost
- Poor distribution

Weakness may be the antithesis of strength.

Opportunities come from an evaluation of the marketplace, customer desires, and competitor strengths, which will give indications for profit and growth opportunities. Some examples are:

- A customer-indicated need that is not met
- Implementation of new technologies
- International trade barriers
- Regulations

Threats may consist of changes in the marketplace that threaten the company's products or services. Some examples are:

- Changes in consumer desires and wants
- Introduction of competitor products
- Changes in regulations
- Changes in trade barriers

Additional Reading

Bradford, Robert W., Peter J. Duncan, and Brian Tarcy. *Simplified Strategic Planning: A No-Nonsense Guide for Busy People Who Want Results Fast!* (Worcester, MA: Chandler House Press, 2000)

Picard, Daniel (ed.). *The Black Belt Memory Jogger* (GOAL/QPC and Six Sigma Academy, 2002).

Wortman, Bill. *The Certified Six Sigma Black Belt Primer* (Quality Council of Indiana, 2001).

TAKT TIME

When people are out of balance, they fall. When a process is out of balance, it fails.

Definition

Takt is a German word meaning "beat." It refers to matching the rate of production to customer demand. This is a starting point to balancing the work flow. Takt time can be first determined with the formula:

$$\frac{T = Ta}{Td}$$

where:

T = Takt time, e.g., minutes of work/unit produced
T_a = Net time available to work, e.g., minutes of work/day
T_d = Time demand (customer demand), e.g., units required/day

Net available time is the amount of time available for work to be done. This excludes break times and any expected stoppage time (for example, scheduled maintenance, team briefings, etc.).

Just the Facts

If you have five steps in a process that take different times to complete, then the process is gated by the process step that takes the longest time. For example, if the process steps take A = 13, B = 9, C = 10, D = 12, and E = 18 minutes, respectively, then process step B will be waiting for process step A, and the entire process is gated by process step E, which requires the longest time. This means that the work-in-process (WIP) will increase and the overall flow is slowed.

Takt time is used to balance work flow across work steps and to meet demand. Roughly translated, *Takt* is a German word meaning "beat." Producing to the pace of customer demand (no faster, no slower), when the work steps are in balance, this produces maximum productivity and saves cost. When you have total processing time, all you need is customer demand to calculate Takt time. Takt gives you a starting point to think about balancing the workload across work steps. (See Figure 9.22.)

Each work step in a process should operate as close to Takt time as possible without going over. If your process has four operations that take 10, 13, 12, and 17 minutes in that order, work step 4 is holding up the process. As seen above, when you have work steps that are faster than the slowest

$$\text{Takt time} = \frac{\text{Available time}}{\text{Customer demand}}$$

FIGURE 9.22
Takt time formula.

step, and they precede the slowest step, WIP is increased and overall production is slowed. A rebalancing of the work in such a way as to have each work step function as close as possible to each other will improve efficiency.

To get the Takt all the time for a process, you start by calculating the two individual elements of the formula above. First, you determine the portal available processing time for the process. If the process has two shifts per day, and each shift is 8 hours long, the total available time is not 16 hours per process. You must deduct any downtime, such as lunch breaks for employees. Each shift may have downtime for breaks and perhaps for tailgate meetings. If each shift has a 1-hour lunch and two breaks, you must deduct 3 hours of downtime from the total shift time. In this example this would be 13 hours of production time.

The other variable that is required is customer demand, and if customer demand is 600 units per month and the organization is working 5-day weeks (most often 21 productive days per month), processes must complete 29 units per day using both shifts. Takt for this process is 0.45 hours per unit.

Once you determine Takt time, you attempt to adjust your process speed so that all work steps are as close to Takt time as possible without exceeding them. Operations vary in the amount of work assigned to them and can also be very inefficient.

Redistribution of the work to come up with a well-spaced balanced flow process is the goal. You should consider two general approaches to achieve Takt time to route your process: move work from one station to another, or change the number of stations or operations in the process.

There are four approaches that should be considered when attempting to balance work steps in a process:

- If the process has a small amount of variation, it may be possible to redistribute the work more evenly by assigning uniform slack time. Slack time is the time not being used by the faster work steps in the process. For instance, if the slowest step in the process is 10 minutes and the fastest step in the process is 7 minutes, the 7-minute process has 3 minutes of slack time. If the average work time (7 minutes) is below Takt time (for example, 8 minutes), you have 1 minute per unit to work with. If production is 60 units per day, then you have 1 hour of time that might be redistributed to the slower work step. This begins balancing the two work steps, speeding up the slower work step, and making the faster work step more efficient.

- When large amounts of variation exist, you can use a relatively simple way to rebalance the process. You can reassign the work so that any slack time comes off the end of the process.
- When there are small amounts of variation, it may be possible to eliminate a work step for work consolidation. For example, instead of four work steps you now have three. If this can be done within Takt time, the savings this represents makes it much better.
- If, when examining the work, it is found that there is significant opportunity for making the work itself more efficient and improving the process, efficiency should be your first choice. Then, reassess your Takt time and, if necessary, begin balancing the work across the process.

Example

To get the Takt time, follow these steps:

1. Determine the available processing time.
    - Count all shifts.
    - Deduct any downtime:
        - Breaks
        - Lunch
        - Clean-up
        - Shift change meetings
    - This will yield the available processing time.
2. Determine the customer demand.
    - If the customer wants 1,000 units
    - If the company works 5 days/week (21 days/month)
        - Production rate should deliver 48 units per day
3. If we assume:
    - Effective production time to be 13 hours/day
    - 48 units required per day
    - Takt time will be 0.62 hour/unit

Additional Reading

Picard, Daniel (ed.). *The Black Belt Memory Jogger* (GOAL/QPC and Six Sigma Academy, 2002).

THEORY OF CONSTRAINTS (TOC)

There always has to be a limiting factor. All you have to do is find and eliminate it.

Definition

Theory of constraints (TOC) is a management philosophy that Dr. Eliyahu M. Goldratt introduced in 1984. It is designed to help organizations continually achieve their goals. The term *TOC* comes from the concept that any manageable system is limited by a very small number of constraints, and that there is always at least one constraint.

Just the Facts

A constraint is anything that prevents a system from achieving its goal. While there are many types of constraints, there are only a few constraints that are the principle constraints.

Constraints may be internal or external to the process. Internal constraints could be when production can't keep up with customer demand. External constraints could manifest in the problem of producing more than the marketplace demands.

Types of (Internal) Constraints

- Equipment: Equipment limitations in its ability to produce more goods or services.
- People: Poorly trained people.
- Processes: Processes that limit the production capability.

TOC focuses on the mechanism for management of the system. In optimizing the system, constraints are specifically expressed in order to limit the scope of implementation.

Example

For *operations*, the goal is to "pull" materials through the processes instead of pushing them.

For *supply chain* the goal is to focus on use of supplies instead of replenishing supplies.

For *finance and accounting*, the goal is focused on the effect each step in the process has on the overall throughput.

For *project management*, the goal is focused on the critical path of the process, also referred to as the critical chain project management (CCPM).

Additional Reading

Cox, Jeff, and Eliyahu M. Goldratt. *The Goal: A Process of Ongoing Improvement* (Croton-on-Hudson, NY: North River Press 1986).

Picard, Daniel (ed.). *The Black Belt Memory Jogger* (GOAL/QPC and Six Sigma Academy, 2002).

Wortman, Bill. *The Certified Six Sigma Black Belt Primer* (Quality Council of Indiana, 2001).

TREE DIAGRAMS

> In QC, we try as far as possible to make our various judgments based on the facts, not on guesswork. Our slogan is *Speak with Facts.*
>
> **—Katsuya Hosotani**

Definition

Tree diagrams are a systematic approach that helps you think about each phase or aspect of solving a problem, reaching a target, or achieving a goal. It is also called the "systematic diagram."

Just the Facts

This approach systematically maps the details of sub- or smaller activities required to complete a project, resolve an issue, or reach a primary goal. This task is accomplished by starting with a key problem or issue and then developing the branches on the tree into different levels of detail. The tree diagram is most often used when a complete understanding of the task is required (i.e., what must be accomplished, how it is to be completed, and the relationships between goals and actions). It can be used at the beginning of the problem-solving process to assist in analyzing a particular issue's root causes prior to data collection. It can also be used in the final stages of the process to provide detail to a complex implementation plan, allowing for a more manageable approach to the individual elements. Tree diagrams are also used in:

- Strategic planning
- Policy development
- Project management
- Change management

The application of this tool is to logically branch out (flowchart) levels of detail on projects, problems, causes, or goals to be achieved. (See Figure 9.23.)

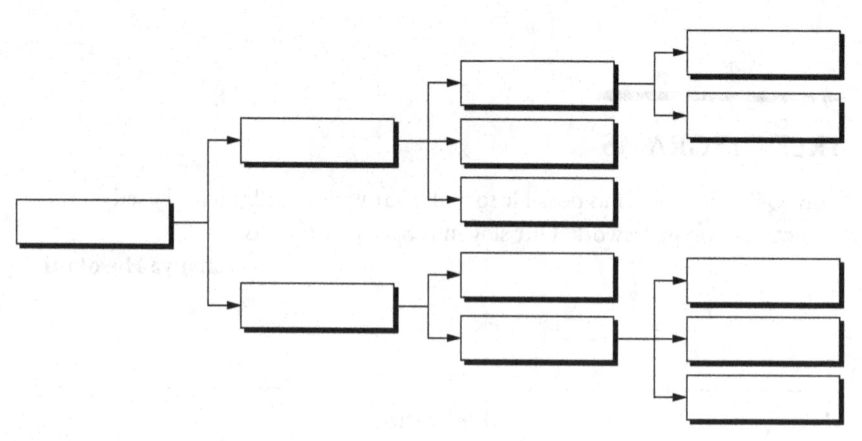

FIGURE 9.23
Typical tree diagram.

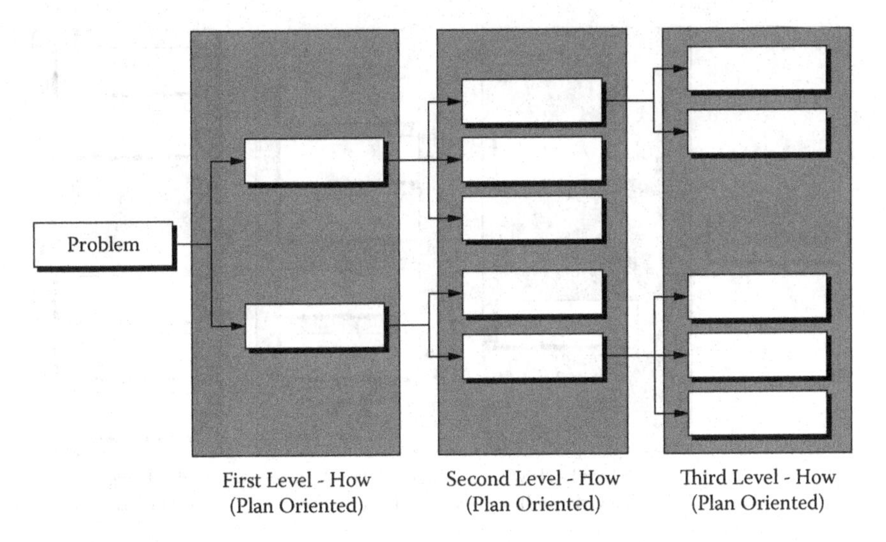

FIGURE 9.24
Example of a components-development tree diagram.

There are basically two types of tree diagrams:

- Components-development tree diagram: Typically used in the early stages of the problem-solving process when analyzing a particular problem and trying to establish the root causes prior to data collection. (This is also known as the "ask why" diagram.) It refers to the components or elements of the work being performed. (See Figure 9.24.)

Steps to completing this type of diagram are relatively simple:

1. State the problem or issue so everyone on the team is in agreement on its meaning. Put that statement in the box on the left of the diagram.
2. By asking why, identify the causes believed to contribute to the problem or issue. Place these causes in a box to the right of the problem or issue. Link them with a line pointing to the cause.
3. Repeat step 2 and continue to develop more causes until a key or root cause is identified.

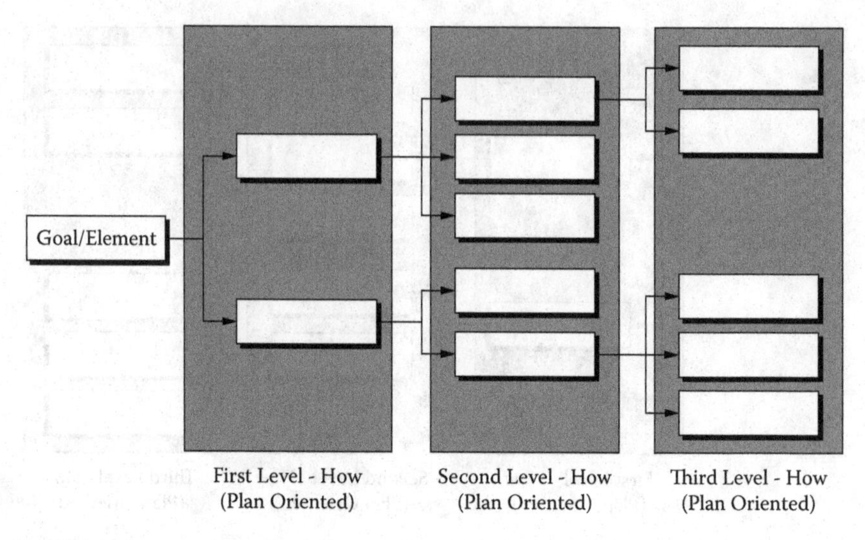

FIGURE 9.25
Example of a plan-development tree diagram.

- Plan-development tree diagram: Used toward the end of the problem-solving process to provide detail to the individual elements of the implementation plan. (This is also known as the "ask how" diagram.) It helps to identify the tasks required to accomplish a goal and the hierarchical relationships between the tasks. (See Figure 9.25.)

  The plan-development tree diagram is developed much the same as the component-development diagram. The difference is here we are trying to provide detail to a particular goal or element of a plan. In step 2, instead of asking "why" you would ask "how" the goal can be achieved. The following are the steps:

  1. State the goal or element so everyone on the team is in agreement on its meaning. Put that statement in the box on the left of the diagram.
  2. By asking "how," identify how the goal, task, or element may be achieved. Place this information in a box to the right of the goal/element. Link them with a line pointing to the cause.
  3. Repeat step 2 and continue to develop more detail on the relationship of the task or element, or until the appropriate level of detail has been provided on achieving the goal.

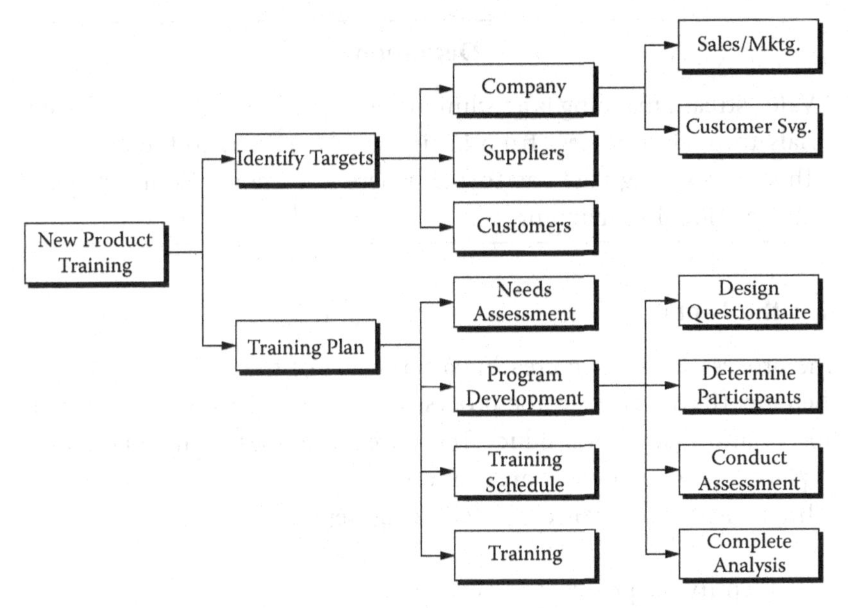

FIGURE 9.26
Example of a semicompleted plan-development tree diagram.

Examples

Figure 9.26 shows a completed plan-development tree diagram, showing how one element of a Total Quality Management (TQM) improvement plan might look.

Additional Reading

Eiga, T., R. Futami, H. Miyawama, and Y. Nayatani. *The Seven New QC Tools: Practical Applications for Managers* (New York: Quality Resources, 1994).

King, Bob. *The Seven Management Tools* (Methuen, MA: Goal/QPC, 1989).

Mizuno, Shigeru (ed.). *Management for Quality Improvement: The 7 New QC Tools* (Portland, OR: Productivity Press, 1988).

VALUE STREAM MAPPING

Produce maximum value to get maximum profits.

Definition

Value stream mapping is a technique used to analyze the flow of materials through a process from beginning to delivery to the customer. This process originated at Toyota, where it is known as material and information flow mapping.

Just the Facts

The process starts with creating a flowchart of the existing process and then identifying areas in the process that cause delays in the continuous flow of information or product. These areas are assessed and plans are put in place to eliminate these areas of waste.

Implementation involves the following steps:

- Identify the process to be evaluated.
- Construct a flowchart of the current process (current state value stream map) to identify:
  - Current steps
  - Delays
  - Information flows required:
    - Raw materials
    - Design flow
    - Handling
    - Storage
- Evaluate the current state value stream map flow to eliminate waste.
- Construct a future state value stream map.
- Implement plans to move toward the future state process flow.

Example

Symbols have been developed to help in the process evaluation. An example from www.valuebasedmanagement.net/methods_value_stream_mapping.html appears in Figure 9.27. An example of a current and future state value stream map from the same source is seen in Figure 9.28.

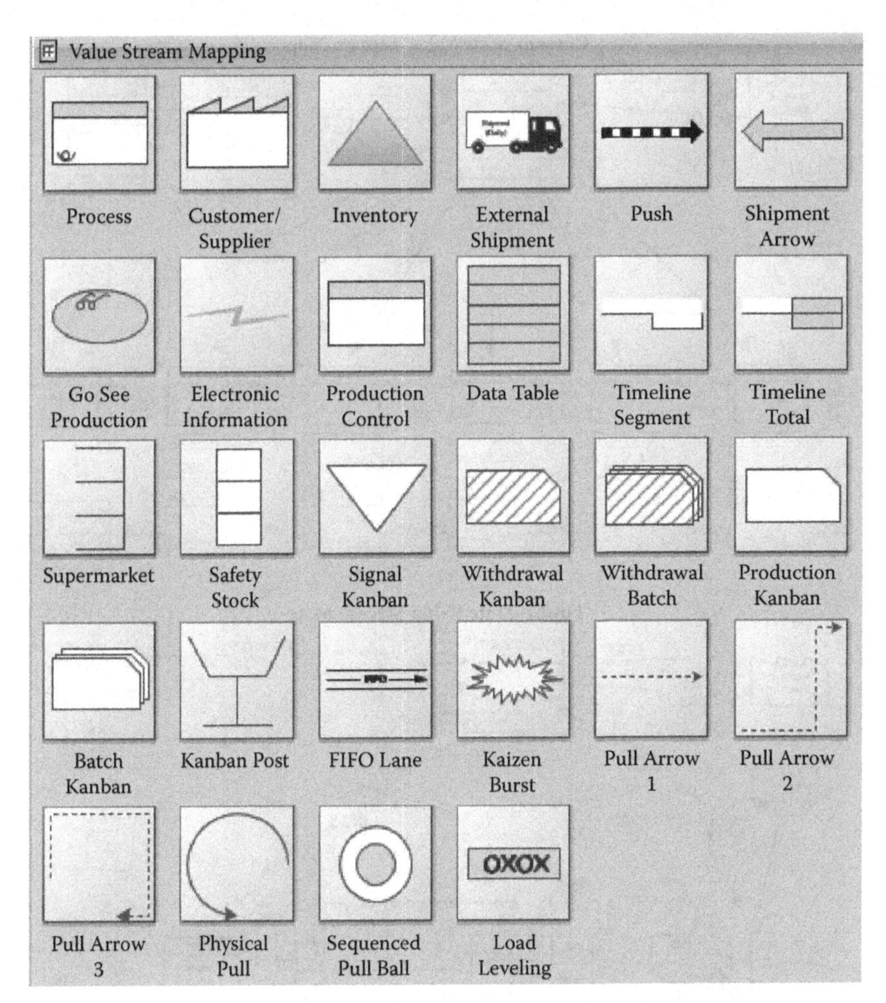

FIGURE 9.27
Value stream map example.

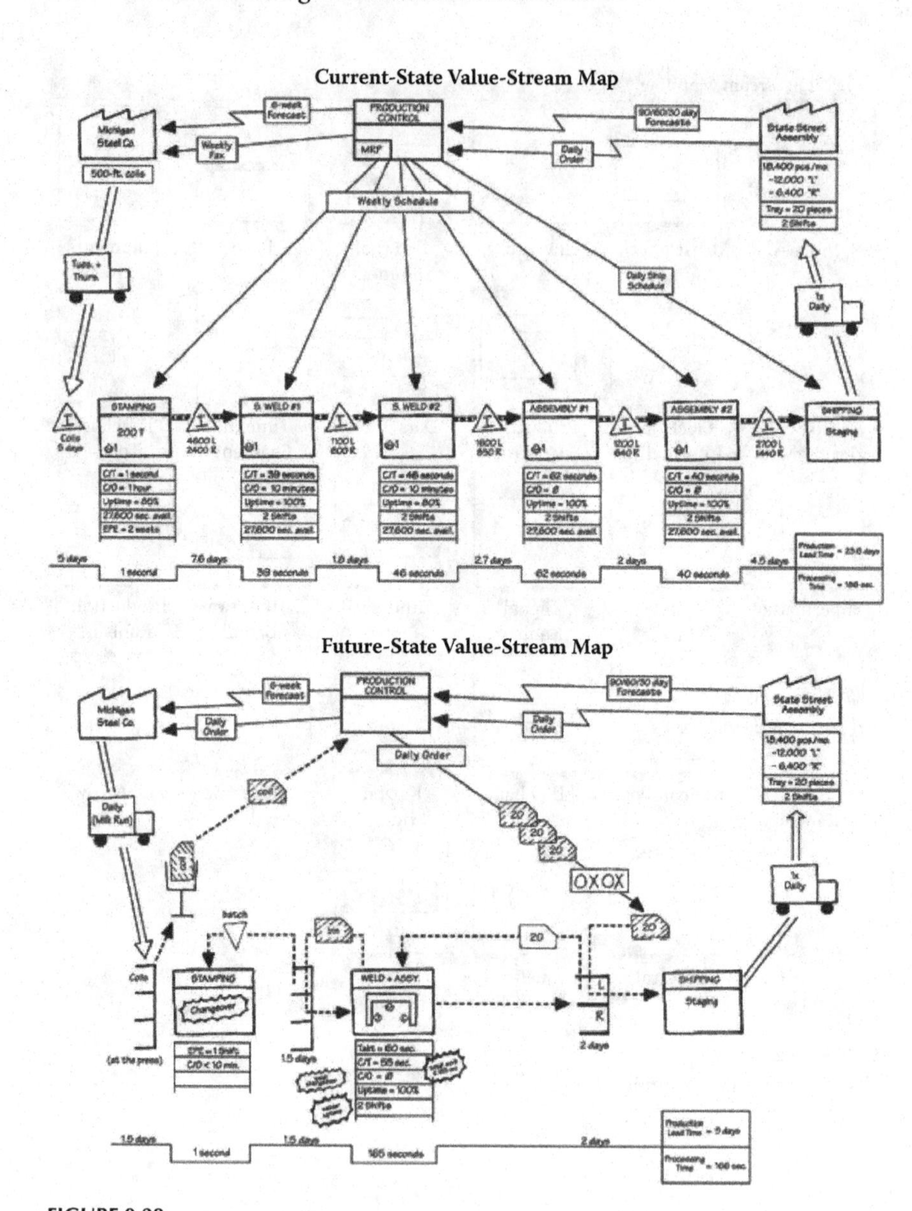

FIGURE 9.28
Current state and future state value stream map.

Additional Reading

Rother, Mike, and John Shook. *Learning to See: Value-Stream Mapping to Create Value and Eliminate Muda* (Cambridge, Ma: Lean Enterprise Institute, 1999).

Shingo, Shigeo. *Quick Changeover for Operators Learning Package: A Revolution in Manufacturing: The SMED System* (New York: Productivity Press, 1985).

Section 5

LSSBB Advanced Statistical Tools

INTRODUCTION

One of the key things that sets the LSSBB apart from the SSGB is the depth of statistical knowledge the LSSBB has acquired. We recognize that many of the people who want to be able to classify themselves or be certified as a LSSBB are less interested in learning complex statistical tools like design of experiments. They may not even use them in the organizational assignment they have right now. However, at some later date they may change organizations or even jobs within the same organization, and if they include being certified as an LSSBB as one of their skills, they will need to understand these statistical tools or they will be misrepresenting themselves to their new employer.

One of the basic concepts of Six Sigma is collecting data and using hard data to make business and improvement decisions. We call it "managing by facts—not hunch." All too often decisions are made based upon false or unfounded data that are presented to management as facts. Through the use of these statistical tools you will be able to qualify and quantify the information you are using to make business decisions.

Another reason you need to master the use of statistical problem-solving tools is because many, if not all, of the problems we face today are complex in nature. These problems are not caused by one root cause, but by the interaction of many root causes, and correcting only one of the root causes could even make the problem worse. Often taking action to solve each of the root causes independently becomes far more expensive than correcting how they interface and interact with each other. Through the use of tools like design of experiments, the Six Sigma Team can look at the interaction between the root causes and understand which root cause or combination of root causes has the biggest impact upon the problem they are analyzing. Relying on the hit-or-miss approach to solving these complex problems may take two to three times longer to come up with a solution that is only 50% as effective as the one you can get through the use of the statistical tools.

As an SSGB, you should have been introduced to the following statistical tools

- Attributes control charts
- Basic probability concepts
- Basic statistical concepts
- Binomial distribution
- Chi-square test
- Correlation coefficient
- Cpk—Operational (long-term) process capability
- Cpk—Using the *K*-factor method
- CPU and CPL—upper and lower process capability indices
- Data accuracy
- Data, scale, and sources
- Histograms
- Hypothesis testing
- Mean
- Median
- Mode
- Mutually exclusive events
- Normal distribution

- Poisson distribution
- Probability theory
- Process capability analysis
- Process capability study
- Process elements, variables, and observations concepts
- Range
- Sampling
- Six Sigma measures
- Standard deviation
- Statistical process control
- Statistical thinking
- Variables control charts
- Variance

At this point you should review Table S5.1 to understand how each of the statistical tools can be used in the two major LSS methodologies.

TABLE S5.1

How the Statistical Tools Can Be Used in the Two Major LSS Methodologies

Tool	D	M	A	I	C	D	M	A	D	V
Analysis of variance (ANOVA)		X			X		X			X
Attributes control charts		X			X		X			X
Basic probability concepts			X					X		X
Basic statistical concepts			X					X		X
Binomial distribution			X		X			X		X
Chi-square test			X		X			X		X
Correlation coefficient			X		X			X		X
Cpk—Operational (long-term) process capability		X			X		X			X
Cpk—Using the *K*-factor method		X	X		X		X	X		X
CPU and CPL—Upper and lower process capability indices		X	X		X		X	X		X
Data accuracy	X	X			X	X	X		X	X
Data, scale, and sources	X	X			X	X	X		X	X

(continued)

TABLE S5.1 (CONTINUED)

How the Statistical Tools Can Be Used in the Two Major LSS Methodologies

Tool	D	M	A	I	C		D	M	A	D	V
Histograms		X			X			X			
Hypothesis testing			X		X				X	X	X
Mean		X	X		X			X	X		X
Median		X	X		X			X	X		X
Mode		X	X		X			X	X		X
Mutually exclusive events	X				X		X				
Normal distribution			X		X				X		X
Poisson distribution			X		X				X		X
Probability theory			X						X	X	X
Process capability analysis			X		X				X	X	X
Process capability study		X	X					X	X		
Process elements, variables, and observations concepts		X			X			X		X	X
Range		X	X					X	X	X	X
Sampling		X			X			X			X
Six Sigma measures	X	X	X		X		X	X	X		X
Standard deviation	X	X	X		X		X	X	X		X
Statistical process control		X	X		X			X	X		X
Statistical thinking	X	X	X				X	X	X		X
Variables control charts		X	X		X			X	X		X
Variance	X	X	X				X	X	X	X	X

To upgrade you to the LSSBB level in statistical knowledge, we will present the following statistical tools:

- ANOVA—one-way
- ANOVA—two-way
- Box plots
- Confidence intervals
- Data transformation
- Design of experiments
- Measurement system analysis
- Method of least squares
- Multivari charts

- Nonparametric statistical tests
- Population and samples
- Regression analysis
- Rolled-throughput yield
- Taguchi methods
- Validation

For ease of reference, the tools are presented in alphabetic order rather than in the order they would be used in solving a problem.

10

Advanced Statistical Tools

In this chapter we present the following statistical tools:

- ANOVA—one-way
- ANOVA—two-way
- Box plots
- Confidence intervals
- Data transformation
- Design of experiments
- Measurement system analysis
- Method of least squares
- Multivari charts
- Nonparametric statistical tests
- Population and samples
- Regression analysis
- Rolled-throughput yield
- Taguchi methods
- Validation

ANALYSIS OF VARIANCE (ANOVA)—ONE-WAY

Less is always better.

Definition

ANOVA (analysis of variance) tests the null hypothesis that the samples in two or more groups are drawn from the same population by comparing their variations.

Just the Facts

When we don't know the population means, an analysis of data samples is required. ANOVA is usually used to determine if there is a statistically significant change in the mean of a Critical To Quality (CTQ) characteristic under two or more conditions introduced by one factor.

The hypothesis statement:

H_0: $\mu_1 = \mu_2 = \ldots = \mu_n$
Ha: $\mu_1 \neq \mu_1$ for at least one pair (i, j)

This is the model where Y_{ij} is the (ij)th observation of the CTQ characteristic ($i = 1, 2, \ldots, g$ and $j = 1, 2, \ldots, n$, for g levels of size n), μ is the overall mean, τ_i is the ith-level effect, and ε_{ij} is an error component.

$$Y = \mu + \tau_i + \varepsilon_i$$

Descriptive Statistics

Level	N	Mean	Standard Deviation
1	15	60.038	2.760
2	15	79.300	3.217
3	15	74.584	3.658

Pooled standard deviation = 3.233

One-Way Analysis of Variance

Source	*DF*	*SS*	*MS*	*F*	*P*
Factor	2	1355.7	677.9	64.87	0.000
Error	42	438.9	10.4		
Total	44	1794.6			

"Source" indicates the different variation sources decomposed in the ANOVA table. "Factor" represents the variation introduced between the factor levels. "Error" is the variation within each of the factor levels. "Total" is the total variation in the CTQ characteristic.

DF: The number of degrees of freedom related to each sum of square (*SS*). They are the denominators of the estimate of variance.

SS: The sums of squares measure the variability associated with each source. They are the variance estimate's numerators. Factor *SS* is due to the change in factor levels. The larger the difference between the means of factor levels, the larger the factor sum of squares will be. Error *SS* is due to the variation within each factor level. Total *SS* is the sum of the factor and error sum of squares (see tool Sum of Squares).

MS: Mean square is the estimate of the variance for the factor and error sources. Computed by $MS = SS/DF$.

F: The ratio of the mean square for the factor and the mean square for the error.

p value: This value has to be compared with the alpha level (α), and the following decision rule is used: If $P < \alpha$, reject H_0 and accept Ha with $(1 - P)100\%$ confidence; if $P \geq \alpha$, don't reject H_0.

The assumption for using this tool is that the data come from independent random samples taken from normally distributed populations, with the same variance. When using ANOVA, we are using a model where its adequacy has to be verified using residual analysis.

Example

1. Analyze the data with Minitab:
    - Verify the assumption of equality of variance for all the levels with the function under Stat > ANOVA > Homogeneity of Variance (to interpret this analysis, see tool Homogeneity of Variance Tests).
    - Use the function under Stat > ANOVA > One Way.
    - Input the name of the column, which contains the measurement of the CTQ characteristic into the "Response" field, and the name of the column that contains the level codes into the "Factor" field.
    - In order to verify the assumption of the model, select the options "Store Residuals" and "Store Fits." Select the "graphs" option and highlight all the available residual plots (to interpret these plots see tool Residual Plots).

- In the event of noncompliance with either of these assumptions, the results of the analysis of variance may be distorted. In this case, the use of graphical tools, such as a Box plot, can be used to depict the location of the means and the variation associated with each factor level. In the case of outliers, these should be investigated.

2. Make a statistical decision from the session window output of Minitab. Either accept or reject H_0. When H_0 is rejected, we can conclude that there is a significant difference between the means of the levels.
3. Translate the statistical conclusion into a practical decision about the CTQ characteristic.

ANALYSIS OF VARIANCE (ANOVA)—TWO-WAY

Definition

An extension to the one-way analysis of variance where there are two independent variables.

Just the Facts

Two-way ANOVA is used to analyze the effect of two random factors on a CTQ. A factor is said to be random when levels are randomly chosen from a population of possible levels and we wish to draw conclusions about the entire population of levels, not just those used in the study. For example, this type of analysis is generally used in gauge R&R studies.

$$Y_{ijk} = \mu + \tau_i + \beta j + (\tau\beta)_{ij} + \varepsilon_{ijk}$$

$$V(Y_{ijk}) = \sigma_\tau^2 + \sigma_\beta^2 + \sigma_{\tau\beta}^2 + \sigma^2$$

$$H_0: \sigma_\tau^2 = 0 \quad Ha: \sigma_\tau^2 > 0$$

$$H_0: \sigma_\beta^2 = 0 \quad Ha: \sigma_\beta^2 > 0$$

$$H_0: \sigma_{\tau\beta}^2 = 0 \quad Ha: \sigma_{\tau\beta}^2 > 0$$

where Y_{ijk} is the (ijk) observation of the characteristic ($i = 1, 2, \ldots, a, j = 1, 2, \ldots, b, k = 1, 2, \ldots, n$) for a level of factor A, b levels for factor B, and n the number of observations in each of the combination of the factor levels.

μ represents the overall mean, τ_i (tau *i*) the effect of factor A, βj (beta *j*) the effect of factor B, $(\tau\beta)_{ij}$ the interaction effect between A and B, and ε_{ijk} (epsilon *ijk*) the error component. For example, with a gauge R&R study factor A can be the operator, factor B the parts, and the interaction is the interaction between the operators and the parts. In this case, *a* will be the number of operators, *b* the number of parts, and *n* the number of repeated measurements. The variance (*V*) of any observation (Y_{ijk}) where σ_τ^2, σ_β^2, $\sigma_{\tau\beta}^2$, and σ^2 are called variance components. The ratio of the variances of each sample are compared against a table based on each sample size to determine if the two variances are significantly different at a given confidence level. For each effect, if H_0 is true, the levels are similar, but if Ha is true, variability exists between the levels

Major Considerations

For a random effect model, we assume that the levels of the two factors are chosen at random from a population of levels. The data should come from independent random samples taken from normally distributed populations. The adequacy of this model has to be verified using residual analysis.

Example

1. Define problem and state the objective of the study.
2. Identify the two factors for study, and the levels associated with these factors. For example, in a gauge R&R study, the factors are parts and operators, and generally the number of levels for the parts is 10 and the number of levels for the operators is 3.
3. Establish sample size.
4. Measure the CT characteristic.
5. Analyze data with Minitab.

BOX PLOTS

This is one box that you will want to stay inside of.

Definition

A box plot, also known as a box-and-whisker diagram or plot, is a graphical way of showing groups of data using five numerical summaries. They are:

- The smallest observation (sample minimum)
- The lower 25% (quartile)
- The median
- The upper 25% (quartile)
- The largest observation (sample maximum)

A box plot may also indicate which observations, if any, might be considered outliers.

Just the Facts

Box plots are one of the more effective graphical summaries of a data set; the box plot generally shows mean, median, 25th and 75th percentiles, and outliers. A standard box plot is composed of the median, upper hinge, lower hinge, higher adjacent value, lower adjacent value, outside values, and far-out values. An example is shown in Figure 10.1. Parallel box plots are very useful for comparing distributions.

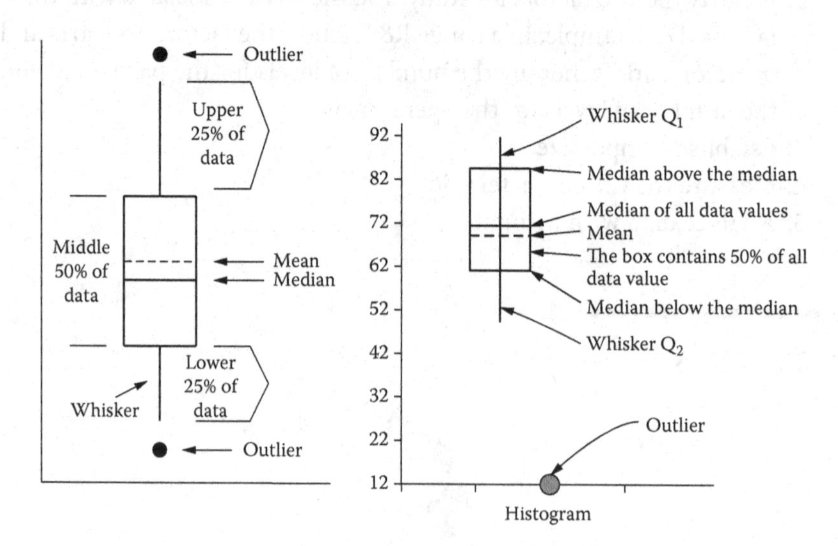

FIGURE 10.1
Box plots.

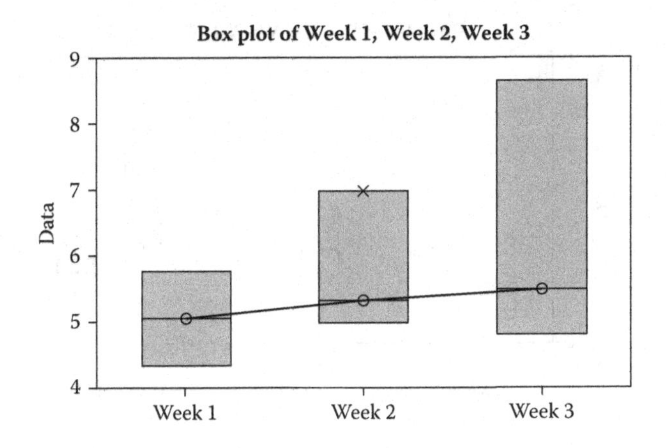

FIGURE 10.2
Box plot example.

Example

Box plots provide a plot tool for viewing the behavior of a data set or for comparing multiple data sets. They are very good for viewing smaller data sets where histograms tend not to do as well. The chart in Figure 10.2 plots the same data set over time week by week. It is easy to see the process output is not the same week to week. Here we observe that the mean is not moving as much as the data are skewing to the right. While we see the skewing taking place, it does not appear that it is related to one of the two operators. Next, we see box plots by the subgroup of machine (Figures 10.3 and 10.4). It is now visually clear that Machine 1 is "drifting" right, explaining the skewing. Also, Machine 1 is responsible for the increase in variance.

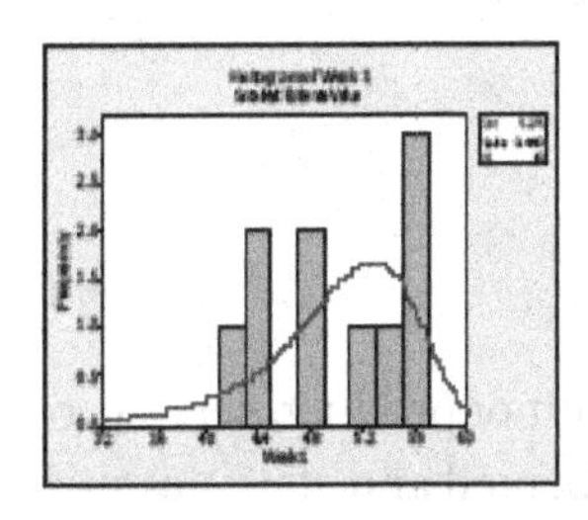
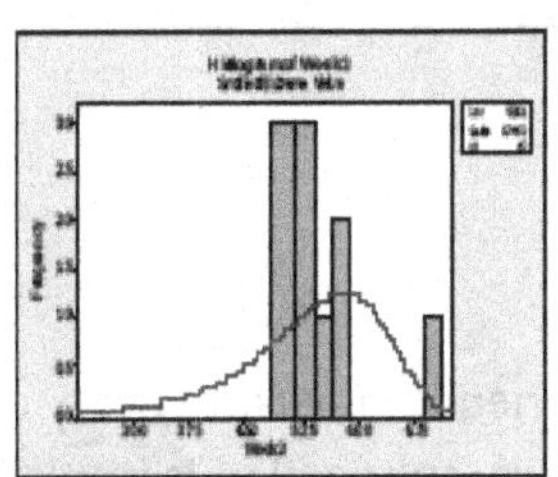
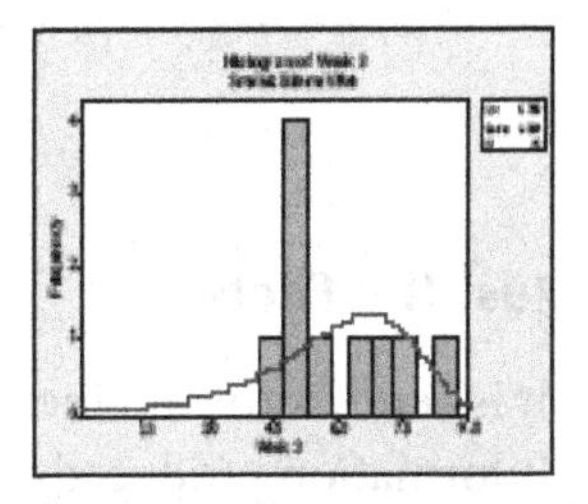

FIGURE 10.3
Box plot.

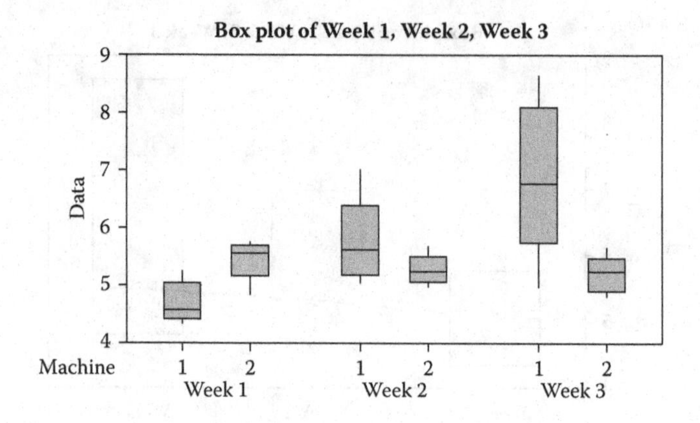

FIGURE 10.4
Box plot.

CONFIDENCE INTERVALS

If you do not have confidence in the data, don't use it.

Definition

The confidence interval allows the organization to estimate the true population parameter with a known degree of certainty.

Just the Facts

It is very costly and inefficient to measure every unit of product, service, or information produced. A sampling plan is implemented and statistics such as the average, standard deviation, and proportion are calculated and used to make inferences about the population parameters. Unfortunately, when

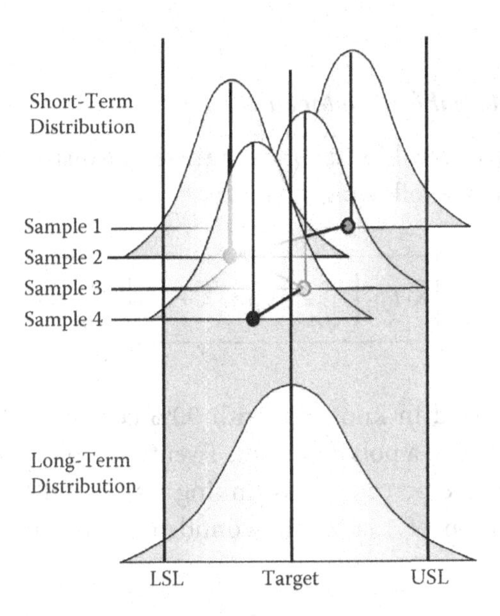

FIGURE 10.5
Calculated sample averages.

a known population is sampled many times, the calculated sample averages can be different even though the population is stable. (See Figure 10.5.)

The differences in these sample averages are simply due to the nature of random sampling. Given that these differences exist, the key is to estimate the true population parameter. The confidence interval allows the organization to estimate the true population parameter with a known degree of certainty.

The confidence interval is bounded by a lower limit and upper limit that are determined by the risk associated with making a wrong conclusion about the parameter of interest. For example, if the 95% confidence interval is calculated for a subgroup of data of sample size *n*, and the lower confidence limit and the upper confidence limit are determined to be 85.2 and 89.3, respectively, it can be stated with 95% certainty that the true population average lies between these values. Conversely, there is a 5% risk (alpha (α) = 0.05) that this interval does not contain the true population average. The 95% confidence interval could also show that:

- Ninety-five of 100 subgroups collected with the same sample size *n* would contain the true population average.
- If another 100 subgroups were collected, 95 of the subgroups' averages would fall within the upper and lower confidence limits.

Example

Confidence Interval for the Mean

The confidence interval for the mean utilizes a t-distribution and can be calculated using the following formula:

$$\overline{X}\text{-}t_{\alpha/2}\left(\frac{s}{\sqrt{n}}\right) \leq \mu \leq \overline{X}+t_{\alpha/2}\left(\frac{s}{\sqrt{n}}\right)$$

We are interested in knowing, with 90% certainty, the average percent cross-linking of a polymer resin. Twenty samples were selected and tested. The average percent cross-linking was found to be 68% with a standard deviation of 15.6%. The confidence interval for the mean μ would be:

$$68 - 1.73(15.6/\sqrt{20}) \leq \mu \leq 68 + 1.73(15.6/\sqrt{20})$$

$$68 - 1.73(3.488266) \leq \mu \leq 68 + 1.73(3.488266)$$

$$68 - 8.0347 \leq \mu \leq 68 + 6.0347$$

$$61.97 \leq \mu \leq 74.03$$

Confidence Interval for the Standard Deviation

The confidence interval for the standard deviation subscribes to a chi-square distribution and can be calculated as follows:

$$s\sqrt{\frac{(n-1)}{\chi^2\alpha/2,\, n-1}} \leq \sigma \leq s\sqrt{\frac{(n-1)}{\chi^2 1-\alpha/2,\, n-1}}$$

S is the standard deviation of the sample and χ^2 is the statistical distribution from a statistical table.

Example

We are interested in knowing, with 95% certainty, the amount of variability in the elasticity of a material being produced. Fourteen samples were measured and the average elasticity was found to be 2.830 with a standard deviation of 0.341.

$$0.341\sqrt{(14.1/24.7)} \leq \sigma \leq \sqrt{(14.1/5.01)}$$

$$0.247 \leq \sigma \leq 0.549$$

Caution: Some software and texts will reverse the direction of reading the table; therefore, $\chi^2_{\alpha/2,\, n-1}$ would be 5.01, not 24.74.

Confidence Interval for the Proportion Defective

For proportion defective (p) we use the binomial distribution; however, in this example the normal approximation will be used. The normal approximation to the binomial may be used when np and $n(1 - p)$ are greater than or equal to 5. A statistical software package will use the binomial distribution.

$$p\text{-}Z_{\alpha/2}\sqrt{\frac{p(1-p)}{n}} \leq P \leq p + Z_{\alpha/2}\sqrt{\frac{p(1-p)}{n}}$$

Example

We want to know, with 95% certainty, the current proportion defective for product measurement forms. Twelve hundred forms were sampled and 14 of these were deemed to be defective. The 95% confidence interval for the proportion defective would be:

$$0.012 - 1.96\sqrt{((0.012(1 - 0.012)/1{,}200)} \leq P \leq 0.012 + 1.96$$

$$\sqrt{((0.012(1 - 0.012)/1{,}200)}$$

$$0.012 - 0.006 \leq P \leq 0.012 + 0.006$$

$$0.006 \leq P \leq 0.018$$

$$0.60\% \leq P \leq 1.80\%$$

Note: np = 1,200 (0.12) = 14.4, which is > 5, and $n(1 - p)$ = 1,200 (0.988) = 1,185.6, which is > 5, so the normal approximation to the binomial may be used.

DATA TRANSFORMATIONS

Data is worthless until it is analyzed.

Definition

When data do not follow a normal distribution, thus allowing us to calculate basic statistics and valid probabilities related to the population (mean, standard deviation, *Z* values, probabilities for defects, yield, etc.), the data need to be transformed into a normal distribution prior to analysis.

Just the Facts

If the distribution is non-normal, then the analysis may be misleading or incorrect due to the violation of the normality and equal variance assumptions. A transformation function can often be used to convert the data closer to a normal distribution to meet the assumptions and allow for a determination of statistical significance. Transformation use may be indicated by nonrandom patterns in the residuals of a regression, ANOVA, or DoE analysis.

A transformation function converts a non-normal distribution to a normal distribution. Caution must be used by the analyst in interpreting transformed data that have no physical significance

Data Transformation Types

It is often difficult to determine which transformation function to use for the data given. Many people decide which function to use by trial and error, using the standard transformation functions. For each function or combination of functions, the transformed data can be checked for normality using the Anderson-Darling test.

A transformation function can incorporate any one or combination of the following equations, or may use additional ones not listed.

Standard Transformation Functions

Moderately positive skewness	Square root:	$x = \text{SQRT}(X)$
Substantially positive skewness	Logarithmic:	$x = \log10\, x$, $\ln x$, etc.
Substantially positive skewness (with zero values)	Logarithmic:	$x = \log10(x + c)$
Moderately negative skewness	Square root:	$X = \text{SQRT}(K - X)$
Substantially negative skewness	Logarithmic (log 10)	$X = \log10(K - X)$

Reciprocal of data: .... $x = 1/x$

Square of data: .... $x = X^2$

There is also benefit in transforming specific distributions such as count data. This type of data often has a nonconstant variance (Poisson distribution). Recommended transformations are listed below:

$$X = \text{SQRT}(X)$$

Freeman-Tukey modification to the square root:

$$X = [\text{SQRT}(x) + \text{SQRT}(c + 1)]/2$$

Another frequently encountered situation requiring transformation occurs when dealing with attribute data. Recommended transformations include:

$$X = \text{Arcsin}\ [\text{SQRT}(p)]$$

Freeman-Tukey modification to arcsin:

$$X = [\arcsin(\text{SQRT}((np)/(n + 1)) + \arcsin(\text{SQRT}((np + 1)/n + 1)]/2$$

By knowing the physics behind a process, the LSSBB can determine the appropriate function that describes the process. This function can be used to transform the data.

TABLE 10.1

Box-Cox Transformations

If the Lambda Value Is:	Then the Transformation Value Will Be:
Use y^{λ} when $\lambda \neq 0$ and ln (*y*) when $\lambda = 0$.	
2	y^2
0.05	$y^{0.05}$
0	ln (*y*)
−0.5	$1/y^{0.5}$
−1	$1/y$
1	No transformation required

Once the transformation function is found for a given set of data, it can be used for additional data collected from the same process. Once a process has been modified, however, the distribution may change and become normal, or it may become a different type of non-normal distribution. A new transformation function may then be needed.

If choosing a transformation is difficult, many software programs will perform a Box-Cox transformation on the data. Some common Box-Cox transformations are listed in Table 10.1. The table can be used to assess the appropriateness of the transformation. For example, the 95% confidence interval for lambda can be used to determine whether the optimum lambda value is close to 1, because a lambda of 1 indicates that a transformation should not be done. If the optimum lambda is close to 1, very little would be gained by performing the transformation.

As another example, if the optimum lambda is close to 0.5, the square root of the data could simply be calculated, because this transformation is simple and understandable. (Note: In some cases, one of the closely competing values of lambda may end up having a slightly smaller standard deviation than the estimate chosen.)

It is highly recommended that a statistical evaluation program such as Minitab or JMP be used to make these calculations.

Example

A data set was evaluated and discovered to be non-normally distributed. The distribution was positively skewed because the exaggerated tail was in the positive direction. Using Minitab, the results are as shown in Figure 10.6.

The same data set was evaluated using a Box-Cox transformation analysis. The Box-Cox transformation plot is shown in Figure 10.7. In this example, the best estimate for lambda is zero or the natural log. (In any practical situation, the LSSBB would want a lambda value that corresponds to an understandable transformation, such as the square root (a lambda of 0.5) or the natural log (a lambda of 0).) Zero is a reasonable choice because it falls within the 95% confidence interval and happens to be the best estimate of lambda. Therefore, the natural log transformation will be used. After performing the recommended transformation on the data set, a test for normality was done to validate the transformation.

The normal probability plot (Figure 10.8) of the transformed data can be graphically interpreted by the fit of the data to a straight line, or analytically interpreted from the results of the Anderson-Darling test for normality. Since the p value (0.346) is greater than the α risk (0.5) we are willing to take, there is sufficient evidence that the transformed distribution is normally distributed.

The original lognormal capability data have been transformed into the normal distribution shown in Figure 10.9. (The * in the graphic indicates transformed data. For example, the actual USL is 1.50; however, the transformed USL is 0.405.) The potential and overall capabilities are now more accurate for the process at hand. The use of non-normal data may result in an erroneous calculation of predicted proportion out of the specified units. When transforming the capability data, remember that any specification limit must also be transformed. (Most software packages will perform this automatically.)

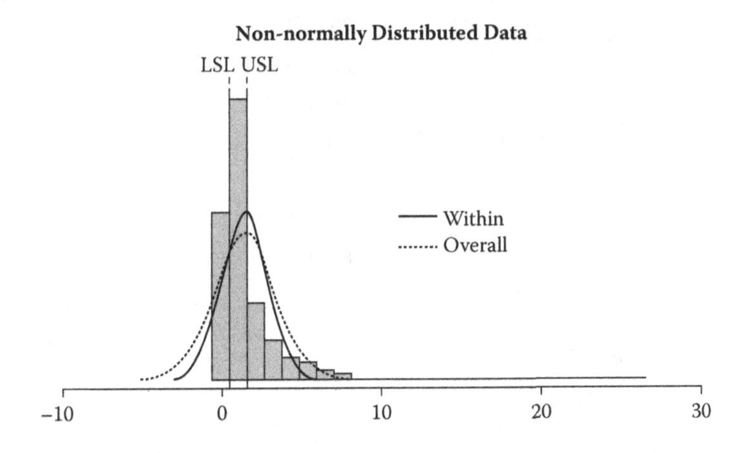

FIGURE 10.6
Non-normally distributed data. (*continued*)

Process Data	
USL	1.50000
Target	*
LSL	0.50000
Mean	1.62833
Sample N	500
StDev (Within)	1.56770
StDev (Overall)	2.18672
Potential (Within) Capability	
C_p	0.11
C_{pu}	−0.03
C_{pl}	0.24
C_{pk}	−0.03
C_{pm}	*
Overall Capability	
P_p	0.08
P_{pu}	−0.02
P_{pl}	0.17
P_{pk}	−0.02
Observed Performance	
P_{pm}<LSL	260000.00
P_{pm}>USL	300000.00
P_{pm} Total	560000.00
Exp. "Within" Performance	
P_{pm}<LSL	235844.41
P_{pm}>USL	532619.36
P_{pm} Total	768463.77
Exp. "Overall" Performance	
P_{pm}<LSL	302930.55
P_{pm}>USL	523398.10
P_{pm} Total	826328.65

FIGURE 10.6 (*continued*)
Nonnormally distributed data.

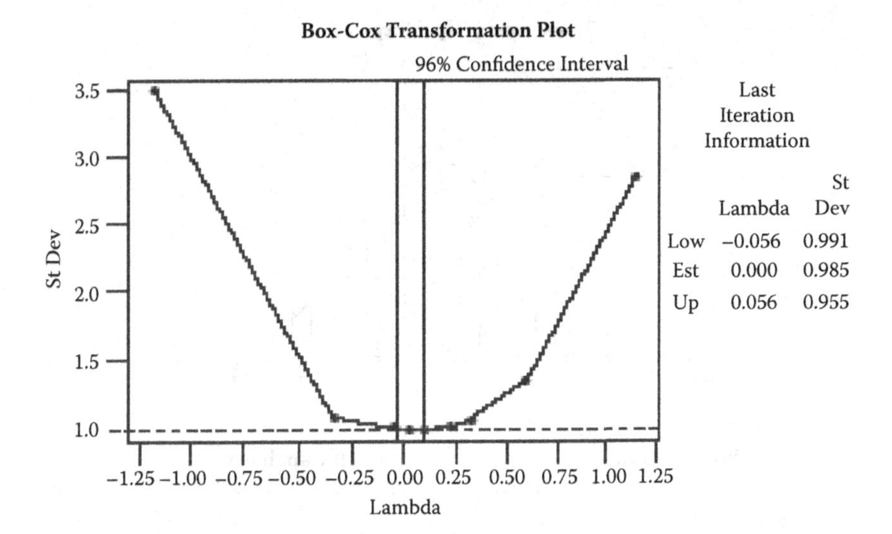

FIGURE 10.7
Box-Cox transformation plot.

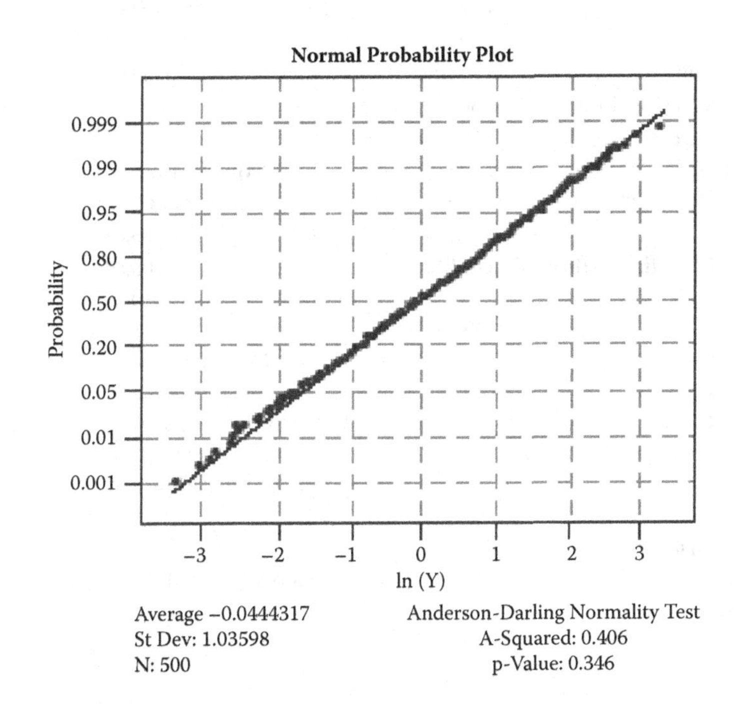

FIGURE 10.8
Normal probability plot.

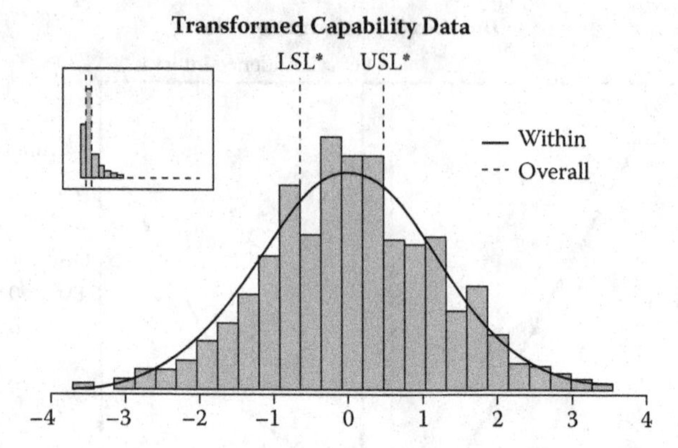

Process Data

USL	1.50000
USL*	0.40547
Target	*
Target*	*
LSL	0.50000
LSL*	−0.69315
Mean	1.62833
Mean*	−0.04443
Sample *N*	500
StDev (within)	1.56770
StDev* (within)	1.02145
StDev (overall)	2.18672
StDev* (overall)	1.03650

Potential (Within) Capability

C_p	0.18
C_{pu}	0.15
C_{pl}	0.21
C_{pk}	0.15
C_{pm}	*

Overall Capability

P_p	0.18
P_{pu}	0.14
P_{pl}	0.21
P_{pk}	0.14

Observed Performance

P_{pm} < LSL	260000.00
P_{pm} > USL	300000.00
P_{pm} total	560000.00

Exp. "Within" Performance

P_{pm} < LSL	262683.84
P_{pm} > USL	329805.84
P_{pm} total	592489.68

Exp. "Overall" Performance

P_{pm} < LSL	265699.92
P_{pm} > USL	332124.86
P_{pm} total	597824.78

FIGURE 10.9
Transformed capability data. (* = calculated for the transformed data the actual values need to be converted)

Notice from a comparison of the before and after capability outputs that we are only slightly more capable than our original estimate with the lognormal data. There is excessive variation in the process. The primary focus of the LSSBB should be to identify and correct the sources of this variation.

If you do not have a program such as Minitab or JMP, and the data fail a normality test, then use the following Application Cookbook.

Application Cookbook

Before using a mathematical transformation, apply steps 1 and 2 of the Application Cookbook. If the data are not normally distributed (e.g., fail the normality test) conduct the following steps:

1. Examine the data to see if there is a nonstatistical explanation for the unusual distribution pattern. For example, if data are collected from various sources (similar machines or individuals performing the same process) and each one has a different mean or standard deviation, then the combined output of the sources will have an unusual distribution, such as a mixture of the individual distributions. In this case, separate analyses could be made for each source (individual, machine, etc.).
2. Analyze the data in terms of averages instead of individual values. Sample averages closely follow a normal distribution even if the population of individual values from which the sample averages came is not normally distributed. If conclusions on a characteristic can be made based on the average value, proceed, but remember these only apply to the average value and not to the individual values in the population.
3. If steps 1 and 2 do not provide reliable estimates, use the Weibull distribution. Consult the Application Cookbook of the tool Distribution—Weibull. The resulting straight line can provide estimates of the probabilities for the population.
4. If all above steps fail in providing reliable estimates, use one of the most common mathematical transformations, which include:

x^2	e^x	$\ln x$	$\ln\frac{x}{1-x}$	$\log\frac{1+x}{1-x}$
$\frac{1}{x}$	e^{-x}	$\log\frac{x}{1-x}$	$\frac{1}{2}\ln\frac{1+x}{1-x}$	
$\sqrt{x}$	$\log x$			

DESIGN OF EXPERIMENTS

The purpose of an experiment is to make nature speak intelligently: All that then remains is to LISTEN.

—Bill Diamond (1978)

Definition

Design of experiments (DoE), or experimental design, is the design of any information-gathering exercises where variation is present, whether under the full control of the experimenter or not. However, in statistics, these terms are usually used for controlled experiments. Other types of study and their design are discussed in published articles on opinion polls and statistical surveys (which are types of observational study), natural experiments, and quasi-experiments (for example, quasi-experimental design). In the design of experiments, the experimenter is often interested in the effect of some process or intervention (the treatment) on some objects (the experimental units), which may be people, parts of people, groups of people, plants, animals, materials, etc. There are many tools used in design of experiments.

Just the Facts

Statistical experimental design usually means multivariable experimental matrices. The primary reason for using statistically designed experiments is to obtain a maximum amount of information for a minimum amount of expenditure. The fundamental purpose of a designed experiment is to determine a course of action. This action results from conclusions drawn from an experiment. Many aspects of designing and optimizing a successful process require efficient, accurate experiments:

1. The scientific method, hence all scientific investigation, is built around experimentation.

2. Optimization of process parameters such as material compositions or environmental parameters is a duty in setting up a process.
3. Cost savings resulting from fewer experiments, increased investigator efficiency, and increased output from an optimal process setup are usually substantial.
4. Statistical methods for designing experiments allow information to be gathered on all factors simultaneously, leading to more economies over single-factor experiments.

Steps in Designing an Experiment

An eight-stage procedure formally details the steps to be taken in designing any experiment:

1. Statement of the problem
2. Formulation of hypotheses
3. Planning of the experiment
   a. Choosing an appropriate experimental technique
   b. Examination of possible outcomes to make certain that the experiment provides the required information
   c. Consideration of possible results from the point of view of the statistical analysis
4. Collection of data (performing the experiment according to the plan)
5. Statistical analysis of the data
6. Drawing conclusions with appropriate significance levels
7. Verification or evaluation of results (conclusions)
8. Drawing final conclusions and recommendations

These eight steps are present in any experiment whether or not it has been effectively designed. They indicate only a logical framework for any experimental process. To go a step further, a *good* experiment has several requirements. It must possess no systematic error, have small standard or experimental error, and a wide range of validity, and allow proper statistical analysis to apply.

The standard and systematic errors that we wish to minimize can be caused by tolerance in the actual experimentation, observation of responses, measurement, variations in experimental materials, or other extraneous factors. A good experiment minimizes these by:

1. Using more homogeneous materials
2. Using information provided by related variables
3. Careful experimentation
4. Using more efficient design

Our matrix experiments, since they tend to be large undertakings, demand adherence to these general experimental practices.

Principles of an Experimental Design

A requirement list for good experiments therefore includes measurements whose accuracy is maximized (as detailed in Chapter 9), sound engineering, and an ability to design our investigation efficiently. There are three general principles also essential to a statistically designed matrix experiment:

- Randomization: If separate experimental runs are run in random order, with no pattern to the factor settings, statistical analysis of responses can be conducted without danger of any bias to the experimental unit. The danger if this is not done could be exemplified by considering an experiment run in a tube furnace. Suppose the experimenter has not considered temperature a factor, though in reality, it plays a part. If one level of material were run in totality first and another totally later (instead of randomly ordering the two levels), any effect of the materials might be biased by a slow heat-up of the furnace.
- Replication: Experimental error—the failure of two identically treated experimental units to give the same value. Replication provides a means of estimating the experimental error, which in turn provides a means of testing interactions. Significant process factors will produce large changes compared to the difference between replicates. Replication is a repetition of the whole experiment in order to estimate experimental error, increase precision (detect smaller changes), and increase sensitivity.

- Local controls: Local controls means experimental planning, referring to the amount of blocking or balancing of grouped experimental runs among different settings of variables being studied. The purpose of local controls is to make the experimental design more efficient, and guard against unforeseen surprises nullifying experimental adjustments.

In summary, replication makes statistical tests possible, randomization makes statistical tests valid, and local controls make the experiment more efficient. Together these principles can be applied to produce good experiments that can be analyzed using statistical analysis of variance methods. We will complete our background with a brief introduction to these methods, and then study the total process of matrix experimental design in various forms through examples.

Setting Up the Appropriate Experiment

The question will lead to an examination of the process with emphasis on the elements of the process that have an effect on the resultant product. Sometimes the answer can be obtained by simply collecting historical data and analyzing them. Sometimes a process control study is appropriate, and sometimes a designed experiment is required. The purpose of the experiment is to determine how the independent variable of the process affects the dependent parameters of the resultant product. The basic questions to be asked are:

1. What is the purpose of this experiment?
   a. What do I expect to accomplish?
   b. What questions do I expect to answer?
2. What are the independent variables (factors)?
   a. (The variables to be changed or controlled during the experiment)
3. What variables are to be held constant?
4. What are the dependent variables (response(s))?
5. Are there any interactions? How many? (known and potential)
6. Is there a process capability study?
7. What stage of development is the process in?
   a. Fresh from research (feasibility, sandbox)
   b. First given to development

   c. Refining the development process
   d. Making the development process manufacturable
   e. The initial manufacturing process
   f. The manufacturing process capability
      i. Establish its initial limits
      ii. Identify the significant variables
   g. Refining the manufacturing process
   h. Stabilizing the manufacturing process
8. How do the dependent variables (responses) relate to answering the question proposed in the purpose for doing the experiment?
9. What measurement tools are you going to use?
   a. How do they relate to the function?
   b. How do they relate to what you want to know?
   c. How do they relate to the specifications?
   d. How good is the measurement?
      i. Precision
      ii. Accuracy
      iii. Repeatability
10. To what degree of precision and accuracy are the measurements going to be made? (Is this sufficient for the purpose of the experiment?)
11. What information can be obtained from existing data?
12. What information needs to be collected?
13. Based on what is known and obtained in 1 through 12 above, what is the appropriate design to use?
   a. Process capability study
   b. Random strategy
   c. Plackett-Burman design
   d. Central composite rotatable design (CCRD)
   e. Fractional factorial (¼, ½, ¾, etc.)
   f. Simple comparison
   g. Paired comparison
   h. Sandbox (one factor at a time)
14. How are the data to be analyzed?
   a. What type of statistical analysis will be used?
      i. Analysis of means
      ii. Analysis of variance (ANOVA)
      iii. Signal to noise
      iv. Control chart
   b. How will the data be analyzed?

i. By hand
ii. By computer program
iii. Existing program
iv. Self-generated program

Analysis (of Means and Variance) Methodologies

The four basic assumptions of the analysis of means and the analysis of variance of data from a statistically designed experiment are normality of the response variable, additivity of the effects, homogeneity of the variances, and statistical independence of each response value. The normality assumption requires that the underlying distribution of the response variable outcomes be a normal distribution. However, the analysis of variance is a "robust" test. This means that false experimental conclusions are not generated by departures from normality. Thus, the normality of the response variable is not a strict requirement.

The assumption of additivity of the effects means that each response value is comprised of the sum of an overall mean plus the sum of all effects due to each factor and the sum of all interaction effects among the factors and the effects due to experimental error.

The homogeneity of distribution assumption requires that the underlying distributions of replicate readings each have the same variance. This is a very strict requirement. The analysis of variance test doesn't apply if this assumption does not hold. Finally, the statistical independence of the response variable outcomes assumption states that each response value is independent of all previous values. This is guaranteed by randomization of the treatment combinations to the test specimens.

When these assumptions are followed, results of a matrix of experimental runs can be economically analyzed together using a statistical test known as the Student's *t*-test and the *F*-ratio. These tests compare variation (in final result) and mean differences caused by a matrix variable to normal noise.

Analysis of Means

In the analysis of means, the means are calculated for each set of response data, and then these means are compared to each other to determine differences. Where a statistical test is required, the Student's *t*-test may be used to determine if there is a significant difference. The Student's *t*-test can be used with distributions that are the same or different. In a multivariate experiment, it is assumed that all distributions are the same. However, there are times when we simply want to evaluate the difference between two sets of readings, or to qualify two different measurement techniques, or laboratory results, etc. For these "simple comparison" tests, there are two analysis techniques that are used: simple comparison and paired comparison. In the simple comparison samples are taken and measured by two or more procedures and the resultant distributions and means are compared using the Student's *t*-test to determine if there is a significant difference. In the paired comparison test, each sample taken from the population is divided and tested by the different testing procedures. These results are then paired together and the differences are used to calculate the *t* value.

The paired comparison test is much more sensitive than the simple comparison test because it removes the variation between samples within the population and shows only the differences between the test results. Consider a test of two chemical mixes for latex rubber. We wish to determine if there is any difference in the average cure time between the two mixtures.

Mix	
A	**B**
8	12
16	17
12	15
6	8
10	14

Comparing two populations (completely randomized experiment):

Mix			Hypothesis
A	**B**		$H_0: \mu_1 = \mu_2$
8_1[a]	12_8		$H_1: \mu_1 \neq \mu_2$
16_7	17_5		$\alpha = 0.05$
12_{10}	15_2		
6_3	8_9		**Pooled Estimate of Standard Deviation**
10_4	14_6		
n 5	5		$Sp = \sqrt{\dfrac{(n_1-1)s_1^2+(n_2-1)s_2^2}{n_1-n_2-2}}$

$\bar{x}$	10.40	13.20	$Sp = \sqrt{\frac{4(3.85)^2 + 4(3.42)^2}{8}}$
s	3.85	3.42	$Sp = \sqrt{13.26} = 3.64$

[a] Subscripts on data values represent the order of randomization for preparation and pouring mixes.

Student's t Value

$$t_8 = \frac{\bar{x}_1 - \bar{x}_2}{Sp\sqrt{\frac{1}{n_1} + \frac{1}{n_2}}} = \frac{10.40 - 13.20}{Sp\sqrt{2/5}} = \frac{-2.80}{3.64\sqrt{2/5}} = 1.22$$

From the t-table we find that $t_{8\,(0.05)} = 2.31$.

Conclusion: There is insufficient evidence to show that there is a difference in the two mixes.

Paired Comparison

	Mix			
Day	**A**	**B**	**Difference**	**Hypothesis**
M	8_1*	12_8	−4	
T	16_7	17_5	−1	$H_0: \mu_1 = \mu_2$
W	12_{10}	15_2	−3	$H_1: \mu_1 \neq \mu_2$
Th	6_3	8_9	−2	$\alpha = 0.05$
F	10_4	14_6	−4	
				Student's *t* Value
n	5	5	5	$\frac{\bar{d}}{t_{n-1} = S_d\sqrt{1/n}}$
$\bar{x}$	10.40	13.20	$\bar{d} = 2.80$	
s	3.85	3.42	$s_d = 1.30$	$\frac{-2.80}{t_4 = 1.30\sqrt{1/5} = 4.802}$

From the t-table we find that $t_{4\,(0.05)} = 2.78$.

Conclusion: There is a statistically significant difference between Mixes A and B.

Analysis of Variance Methodology

The amount of variation (in experimental response) contained in the total experimental matrix is first calculated by summing the squares of each observation and subtracting a correction factor for the overall average response level. This is not unlike the type of calculation performed in finding the standard deviation.

$$s^2 = \frac{\Sigma(x-x)^2}{n-1} = \frac{1}{(n-1)}(\Sigma X^2 - nx^2)$$

$$= \frac{1}{(n-1)}((\Sigma x^2) - (\text{Correction factor}))$$

The amount of variation for each variable in the experiment can be similarly sized by squaring the totals of each level of that variable, and subtracting the same correction factor for overall average. This variation is proportioned for sample size by dividing our sizing (by $n - I$ in the case of the sample variance). The number of measurements in any variation sizing is kept track of by this divisor, called degrees of freedom. For our purposes this can be considered the number of measurements (in the total group) one could vary while still keeping the same average response level. Accordingly, degrees of freedom can be calculated as one less than the number of readings squared in getting the variation total. This is necessary since variation sizings based on little data are less secure than those based on extensive study, and this must be accounted for in our statistical evaluation.

Since all variation sizing of this type must mathematically add up to the total experimental variation, variation due to experimental error can be found by subtracting variation sizing for individual variables from the total experiment's variation. The analysis of variance, or statistical F-ratio test, consists of comparing the variation caused by any specific variable to variation due to experimental error. If the particular variable causes a lot of variation in the final result (compared to normal fluctuations seen in experimental error), it brands itself a major process impactor. These significant process impactors can then be carefully specified for optimum process setup.

The quantified test for significance is conducted by calculating a sample F-ratio, which is the comparison between a variable's variance components and experimental error (as we will work out in examples). This value is compared with critical values of *F* having the same degrees of freedom settings as the sample *F*-ratio, and an error exposure (alpha) appropriate to the text. Sample *F* values exceeding this critical value indicate variation in the end result is significantly large for the studied variable when compared with random variation seen in the experiment. As we have mentioned, the analysis of variance technique is fairly reliable even when assumptions are not 100% valid. Since the analysis method lets us reuse the same experimental variations, and since the same experimental runs can be reused in calculating variances for several variables, these matrix experiments are extremely economical and efficient ways to study several variables at once.

One-Way and Two-Way ANOVA

One-Way ANOVA

Let's consider an example in the simplest case of a one-factor experiment:

$$SST = SSA + SSE$$

In the more complicated case of a two-factor experiment with replication we would have:

$$SST = SSA + SSB + SSAB + SSE$$

where:

SST = Total sum of squares for the entire experimental matrix

SST = Sum of squares of all items minus the correction factor

SSA or *SSB* = Treatment sum of squares or the sum of squares for factor A or B

SSE = Error sum of squares (noise)

SSE = Total sum of squares minus the treatment sum of squares

SSAB = Sum of squares for interaction between A and B

SSAB = Cell totals squared minus *SSA* and *SSB*, and we will also use (later) *CF*

CF = Correction factor (for the mean) = grand total squared – number of items summed to make up the grand total

Notice we have just broken up the total variation (in our experimental results) into pieces attributable to each experimental variable, plus "noise" (normal experimental error variations).

Example Experiment 1

Response Data

	I	II	III	IV	V
	58	60	49	59	52
	55	58	42	64	49
	63	60	48	64	58
	60	55	46	63	50
	59	56	46	69	52
Total	295	289	231	319	261

Grand total = 295 + 289 + 231 + 319 + 261 = 1,395

X	59.0	57.8	46.2	63.8	52.2

$$X = \frac{59.0+57.8+46.2+63.8+52.2}{5} = \frac{279}{5} = 55.8$$

Analysis of Variance (ANOVA)

1. Total sum of squares (SST or SS total):

$$\text{SS total} = \frac{\Sigma(\text{each measurement})^2}{1} - \frac{(\Sigma \text{ all measurements})^2}{\text{total number of measurements}}$$

$$= \text{sum of square readings} - \text{correction factor}$$

$$= 58^2 + 55^2 + 63^2 + \ldots 52^2 - \frac{(1395)^2}{25} = 78{,}941 - 77{,}841 = 1100.00$$

2. Sum of squares for coating-type totals (SSA or SS type):

$$\text{SS type} = \frac{\Sigma(\text{type total})^2}{\text{type sample size}} - \frac{(\Sigma \text{all measurements})}{\text{total number of measurements}}$$

$$= \frac{295^2 + 289^2 + \cdots 261^2}{5} - \frac{(1395)^2}{25} = 78{,}757.8 - 77{,}841 = 916.8$$

Source of Variation	Sum of Squares (SS)	Degrees of Freedom (DF)	Mean Square (MS)	*F*-Ratio
Among five types (treatment)	916.8 (SSA)	$5 - I = 4$	$\frac{916.8}{4} = 229.2$ (MSA)	$F_{calc} = \frac{229.2}{9.16} = 25.02$
Within type (error)	183.2 (SSE)	$25 - 5 = 20$	9.16 (MSE)	$F_{.05}\left(\frac{4}{20}\right) = 2.87$
Total	1,100	$25 - I = 24$		

Since F calculated is larger than $F_{.05}(\frac{4}{20})$ we reject the null hypothesis that the coating-type means are equal.

The table on conductivity of coating contains data from a real experimental matrix that studied several conductive coating application processes. The response number is the conductivity of coating measured in mhos. The single factor is the coating process. The five levels are the five different coating processes. There are five replications per each level.

The calculation is conveniently arranged in the following ANOVA table:

Source	*DF*	*SS*	*MS*	F_{calc}	F_{tab}
Total	$N - 1$	*SST*			
Treatments	$K - 1$	*SSA*	$\frac{SSA}{K-1}$	$\frac{MSA}{MSE}$	$F\alpha\,(K - 1, N - K)$
Error	$N - K$	*SSE*	$\frac{SSE.}{N-K}$		

N = Total number of measurements
K = Number of levels of the one factor
SS = Sum of squares appropriate to the source, a variance estimate
DF = Degrees of freedom, an indexing count determined by number of readings used in invariance estimate
MS = Mean square = SS divided by DF
MSA = Measure of the variation among the K sample means of the different variable (here, coating process) settings
MSE = Measure of the variation that is due to chance

F is a ratio of the variation among the sample means to the variation due to chance. If *F* is large, then the sample means vary widely and thus are significant result changers. Stated another way, processes significantly differ. If *F* is small, then the differences among the sample means can be attributed to chance. Notice both the total degrees of freedom and the total sum of squares are divided into pieces appropriate to our treatments and error.

ID	Average	IV	I	II	V	III
IV	63.8	0	4.8	6.0	11.6	17.6
I	59.0		0	1.2	6.8	12.8
II	57.8			0	5.6	11.6
V	52.2				0	6.0
III	46.2					0

From the data, $MSE = 9.16$ and $n = 5$; thus,

$$SEM = \left(\frac{9.16}{5}\right) = 1.832 = 1.35$$

$$DF = 20$$

$$t_{0.025(20)} = 2.086$$

$$LSD = 2.086 \times 1.4142 \times 1.35 = 3.98$$

$$LSD = \text{least significant difference}$$

Thus, process IV would be recommended if high conductivity is desired.

If the difference between two means exceeds the LSD value, then they come from different distributions.

Since the mean of process IV differs by 4.8 from the mean of process I, processes I and IV have separate distributions. Also, process IV is significantly larger than all other processes.

Since the *F*-ratio is significant, we should like to go further and determine where the separate distributions are located. This can be accomplished by using the LSD method to separate the process distributions. First calculate the critical difference.

$$LSD = t_{\alpha/2}, DF = (N - K) \times (\sqrt{2}) \times (SEM)$$

For this formula: $SEM = \sqrt{(MSE/n)}$

t = value that traps the desired area in the table of the t-distribution (see t-table).

Next average the treatment means in ascending order and compute the pairwise positive differences, then arrange them in a table as follows: where:

MSE = mean square due to error (noise)
n = number of replicates
k = number of populations being compared

	(1)	(2)	(3)	(4)	(5)
$\overline{X}_{(1)}$	0	$\overline{X}_{(1)} - \overline{X}_{(2)}$	$\overline{X}_{(1)} - \overline{X}_{(3)}$	$\overline{X}_{(1)} - \overline{X}_{(4)}$	$\overline{X}_{(1)} - \overline{X}_{(5)}$
$\overline{X}_{(2)}$		0	$\overline{X}_{(2)} - \overline{X}_{(3)}$	$\overline{X}_{(2)} - \overline{X}_{(4)}$	$\overline{X}_{(2)} - \overline{X}_{(5)}$
$\overline{X}_{(3)}$			0	$\overline{X}_{(3)} - \overline{X}_{(4)}$	$\overline{X}_{(3)} - \overline{X}_{(5)}$
$\overline{X}_{(4)}$				0	$\overline{X}_{(4)} - \overline{X}_{(5)}$
$\overline{X}_{(5)}$					0

As indicated above, analysis of our example showed process IV to produce higher conductivities than I and II, which were in turn higher than V followed by III.

Example Experiment 2

	Vendor				
Material	**A**	**B**	**C**	**D**	**Totals**
I	96	28	106	88	
	78	36	88	104	
	174	64	194	192	624
II	57	62	71	78	
	69	50	89	68	
	126	112	160	146	544
III	82	38	88	102	
	98	32	110	82	
	180	70	198	184	632
Totals	480	246	552	522	1,800

1. Total sum of squares:

$$SS_{total} = \sum(\text{each measurement})^2 - \left(\frac{(\Sigma \text{ all measurements})^2}{\text{total number of measurements}}\right)$$

$$= 96^2 + 78^2 + 57^2 + \cdots + 82^2 - \left(\frac{(1800)^2}{24}\right) = 148{,}712 - 135{,}000 = 13{,}712$$

2. Sum of squares for 12 cells:

$$SS_{12\text{cells}} = \frac{\Sigma(\text{cell total})^2}{\text{cell sample size}} - \frac{(\Sigma \text{ all measurements})^2}{\text{total numbers of measurements}}$$

$$= \frac{174^2}{2} + \frac{126^2}{2} + \frac{184^2}{2} + \cdots + \frac{(1800)^2}{24} = 147{,}284 - 135{,}000 = 12{,}284$$

3. Sum of squares within cell (errors):

$$SS_{within} = SS_{total} - SS_{cell} = 13{,}712 - 12{,}284 = 1{,}428$$

Two-Way ANOVA

When two factors are involved we are concerned about the main effects of each of the two factors and the effect of their interaction. If the experiment has been replicated, then we have an independent estimate of experimental error and can thus test the interaction effect. The following example illustrates these ideas. Notice that this interaction could not be studied if we ran simple experiments on each variable individually.

In this experiment, four different potential plastics suppliers each provided samples of three different polymers for evaluation. Tensile strengths of two samples of each material from each vendor were randomly tested with results as shown. The experiment will allow us to discover if different materials or different vendors produce stronger material. In addition, we will be able to notice vendor-material interactions if they are significant, that is, whether some vendors are better with certain materials.

As indicated above, the total variation in the experimental matrix is found as before. The portion of this variation due to error can be found by subtracting the variation estimate from all cells in the experiment (which includes both variables and their interaction) from the total variation. Variation due to each variable can be found using that variable's level totals as before. Since the cell variation was due to the variable-caused variation plus interaction, we can find it by subtraction.

Degrees of freedom (used to choose significance cutoffs for F-ratio tests) can be calculated using the basic definition given earlier plus subtraction methods similar to those we just used in variance estimates. Since 24 squares were used in getting total variation, there should be a total of (24 – 1) = 23 degrees of freedom. Three different material levels and four vendor totals (used in those variance estimates) give 2 and 3 degrees of freedom for those variables. Since there were 12 cell totals, and thus 12 – 1 = 11 degrees of freedom for variables plus interaction, there must then be 11 – 2 – 3 = 6 degrees of freedom left for interaction. This also leaves 23 – 11 = 12 degrees of freedom for error. Putting all this analysis in table form produces the following chart.

Sum of squares for materials:

$$SS_m = \frac{\Sigma(\text{material total})^2}{\text{material sample size}} - \frac{(\Sigma\text{ all measurements})^2}{\text{total number of measurements}}$$

$$= \frac{624^2}{8} + \frac{544^2}{8} + \frac{632^2}{8} - \frac{(1800)^2}{24} = 135{,}592 - 135{,}000 = 592$$

Sum of squares for vendors:

$$SS_v = \frac{\Sigma(\text{vendor total})^2}{\text{vendor sample size}} - \frac{(\Sigma\text{ all measurements})^2}{\text{total number of measurements}}$$

$$= \frac{480^2}{6} + \frac{246^2}{6} + \frac{552^2}{6} + \frac{552^2}{6} - \frac{(1800)^2}{24} = 144{,}684 - 135{,}000 = 9{,}684$$

Sum of squares for interaction:

$$SS_1 = SS_{12\,cells} - [SS_m + SS_v] = 12{,}284 - [592 + 9{,}684] = 2{,}008$$

Source of Variation	Sum of Squares	Degrees of Freedom	Mean Square	Calculated F-Ratio	Significance
Materials	592	2	296	$\frac{296}{119} = 2.49$	$F_{.05}\binom{2}{12} = 3.89$
Vendors	9,684	3	3,228	$\frac{3228}{119} = 27.1$	$F_{.05}\binom{3}{12} = 3.49$
Interaction	2,008	6	334.67	$\frac{334.67}{119} = 2.81$	$F_{.05}\binom{6}{12}$ 3.00
Error	1,428	12	119		
Total	13,712	23			

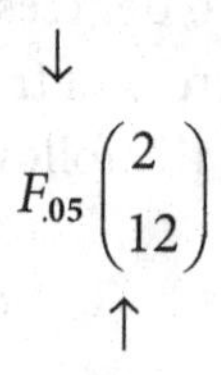

There is significant difference between vendors at the 95% confidence level. At $\alpha = 0.05$, there is no evidence of significant difference between materials or of a significant material-vendor interaction. Accordingly, choice of vendors is an important consideration in the strengths of materials to be received. Material type should not have a major impact, nor should any particular vendor's skill with a particular material.

The LSD test could next be applied to the vendor mean values to separate significant contributors into significantly different or roughly equal classes.

You may have observed that our estimate of error variation did triple duty as a comparison standard for materials and vendors individually, as well as their interaction. Going to the cells themselves, the cell total for material I and vendor A (for example) helped estimate variation for materials, was reused to estimate vendor variation, and helped in interaction evaluation. These economies enabled both variables and their combination to be closely examined with only 24 well-chosen runs.

Types of Experimental Designs

Residual error: The difference between the observed and the predicted value (*C* or *E*) for that result, based on an empirically determined model. It can be variation in outcomes of virtually identical test conditions.

Residuals: The difference between experimental responses and predicted model values.

Resolution I: An experiment in which tests are conducted, adjusting one factor at a time, hoping for the best. This experiment is not statistically sound (definition totally fabricated by the authors).

Resolution II: An experiment in which some of the main effects are confounded. This is *very* undesirable.

Resolution III: A fractional factorial design in which no main effects are confounded with each other, but the main effects and two-factor interaction effects are confounded.

Resolution IV: A fractional factorial design in which the main effects and two-factor interaction effects are not confounded, but the two-factor effects may be confounded with each other.

Resolution V: A fractional factorial design in which no confounding of main effects and two-factor interactions occurs. However, two-factor interactions may be confounded with three-factor and higher interactions.

Response: The graph of a system response plotted against one or more system factors.

Surface: Response surface methodology (RSM). Employs experimental design to discover the "shape" of the response surface and then uses geometric concepts to take advantage of the relationships discovered.

Robust design: A term associated with the application of Taguchi experimentation in which a response variable is considered robust or immune to input variables that may be difficult or impossible to control.

Screening: A technique to discover the most (probable) important factors.

Experiment: In an experimental system. Most screening experiments employ two-level designs. A word of caution about the results of screening experiments: If a factor is not highly significant, it does not necessarily mean that it is insignificant.

Sequential: Experiments are done one after another, not at the same time.

Experiments: This is often required by the type of experimental design being used. Sequential experimentation is the opposite of parallel experimentation.

Simplex: A geometric figure that has a number of vertexes (corners) equal to one more than the number of dimensions in the factor space.

Simplex design: A spatial design used to determine the most desirable variable combination (proportions) in a mixture.

Test coverage: The percentage of all possible combinations of input factors in an experimental test.

Treatments: In an experiment, the various factor levels that describe how an experiment is to be carried out. The level of pH at 3 and the level of temperature at 37°C describe an experimental treatment.

Applications of DoE

Situations where experimental design can be effectively used include:

- Choosing between alternatives
- Selecting the key factors affecting a response
- Response surface modeling to:
  - Hit a target
  - Reduce variability
  - Maximize or minimize a response
  - Make a process robust (i.e., the process gets the "right" results even though there are uncontrollable "noise" factors)
- Seek multiple goals

DoE Steps

Getting good results from a DoE involves a number of steps:

- Set objectives
- Select process variables
- Select an experimental design
- Execute the design
- Check that the data are consistent with the experimental assumptions
- Analyze and interpret the results
- Use/present the results (may lead to further runs or DoEs)

Important practical considerations in planning and running experiments are:

- Check the performance of gauges/measurement devices first
- Keep the experiment as simple as possible
- Check that all planned runs are feasible

- Watch out for process drifts and shifts during the run
- Avoid unplanned changes (e.g., switching operators at half-time)
- Allow some time (and backup material) for unexpected events
- Obtain buy-in from all parties involved
- Maintain effective ownership of each step in the experimental plan
- Preserve all the raw data—do not keep only summary averages!
- Record everything that happens
- Reset equipment to its original state after the experiment

Experimental Objectives

Choosing an experimental design depends on the objectives of the experiment and the number of factors to be investigated. Some experimental design objectives are discussed below:

1. **Comparative objective:** If several factors are under investigation, but the primary goal of the experiment is to make a conclusion about whether a factor (in spite of the existence of the other factors) is "significant," then the experimenter has a comparative problem and needs a comparative design solution.
2. **Screening objective:** The primary purpose of this experiment is to select or screen out the few important main effects from the many lesser important ones. These screening designs are also termed main effects or fractional factorial designs.
3. **Response surface (method) objective:** This experiment is designed to let an experimenter estimate interaction (and quadratic) effects, and therefore give an idea of the (local) shape of the response surface under investigation. For this reason they are termed response surface method (RSM) designs. RSM designs are used to:
   - Find improved or optimal process settings
   - Troubleshoot process problems and weak points
   - Make a product or process more robust against external influences
4. **Optimizing responses when factors are proportions of a mixture objective:** If an experimenter has factors that are proportions of a mixture and wants to know the "best" proportions of the factors, to maximize (or minimize) a response, then a mixture design is required.
5. **Optimal fitting of a regression model objective:** If an experimenter wants to model a response as a mathematical function (either known or empirical) of a few continuous factors, to obtain "good" model parameter estimates, then a regression design is necessary.

It should be noted that most good computer programs will provide regression models. The best design sources are full factorial (in some cases with replication) and response surface designs. Designs for the first three objectives above are summarized in the following.

Select and Scale the Process Variables

Process variables include both inputs and outputs, i.e., factors and responses. The selection of these variables is best done as a team effort. The team should:

- Include all important factors (based on engineering and operator judgments)
- Be bold, but not foolish, in choosing the low and high factor levels
- Avoid factor settings for impractical or impossible combinations
- Include all relevant responses
- Avoid using responses that combine two or more process measurements

When choosing the range of settings for input factors, it is wise to avoid extreme values. In some cases, extreme values will give runs that are not feasible; in other cases, extreme ranges might move the response surface into some erratic region.

The most popular experimental designs are called two-level designs. Why only two levels? It is ideal for screening designs, simple and economical; it also gives most of the information required to go to a multilevel response surface experiment if one is needed. However, two-level design is something of a misnomer. It is often desirable to include some center points (for quantitative factors) during the experiment (center points are located in the middle of the design "box").

Design Guidelines

Number of Factors	Comparative	Screening Objective	Response Surface Objective
1	One-factor completely randomized design		
2–4	Randomized block design	Full or fractional factorial	Central composite or Box-Behnken
5 or more	Randomized block fractional factorial screen first to reduce design or Plackett-Burman number of factors		

The choice of a design depends on the amount of resources available and the degree of control over making wrong decisions (Type I and Type II hypothesis errors) that the experimenter desires. It is a good idea to choose a design that requires somewhat fewer runs than the budget permits, so that additional runs can be added to check for curvature and to correct any experimental mishaps.

A Typical DoE Checklist

Planning an experiment can be as simple as answering some basic questions. Such questions include:

- What do I want to accomplish?
  - What results am I looking for?
  - What relationships am I looking for?
  - What is the final objective of the experiment?
- What are the dependent variable(s)?
  - How am I going to measure the dependent variable?
  - What changes do I expect to see?
  - How am I going to measure these changes?
  - Measuring instrumentation:
    - What is its precision?
    - What is its accuracy?
  - Based on my ability to measure and the changes expected:
    - How many replicas are required to guarantee a valid result?
    - Are there any systematic relationships that might affect the order in which the experiment is run?
  - How do the dependent variable(s) relate to the question being asked?
  - Can the question being asked be answered by these observations?
  - Can inference be drawn from these observations concerning the question?
- What is the independent variable?
  - Which variables affect the dependent variable(s)?
  - How much do you know about the independent variables?
  - Which are the most significant variables?
  - Are there any known or suspected interactions between the variables?
  - How many independent variables will be used?
- Which experimental design is best given the answers to the above questions?
  - What are the advantages of each design?

What are the disadvantages of each design?
How many trials are required for each of the proposed designs?
What order will the experiment be run (how will it be randomized)?
What equipment is required to run the experiment?
What manpower is required to run the experiment?
How long will it take to run the experiment?
What resources are required to evaluate the data?
How will the results be interpreted?
How will the results be presented?

The Iterative Approach to DoE

It is often a mistake to believe that "one big experiment will give the answer." A more useful approach to experimental design is to recognize that while one experiment might give a useful result, it is more common to perform two or three, or perhaps more, experiments before a complete answer is attained. In other words, an iterative approach is best and, in the end, is the most economical. Putting all one's eggs in one basket is not advisable. The reason an iterative approach frequently works best is because it is logical to move through stages of experimentation, each stage supplying a different kind of answer. One author supported an R&D team in the development of a new textile dyestuff in which 33 iterative experiments over 2 years improved colorfastness from 18% to 94%.

Experimental Assumptions

In all experimentation, one makes assumptions. Some of the engineering and mathematical assumptions an experimenter makes include:

- Are the measurement systems capable for all responses?
- Is the process stable?
- Are the residuals (the difference between the model predictions and the actual observations) well behaved?

Is the Measurement System Capable?

It is not a good idea to find, after finishing an experiment, that the measurement devices are incapable. This should be confirmed before embarking on the experiment itself. In addition, it is advisable, especially if the experiment lasts over a protracted period, that a check be made on all measurement devices from the start to the conclusion of the experiment. Strange experimental outcomes can often be traced to "hiccups" in the metrology system.

Is the Process Stable?

Experimental runs should have control runs that are done at the "standard" process set points, or at least at some identifiable operating conditions. The experiment should start and end with such runs. A plot of the outcomes of these control runs will indicate if the underlying process itself drifted or shifted during the experiment. It is desirable to experiment on a stable process. However, if this cannot be achieved, then the process instability must be accounted for in the analysis of the experiment.

Are the Residuals Well Behaved?

Residuals are estimates of experimental error obtained by subtracting the observed response from the predicted response. The predicted response is calculated from the chosen model, after all the unknown model parameters have been estimated from the experimental data. Residuals can be thought of as elements of variation unexplained by the fitted model. Since this is a form of error, the same general assumptions apply to the group of residuals that one typically uses for errors in general. One expects them to be normally and independently distributed with a mean of zero and some constant variance.

These are the assumptions behind ANOVA and classical regression analysis. This means that an analyst should expect a regression model to err in predicting a response in a random fashion; the model should predict values higher than actual and lower than actual with equal probability. In addition, the level of the error should be independent of when the observation occurred in the study, or the size of the observation being predicted, or even the factor settings involved in making the prediction.

The overall pattern of the residuals should be similar to the bell-shaped pattern observed when plotting a histogram of normally distributed data. Graphical methods are used to examine residuals. Departures from assumptions usually mean that the residuals contain structure that is not accounted for in the model. Identifying that structure and adding a term representing it to the original model leads to a better model. Any graph

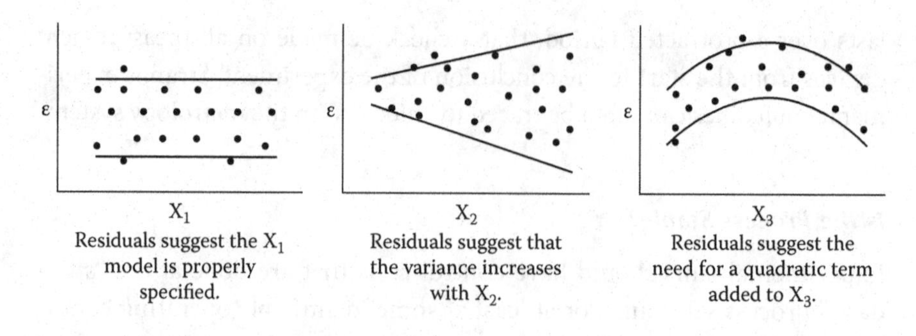

FIGURE 10.10
Residual types.

suitable for displaying the distribution of a set of data is suitable for judging the normality of the distribution of a group of residuals. The three most common types are: histograms, normal probability plots, and dot plots. Shown in Figure 10.10 are examples of dot plot results.

Interactions

An interaction occurs when the effect of one input factor on the output depends upon the level of another input factor. Refer to the diagrams in Figure 10.11.

Interactions can be readily examined with full factorial experiments. Often, interactions are lost with fractional factorial experiments. The preferred DoE approach examines (screens) a large number of factors with highly fractional experiments. Interactions are then explored or additional levels examined once the suspected factors have been reduced. Often, a full factorial or three-level fractional factorial trial (giving some interactions) is used in the follow-up experiment.

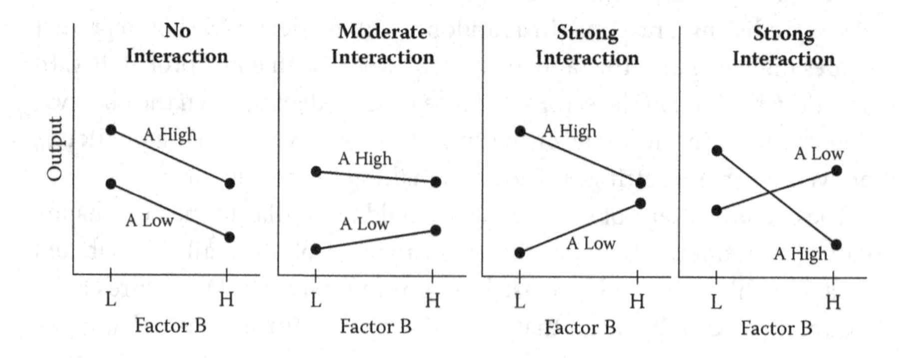

FIGURE 10.11
Examples of interactions.

Interaction Case Study

A simple 2 × 2 factorial experiment (with replication) was conducted in the textile industry. The response variable was ends down/thousand spindle hours (ED/MSH). The independent factors were relative humidity (RH) and ion level (the environmental level of negative ions). Both of these factors were controllable. A low ED/MSH is desirable since fewer thread breaks means higher productivity.

An ANOVA showed the main effects were not significant, but the interaction effects were highly significant. Consider the data table and plots in Figure 10.12.

MAIN EFFECTS	
FACTOR	ED/MSH
RH 37%	7.75
RH 41%	7.85
LOW ION	7.9
HIGH ION	7.7

INTERACTION EFFECTS		
	RH 37%	RH 41%
LOW ION	7.2	8.3
HIGH ION	8.6	7.1

The above interaction plot demonstrates that if the goal was to reduce breaks, an economic choice could be made between low ion/low RH and high ion/high RH.

Categories of Experimental Designs

Three-Factor, Three-Level Experiment

Often a three-factor experiment is required after screening a larger number of variables. These experiments may be full or fractional factorial.

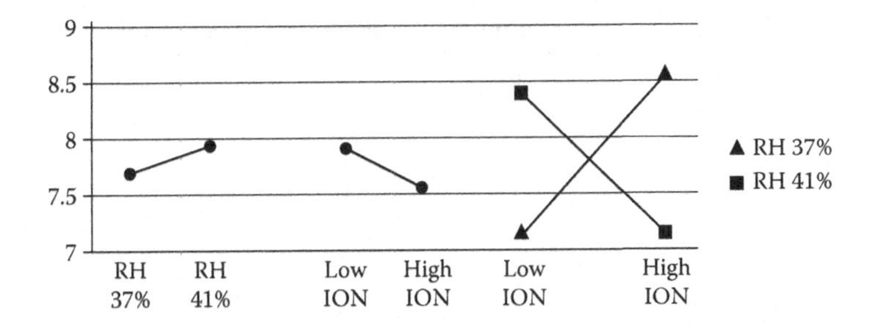

FIGURE 10.12
Relative humidity and ion level effects on ED/MSH.

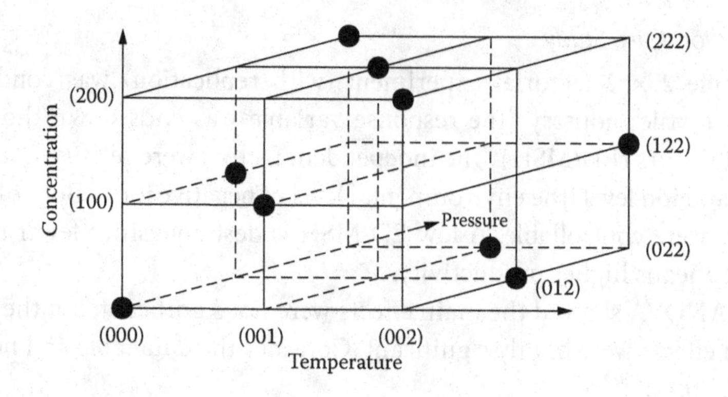

FIGURE 10.13
Example of a 113 fractional factorial design, three factors, three levels.

Shown in Figure 10.13 is a 113 fractional factorial design. Generally the (–) and (+) levels in two-level designs are expressed as 0 and 1 in most design catalogues. Three-level designs are often represented as 0, 1, and 2.

From the design catalogue test plan (plan 3, columns 1, 2, 4), the selected fractional factorial experiment looks like that shown in Table 10.2.

Randomized Block Plans

In comparing a number of factor treatments, it is desirable that all other conditions be kept as nearly constant as possible. The required number of tests may be too large to be carried out under similar conditions. In such cases, we may be able to divide the experiment into blocks, or planned

TABLE 10.2

Fractional Factorial Experiment

Experiment	Concentration	~Pressure	Temperature
1	0	0	0
2	0	1	2
3	0	2	1
4	1	0	1
5	1	1	0
6	1	2	2
7	2	0	2
8	2	1	1
9	2	2	0

TABLE 10.3

Randomized Incomplete Block Design

	Treatment			
Block (Days)	**A**	**B**	**C**	**D**
1	−5	Omitted	−18	−10
2	Omitted	−27	−14	−5
3	−4	−14	−23	Omitted
4	−1	−22	Omitted	−12

homogeneous groups. When each group in the experiment contains exactly one measurement on every treatment, the experimental plan is called a randomized block plan. For example, an experimental scheme may take several days to complete. If we expect some biasing differences among days, we might plan to measure each item on each day, or to conduct one test per day on each item. A day would then represent a block. A randomized incomplete block (tension response) design is shown in Table 10.3.

Only treatments A, C, and D are run on the first day, with B, C, and D on the second day, etc. In the whole experiment, note that each pair of treatments, such as BC, occurs twice together. The order in which the three treatments are run on a given day follows a randomized sequence.

Another randomized block design for air permeability response is shown in Table 10.4.

Latin Square Designs

A Latin square plan is often useful when it is desirable to allow for two sources of nonhomogeneity in the conditions affecting test results. Such designs were originally applied in agriculture when the two sources of

TABLE 10.4

Randomized Block Design for Air Permeability Response

	Fabric Types			
	I	**II**	**III**	**IV**
Chemical Applications	B(15.1)	D(11.6)	A(15.4)	C(9.9)
	C(12.2)	C(13.1)	B(16.3)	D(9.4)
A, B, C, D	A(19.0)	B(17.6)	D(16.0)	B(8.6)
	D(11.5)	A(13.0)	C(10.8)	A(11.5)

nonhomogeneity were the two directions on the field. The square was literally a plot of ground.

In Latin square designs a third variable, the experimental treatment, is then applied to the source variables in a balanced fashion. The Latin square plan is restricted by two conditions:

- The number of rows, columns, and treatments must be the same.
- There should be no expected interactions between row and column factors, since these cannot be measured. If there are, the sensitivity of the experiment is reduced.

A Latin square design is essentially a fractional factorial experiment that requires less experimentation to determine the main treatment results. Consider the 5 × 5 Latin square in Table 10.5.

In the design in Table 10.5, five automobiles and five carburetors are used to evaluate gas mileage by five drivers (A, B, C, D, and E). Note that only 25 of the potential 125 combinations are tested. Thus, the resultant experiment is a one-fifth fractional factorial. Similar 3 × 3, 4 × 4, and 6 × 6 designs may be utilized.

Graeco-Latin Designs

Graeco-Latin square designs are sometimes useful to eliminate more than two sources of variability in an experiment. A Graeco-Latin design is an extension of the Latin square design, but one extra blocking variable is added for a total of three blocking variables. Consider the following 4 × 4 Graeco-Latin design:

The output (response) variable could be gas mileage by four drivers (A, B, C, D). (See Table 10.6.)

TABLE 10.5

Graeco-Latin Carburetor Type

Car	I	II	III	IV	V
1	A	B	C	D	E
2	B	C	D	E	A
3	C	D	E	A	B
4	D	E	A	B	C
5	E	A	B	C	D

TABLE 10.6

Graeco-Latin Design

	Carburetor Type			
Car	**I**	**II**	**III**	**IV**
1	Aα	Bβ	Cy	Dδ
2	Bδ	Ay	Dβ	Cα
3	Cβ	Dα	Aδ	By
4	Dy	Cδ	Bα	Aβ

Hyper-Graeco-Latin Designs

A hyper-Graeco-Latin square design permits the study of treatments with more than three blocking variables. Consider the 4 × 4 hyper-Graeco-Latin design in Table 10.7. The output (response) variable could be gas mileage by four drivers (A, B, C, D).

Plackett-Burman Designs

Plackett-Burman (PB) (1946)[6] designs are used for screening experiments. PB designs are very economical. The run number is a multiple of 4 rather than a power of 2. PB geometric designs are two-level designs with 4, 8, 16, 32, 64, and 128 runs and work best as screening designs. Each interaction effect is confounded with exactly one main effect. All other two-level PB designs (12, 20, 24, 28, etc.) are nongeometric designs. In these designs a two-factor interaction will be partially confounded with each of the other main effects in the study. Thus, the nongeometric designs are essentially "main effect designs," when there is reason to believe any interactions are of little practical importance. A PB design in 12 runs, for

TABLE 10.7

Hyper-Graeco-Latin Designs

	Carburetor Type			
Car	**I**	**II**	**III**	**IV**
1	AαMΦ	BβNX	CyOψ	DδPΩ
2	BδNΩ	AyMΨ	DβPX	CαOΦ
3	CβOX	DαPΦ	AδMΩ	ByNΨ
4	DyPΨ	CδOΩ	BαNΦ	AβMX

TABLE 10.8

Plackett-Burman Nongeometric Design (12 Runs/11 Factors)

	Factors											
Exp.	**X1**	**X2**	**X3**	**X4**	**X5**	**X6**	**X7**	**X8**	**X9**	**X10**	**X11**	**Results**
1	+	+	+	+	+	+	+	+	+	+	+	
2	−	+	−	+	+	+	−	−	−	+	−	
3	−	−	+	−	+	+	+	−	−	−	+	
4	+	−	−	+	−	+	+	+	−	−	−	
5	−	+	−	−	+	−	+	+	+	−	−	
6	−	−	+	−	−	+	−	+	+	+	−	
7	−	−	−	+	−	−	+	−	+	+	+	
8	+	−	−	−	+	−	−	+	−	+	+	
9	+	+	−	−	−	+	−	−	+	−	+	
10	+	+	+	−	−	−	+	−	−	+	−	
11	−	+	+	+	−	−	−	+	−	−	+	
12	+	−	+	+	+	−	−	−	+	−	−	

example, may be used to conduct an experiment containing up to 11 factors. See Table 10.8.

With a 20-run design, an experimenter can do a screening experiment for up to 19 factors. As many as 27 factors can be evaluated in a 28-run design.

Hadamard Matrices of Order 2n

No. of Trials	
8	+ + + − + − −
16	+ + + − + − + + − − + − − −
32	+ + + + + − − + + − + − − + − − − − + − + − + + + − + + − − −
64	+ + + + + + − + − + − + + − − + + − + + + − + + − + − − + − −
	+ + + − − − + − + + + + − − + − + − − − + + − − − − + − − − − −
128	+ + + + + + + − + − + − + − − + + − − + + + − + + + − + − − + −
	+ + − − − + + − + + + + − + + − + − + + − + + − − + − − + − − − +
	+ + − − − − + − + + + + + − − + − + − + + + − − + + − + − − − + −
	− + + + + − − − + − + − − − − + + − − − − − + − − − − −

Other Hadamard Matrices Not of Order 2n

Hadamard matrices other than the 2*n* matrices also exist; however, they can only generate resolution III designs. The following are some useful vectors:

12	+ + − + + + − − − + −
20	+ + − − + + + + − + − + − − − − + + −

```
24    + + + + - + - + + - - + + - - + - + - - - -
44    + + - - + - + - - + + + - + + + + + - - - + - + + + - - - - - + - - - +
      + - + - + + -
```

Setting up the (8 × 8) Hadamard Matrix from the Vector

A. The proper vector is put in a column form.

```
+
+
+
-
+
-
-
```

B. The vector is permuted six (T – 2) times.

```
+  -  -  +  -  +  +
+  +  -  -  +  -  +
+  +  +  -  -  +  -
-  +  +  +  -  -  +
+  -  +  +  +  -  -
-  +  -  +  +  +  -
-  -  +  -  +  +  +
```

C. A set of minus (–) signs is added at the bottom.

```
+  -  -  +  -  +  +
+  +  -  -  +  -  +
+  +  +  -  -  +  -
-  +  +  +  -  -  +
+  -  +  +  +  -  -
-  +  -  +  +  +  -
-  -  +  -  +  +  +
-  -  -  -  -  -  -
```

D. A column of plus (+) signs is added to the left, and the columns are numbered.

0	1	2	3	4	5	6	7
+	+	-	-	+	-	+	+
+	+	+	-	-	+	-	+
+	+	+	+	-	-	+	-
+	-	+	+	+	-	-	+
+	+	-	+	+	+	-	-
+	-	+	-	+	+	+	-
+	-	-	+	-	+	+	+
+	-	-	-	-	-	-	-

Using the Hadamard (8 × 8) Matrix to Evaluate One, Two, and Three Factors

One variable, each treatment combination replicated four times:

		A							Treatment
Trial	**0**	**1**	**2**	**3**	**4**	**5**	**6**	**7**	**Combinations**
1	+	+	−	−	+	−	+	+	a
2	+	+	+	−	−	+	−	+	a
3	+	+	+	+	−	−	+	−	a
4	+	−	+	+	+	−	−	+	(1)
5	+	+	−	+	+	+	−	−	a
6	+	−	+	−	+	+	+	−	(1)
7	+	−	−	+	−	+	+	+	(1)
8	+	−	−	−	−	−	−	−	(1)

Two variables, each treatment combination replicated twice:

		A	B		−AB				Treatment
Trial	**0**	**1**	**2**	**3**	**4**	**5**	**6**	**7**	**Combinations**
1	+	+	−	−	+	−	+	+	a
2	+	+	+	−	−	+	−	+	ab
3	+	+	+	+	−	−	+	−	ab
4	+	−	+	+	+	−	−	+	b
5	+	+	−	+	+	+	−	−	a
6	+	−	+	−	+	+	+	−	b
7	+	−	−	+	−	+	+	+	(1)
8	+	−	−	−	−	−	−	−	(1)

Three variables:

		A	B	C	−AB	−BC	ABC	−AC	Treatment
Trial	**0**	**1**	**2**	**3**	**4**	**5**	**6**	**7**	**Combinations**
1	+	+	−	−	+	−	+	+	a
2	+	+	+	−	−	+	−	+	ab
3	+	+	+	+	−	−	+		- abc
4	+	−	+	+	+	−	−	+	bc
5	+	+	−	+	+	+	−	−	ac
6	+	−	+	−	+	+	+	−	b
7	+	−	−	+	−	+	+	+	c
8	+	−	−	−	−	−	−	−	(1)

Taguchi Designs

The Taguchi philosophy emphasizes two tenets: (1) Reduce the variation of a product or process (improve quality), which reduces the loss to society, and (2) use a proper development strategy to intentionally reduce variation.

One development approach is to identify a parameter that will improve some performance characteristic. A second approach is to identify a less expensive, alternative design, material, or method that will provide equivalent or better performance. Orthogonal arrays (OAs) have been designed to facilitate the development test strategy.

Orthogonal Arrays Degrees of Freedom

Let df = degrees of freedom and k = number of factor levels.

Examples:

For factor A, $df_A = k_A - 1$.
For factor B, $df_B = k_B - 1$.
For A × B interaction, $df_{AB} = df_A \times df_B$.
$df_{min} = Edf$ all factors + Edf all interactions of interest.

Two-Level OAs OAs can be used to assign factors and interactions. The simplest OA is an L4 (four trial runs). (See Table 10.9.)

Factors A and B can be assigned to any two of the three columns. The remaining column is the interaction column. Assume a trial is conducted with two repeat runs for each trial. Let us assign factor A to column 1 and

TABLE 10.9

An L4 OA Design

	Columns		
Trial	**1**	**2**	**3**
1	1	1	1
2	1	2	2
3	2	1	2
4	2	2	1

TABLE 10.10

An L4 OA Design with Data

	Column			Raw Data		*I* Simplified		(Simplified)² *I*	
Trial	**1**	**2**	**3**	(y)		(y, −40)		(y, −40)²	
1	1	1	1	44	47	4	7	16	49
2	1	2	2	43	45	3	5	9	25
3	2	1	2	41	42	1	2	1	4
4	2	2	1	48	49	8	9	64	81
					Totals	39		249	

factor B to column 2. The interaction is then assigned to column 3. (See Table 10.10.)

Factor A $\quad EA_1 = 4 + 7 + 3 + 5 = 19 \quad EA_2 = 1 + 2 + 8 + 9 = 20$

Factor B $\quad EB_1 = 4 + 7 + 1 + 2 = 14 \quad EB_2 = 3 + 5 + 8 + 9 = 25$

A × B interaction $E3_1 = 4 + 7 + 8 + 9 = 28$

$$E3_2 = 3+5+1+2 = 112$$

$$SS_T = (16 + 49 + 9 + 25 + 1 + 4 + 64 + 81) - (^39) = 58.875$$

$ss = (20 - 19)^2 = 0.125 \quad SS = (25 - 14)2 = 15.125\ SS \quad (28 - 11)2 = 6.125$

A8 $\quad$ B – SS_3 8

$$SS_8 = SS_T - SS_A - SS_B - SS_3 = 58.875 - 0.125 - 15.125 - 36.125 = 7.5$$

A standard ANOVA table can now be set up to determine factor significance at a selected alpha value. ANOVA was previously reviewed. The assignment of factors and interactions to arrays is reviewed in the following discussion. We will consider an L4 OA first. The L4 linear graph shows that if the two factors are assigned to columns 1 and 2, the interaction will be in column 3. The L4 triangular table shows that if the two factors are put in columns 1 and 3, the other point of the triangle for the interaction is in column 2. If the two factors are put in columns 2 and 3, the interaction will be found in column 1.

The next level of linear graphs is for an L8 OA. Two linear graphs available for an L8 experiment are shown in Figure 10.14. The linear graph

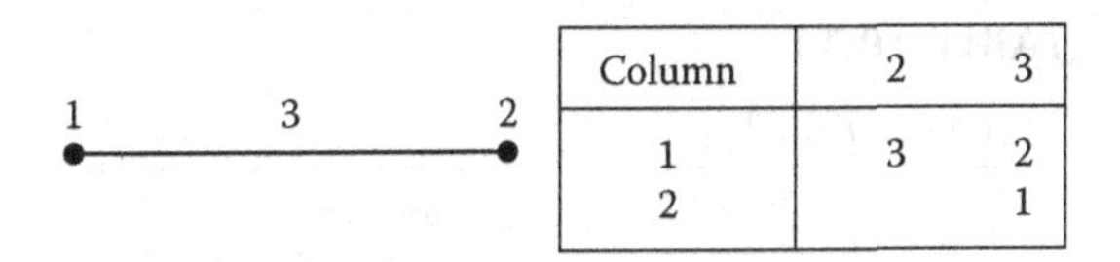

FIGURE 10.14
L8 linear graph.

in Figure 10.14 indicate that several factors can be assigned to different columns and several different interactions may be evaluated in different columns. If three factors (A, B, and C) are assigned, the L8 linear graph indicates the assignment to columns 1, 2, and 4 located at the vertices in the type A triangle. (See Figures 10.15 and 10.16.)

The L8 in Table 10.11 shows the B × C interaction (columns 2 and 4) takes place in column 6. The A× B× C interaction is located by finding the interaction of factor A and the B × C interaction. The B × C interaction is in column 6 (columns 2 and 4 intersection).

Factor	A × B × C = A × B × C
Column	1 6 7

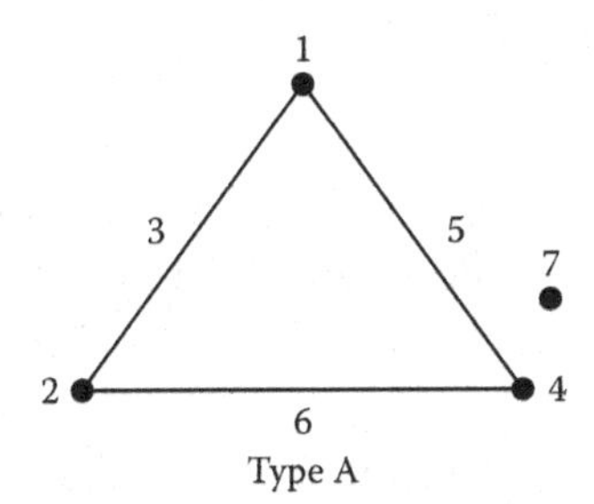

FIGURE 10.15
L4 linear graph.

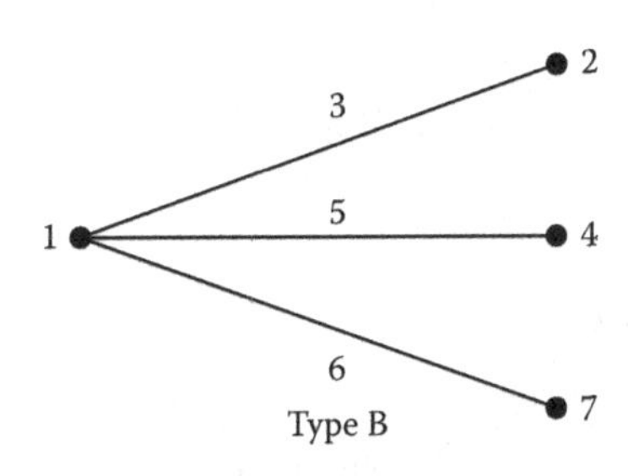

FIGURE 10.16
L4 triangular table.

TABLE 10.11

Triangular *T*-Table

	Column No.					
Column No.	**2**	**3**	**4**	**5**	**6**	**7**
1	3	2	5	4	7	6
2		1	6	7	4	5
3			7	6	5	4
4				1	2	3
5					3	2
6						1

The A × C interaction occurs in column 5 (columns 1 and 4 intersection).

Factor	B × A × C = A× B× C
Column	2 5 7

The A × B interaction occurs in column 3 (columns 1 and 2 intersection).

Factor	C × A × B = A × B × C
Column	4 3 7

The column assignments for the factors and their interactions are shown in Table 10.12. All main effects and all interactions can be estimated, which results in a high-resolution experiment. This is also a full factorial experiment.

The L8 in Table 10.13 shows the B × C interaction (columns 2 and 4) takes place in column 6. The A × B × C interaction is located by finding the interaction of factor A and the B × C interaction. The B × C interaction is in column 6 (columns 2 and 4 intersection).

Factor	A × B × C = A × B × C
Column	1 6 7

TABLE 10.12

Column Assignments for an L8 Linear Graph

Column No.						
1	2	3	4	5	6	7
A	B	A × B	C	A × C	B × C	A × B × C

TABLE 10.13

Triangular *T*-Table

	Column No.					
Column No.	**2**	**3**	**4**	**5**	**6**	**7**
1	3	2	5	4	7	6
2		1	6	7	4	5
3			7	6	5	4
4				1	2	3
5					3	2
6						1

The A × C interaction occurs in column 5 (columns 1 and 4 intersection).

Factor B × A × C = A × B × C
Column 2 5 7

The A × B interaction occurs in column 3 (columns 1 and 2 intersection).

Factor C × A × B = A × B × C
Column 4 3 7

The column assignment for the factors and their interactions is shown in Table 10.14. All main effects and all interactions can be estimated, which results in a high-resolution experiment. This is also a full factorial experiment.

Randomization

The order of running the various trials should include some form of randomization. Randomization protects the experimenter. The pattern of any unknown or uncontrollable effects will be evenly spread across the

TABLE 10.14

Column Assignments for an L8 Linear Graph

Column No.						
1	2	3	4	5	6	7
A	B	A × B	C	A × C	B × C	A × B× C

experiment if a random pattern of run order is established. This will prevent bias of the factors and interactions assigned to the experimental columns. The three most common forms of randomization follow.

Complete Randomization Complete or ideal randomization means that any trial has an equal chance of being selected for the first test run. Each of the remaining trials has an equal chance of being selected for the second test run, and so on for the rest of the trials. Random numbers can be obtained from a random number table or a random number generator. If replications of each trial are required, all random arrangements of the trials should be completed for the first trial observation before the next new random arrangement of trials. In many experiments, an analysis can be performed at the end of each complete set of trials.

Simple Repetition Simple repetition means that each trial has an equal random chance of being selected for the first test run, but once selected, all repetitions are run before the next trial is randomly selected. This approach is used when changing test setups is difficult or expensive.

Complete randomization within blocks is used when changing the setup for say, factor C, and may be difficult or expensive, but changing all other factors is relatively easy. The experiment could be completed in two halves or blocks. All C_1 trials could be randomly selected and run, and then all C_2 trials could be randomly selected.

Analysis of Experiment Data One of the basic properties of OAs is that total variation can be determined by summing the variation from all columns. The total sums of squares for the unassigned columns are equal to the error sums of squares.

Taguchi Robust Concepts

Dr. Genichi Taguchi, an engineer from Japan, learned DoE techniques and modified some of the terminologies and approaches to obtain a dramatic improvement in the quality of products. Since the early 1980s, his approach has found widespread use in achieving "robustness of design." The Japanese say, "An engineer who does not know design of experiments is only a half an engineer."

Dr. Taguchi's basic message is that the consistency of the product is very important. However, it is futile to try to achieve consistent output characteristics by controlling the variation in every variable. There are a few key variables (signal or noise factors) and their interaction in any process. When these factors are fixed at the right levels, they will make the product characteristic robust. That is, the process will be insensitive to variations in other input factors. The Taguchi approach focuses on controlling some inputs stringently to reduce product variation. An example may make the point clearer.

Example of the Taguchi Approach

In a paint shop at a Ford Motor Company plant, paint blistering on an automobile hood led to expensive rework. No one apparently had any clue why the blistering occurred. Engineers, operating personnel, and suppliers gathered to brainstorm and identified seven suspect factors as the likely causes of blistering: paint supplier, undercoat thickness, drying time for undercoat, paint thickness, drying time, temperature, and humidity. Only eight experimental runs were needed to study the effects of the seven variables. However, to have more confidence in the data, the experiment was conducted 5 times, a total of 40 runs. The experiment proved that blistering occurred most frequently when the undercoat thickness was higher. It was not necessary to control other variables as stringently. The experiment proved that aiming for consistency in product output is not only sound judgment but of economic necessity.

Achieving Design Robustness

Three types of design considerations are involved for any product or process:

1. **System design:** Includes the selection of parts, methods, and tentative product parameter values. Engineers and scientists are best equipped to handle system design (also called concept design).
2. **Parameter design:** The selection of nominal product and process operating levels to determine the optimum combinations. The levels should be chosen so as to make the output characteristics insensitive to (continued) variation in environmental factors (noise) over which there is little control. Parameter design is the most neglected aspect of product design. Quality professionals can make an important contribution in achieving quality without increasing cost via parameter design.
3. **Tolerance design:** The establishment of the permissible variation in the product and process to achieve a consistent output. Traditionally, tolerances are made stringent to achieve a better product.

Signal Factors

Signal factors are the factors that strongly influence the mean response. Signal factors normally have minimal influence on variation of the output response and are controllable. We should vary their level to adjust the mean. Choose a correct level for ones that introduce variability.

Noise Factors

Noise factors are those factors that influence the variation in the output. These may be controllable or noncontrollable. The controllable ones are varied during the experimentation to see which combination gives the highest (or lowest as desired) signal-to-noise ratio (S/N). In essence, choose the levels of controllable noise factors such that the S/N ratio is maximized (or minimized) and the output response is insensitive to the variation in uncontrollable noise factors. This is the key to achieving design robustness.

S/N Ratios

The S/N ratio is a calculation to quantify the effects (in dB) of variation in the controllable factors resulting in the variation of output. (There has been some criticism in the West regarding the reliability of S/N ratios.)

When higher results are desired:

$$\text{S/N (in dB)} = -10\ \text{LOG}\frac{1}{n}\left(\frac{1}{(Y_1)^2}+\frac{1}{(Y_2)^2}+\cdots+\frac{1}{(Y_n)^2}\right)$$

When lower results are desired:

$$\text{S/N (in dB)} = -10\ \text{LOG}\frac{1}{n}\left((Y_1)^2+(Y_2)^2+\cdots+(Y_n)^2\right)$$

where *n* refers to the number of observations of controllable factors within an experiment and *Y* denotes the output response for each experiment conducted.

Levels of the Controllable Factor

The input (controllable) factor should be chosen between the medium and high levels, since input variation will cause little output variation. Suggested reading: *The Taguchi Approach to Parametric Design* (Byrne and Taguchi, 1987).[2]

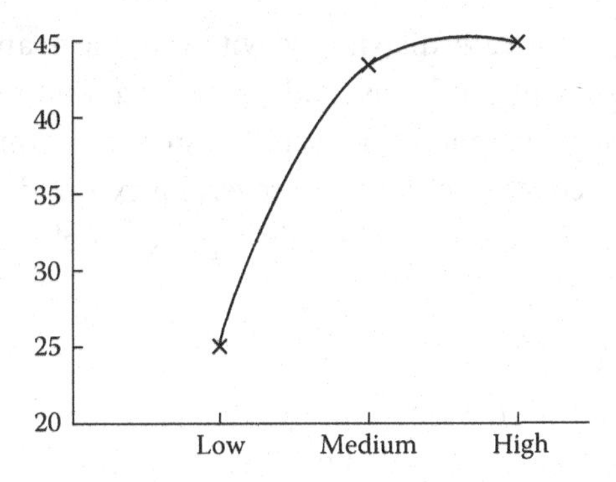

FIGURE 10.17
Signal-to-noise ratio.

Mixture Designs

In a mixture experiment, the independent factors are proportions of different components of a blend. The fact that the proportions of the different factors must sum to 100% complicates the design as well as the analysis of mixture experiments.

When the mixture components are subject to the constraint that they must sum to 1, there are standard mixture designs for fitting standard models, such as simplex-lattice designs and simple-centroid designs. When mixture components are subject to additional constraints, such as a maximum and/or minimum value for each component, designs other than the standard mixture designs, referred to as constrained-mixture designs or extreme-vertices designs, are appropriate.

In mixture experiments, the measured response is assumed to depend only on the relative proportions of the ingredients or components in the mixture and not on the amount of the mixture. The amount of the mixture could also be studied as an additional factor in the experiment. However, this is an example of where mixture and process variables are treated together. In mixture problems the purpose of the experiment is to model the blending surface with some form of mathematical equation so that:

- Predictions of the response for any mixture or combination of the ingredients can be made empirically
- Some measure of the influence on the response of each component singly and in combination with other components can be obtained

The usual assumptions made for factorial experiments are also assumed for mixture experiments. In particular, the errors are assumed to be independent and identically distributed with zero mean and common variance. Another assumption that is made, similar to that made for factorial designs, is that the true underlying response surface is continuous over the region being studied. While there are several mixture design alternatives as noted above, this handbook will only explore simplex-lattice designs.

Simplex-Lattice Designs

A $\{q, m\}$ simplex-lattice design for q components consists of points defined by the following coordinate settings: The proportions assumed by each component take the $m + 1$ equally spaced values from 0 to 1.

$$X = 0, 1/m, 2/m, \ldots, 1 \quad \text{for } I = 1, 2, \ldots, q$$

and all possible combinations (mixtures) of the proportions from this equation are used.

Note that the standard simplex-lattice is a boundary point design; that is, with the exception of the overall centroid, all the design points are on the boundaries of the simplex. When one is interested in prediction in the interior, it is highly desirable to augment the simplex-type designs with interior design points.

Consider a three-component mixture where the number of equally spaced levels for each component is four (i.e., $X1 = 0, 0.333, 0.667, 1$). In this example, then, $q = 3$ and $m = 3$. If one considers all possible blends of the three components with these proportions, then the {3, 3} simplex-lattice contains the 10 blending coordinates listed below. The experimental region and the distribution of design run over the simplex region are shown in Figure 10.18. There are a total of 10 design runs for the {3, 3} simplex-lattice design. The number of design points in the simplex-lattice is $(q + m - 1)!/(m!(q - 1)!)$.

Now consider the form of the polynomial model that one might fit to the data from a mixture experiment. Due to the restriction $X + X2 + \ldots + Xq = 1$, the form of the regression function is somewhat different than the traditional polynomial fit and is often referred to as the canonical polynomial.

The canonical polynomial is derived using the general form of the regression function that can be fit to data collected at the points of a $\{q, m\}$ simplex-lattice design and substituting into this the dependence relationship among the X_i terms. The number of terms in the $\{q, m\}$ polynomial is $(q + m - 1)!/m!(q - 1)!$. This number is equal to the number of points that make up the associated $\{q, m\}$ simplex-lattice design.

X1	X2	X3	Yield
0	0		
0	0.333	0.667	
0	0.667	0.333	
0	1	0	
0.333	0	0.667	
0.333	0.333	0.333	
0.333	0.667	0	
0.667	0	0.333	
0.667	0.333	0	
1	0	0	

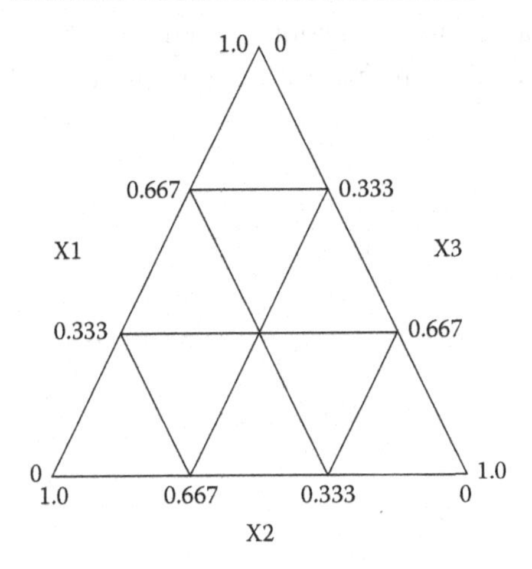

FIGURE 10.18
Configuration of design runs for a {3, 3} simplex-lattice design.

In general, the canonical forms of the mixture models are as follows:

$$E(Y)=\sum_{i=1}^{q}\beta_i x_j$$

$$E(Y)=\sum_{i=1}^{q}\beta_i x_i+\sum_{i=1}^{q}\sum_{i<j}^{q}\beta_{ij}x_i x_j$$

$$E(Y)=\sum_{i=1}^{q}\beta_i x_i+\sum_{i=1}^{q}\sum_{i<j}^{q}\beta_{ij}x_i x_j+\sum_{i=1}^{q}\sum_{i<j}^{q}\delta_{ij}x_i x_j(x_i - x_j+\sum_{k=1}^{q}\sum_{j<k}^{q}\sum_{i=j}^{q}\beta_{ijk}x_i x_j x_k$$

The terms in the canonical mixture polynomials have simple interpretations. Geometrically, the parameter 3_i in the above equations represents the expected response to the pure mixture $X_i = 1$, $X_i = 0$, ij, and is the height of the mixture surface at the vertex $X_i = 1$. The portion of each of the above polynomials given by

$$\sum_{i=1}^{q} \beta_i X_i$$

is called the linear blending portion. When blending is strictly additive, then the linear model form above is an appropriate model.

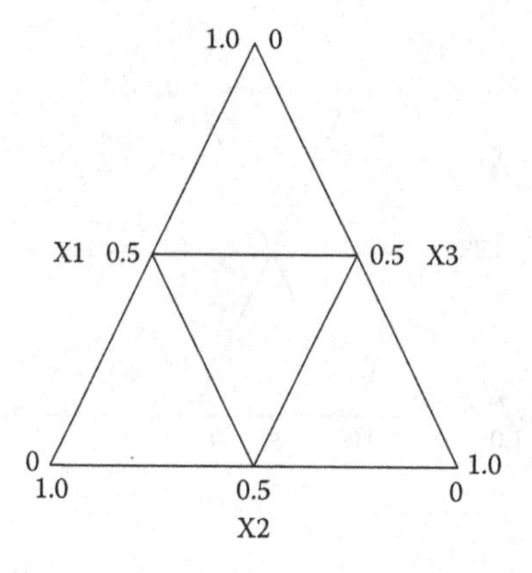

Simplex-Lattice Design Example

The following example is from Cornell (1990)[3] and consists of a three-component mixture problem. The three components are polyethylene ($X1$), polystyrene ($X2$), and polypropylene ($X3$), which are blended together to form fiber that will be spun into yarn. The product developers are only interested in the pure and binary blends of these three materials.

The response variable of interest is yarn elongation in kilograms of force applied. A {3, 2} simplex-lattice design is used to study the blending process. The design and the observed responses are listed in Table 10.15. There were two replicate observations run at each of the pure blends. There were three replicate observations run at the binary blends. There are a total of 15 observations with six unique design runs. A {3, 2} simplex-lattice design is shown below.

TABLE 10.15

Design Runs for the {3, 2} Simplex-Lattice

X1	*X2*	*X3*	Observed Values
0	0	1	16.8, 16.0
0	0.5	0.5	10.0, 9.7, 11.8
0	1	0	8.8, 10.0
0.5	0	0.5	17.7, 16.4, 16.6
0.5	0.5	0	15.0, 14.8, 16.1
1	0	0	11.0, 12.4

The design runs listed in the above table are in standard order. The actual order of the 15 treatment runs was completely randomized.

An analysis was performed using SAS software JMP 3.2 and is summarized in Table 10.16.

Summary of Fit	
R Square	0.951356
R Square Adj.	0.924331
Root Mean Square Error	0.85375
Mean of Response	13.54
Observations (or Sum wgts.)	15

Analysis of Variance

Source	*DF*	Sum of Squares	Mean Square	*F*-Ratio
Model	5	128.29600	25.6592	35.2032
Error	9	6.56000	0.7289	
C total	14	134.85600		

TABLE 10.16

JMP Output for {3,2} Simplex-Lattice Design

Parameter		Estimates	
Term	Estimate	Standard Error	*t*-Ratio
X_1	11.7	0.603692	19.38
X_2	9.4	0.603692	15.57
X_3	16.4	0.603692	27.17
$(X_2 \bullet X_1)$	19.0	2.608249	7.28
$(X_3 - X_1)$	11.4	2.608249	4.37
$(X_3 \bullet X_2)$	-9.6	2.608249	-3.68

Under the parameter estimates section of the output are the individual *t*-tests for each of the parameters in the model. The three cross-product terms are $(X_1 \bullet X_2)(X_3 \bullet X_i)(X_3 \bullet X_2)$, indicating a significant quadratic fit. The fitted quadratic mixture model is:

$$Y = 11.7X_1 + 9.4X_2 + 16.4x_3 + 19.0X_1\ X_2 + 11.4X_1X_3 - 9.6X_2X_3$$

Since $b_3 > b_1 > b_2$, one can conclude that component 3 (polypropylene) produces yarn with the highest elongation. Additionally, since b_{12} and b_{13} are positive, blending components 1 and 2 or components 1 and 3 produces higher elongation values than would be expected just by averaging the elongations of the pure blends. This is an example of "synergistic" blending effects. Components 2 and 3 have antagonistic blending effects because b_{23} is negative.

Steepest Ascent/Descent

Most of the discussion in this element is directed at steepest ascent but is applicable to steepest descent when lower values are desirable. In an experimental problem the contours are usually not known, although in some situations the equations may be known. The object is to move from some initial point *P* in the (*X X*2) space within the contour system. Notice in Figure 10.19 the small circle drawn around *P*. Consider the directional line arrow

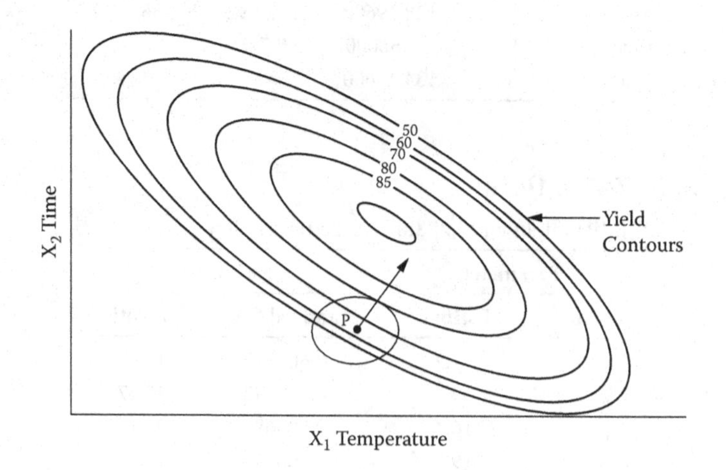

FIGURE 10.19
Illustration of steepest ascent methodology.

obtained by joining the center of the circle (point *P*) to the point on the circumference of the circle where it just touches one of the response contours.

As the diameter of the circle is decreased, the directional line is said to point in the direction of steepest ascent on the surface at point *P*. This direction achieves the greatest rate of increase per unit of distance traveled in contour space. It can be shown that this directional arrow can be followed by making changes that are proportional to partial derivatives calculated at *P*. This method is often used for obtaining the maximum or minimum of known functions.

When the function is unknown, we use an experimental procedure with observations taken, but subject to error, that was first introduced by Box and Wilson (1951).[1] In this procedure, a small experimental design is performed around point *P* and the derivatives are estimated numerically from the experimental observations. (See Figure 10.19.)

In general, the path of steepest ascent is perpendicular to the contour lines if the space is measured in the same relative units chosen to scale the design.

Simplex Approaches to Steepest Ascent

For a simplex approach to steepest ascent, see Figure 10.20.

1. Requires one more point than the number of independent variables.
2. Move away from the lowest response point through the midpoint of the other two points to an equal distance on the other side.
3. Repeat this cycle, dropping the lowest point at each step.

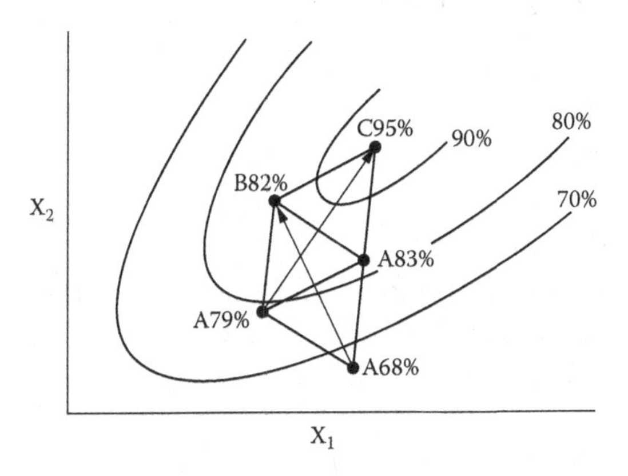

FIGURE 10.20
A simplex approach to steepest ascent.

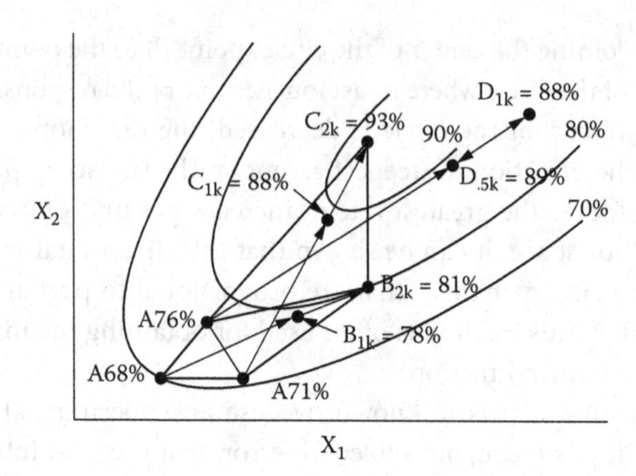

FIGURE 10.21
A modified simplex approach to steepest ascent.

Note: This can be expanded to any number of variables, projecting away from the lowest, through the centroid of the remaining points.

For a modified simplex approach to steepest ascent, see Figure 10.21. Run regular simplex point ($k = 1$), then:

1. If $k = 1$ gives a new best, run $k = 2$ point, use better of $k = 1$ or $k = 2$.
2. If $k = 1$ gives better than only the current worst, run $k = 0.5$ point.
3. If $k = 0.5$ point is even worse than the current worst, then quit or restart a smaller simplex around $k = -0.5$ point.

Exchange only one point at each stage.

Response Surfaces

See Figures 10.22 and 10.23.

Experimental Equations Are Pictures

The equation represents a response line, plane, or surface of the factors being evaluated. (See Figure 10.24.) The *S*, *C*, and *T* values depend on the size of the slopes, curves, and twists, respectively.

Central Composite Designs

A Box-Wilson central composite design, commonly called a central composite design, contains an embedded factorial or fractional factorial

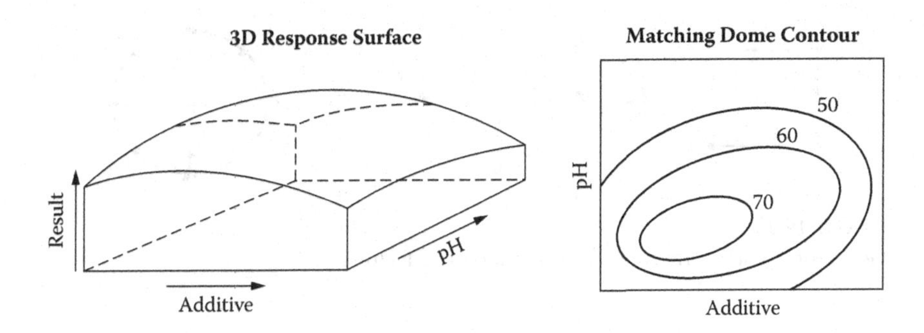

FIGURE 10.22
Comparison of 3D and 2D response surfaces.

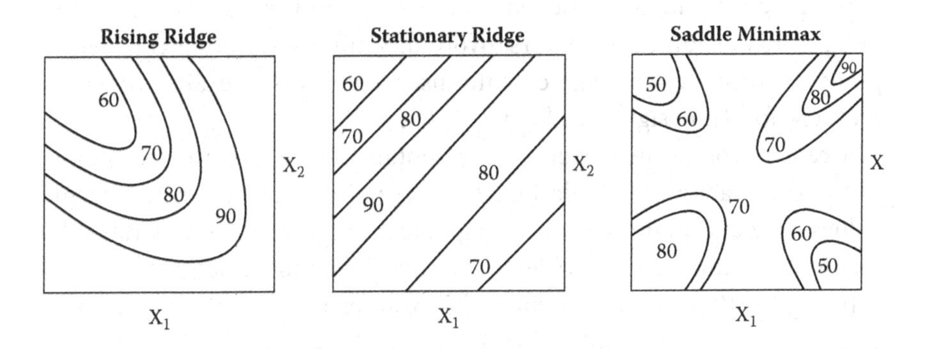

FIGURE 10.23
Various contour examples.

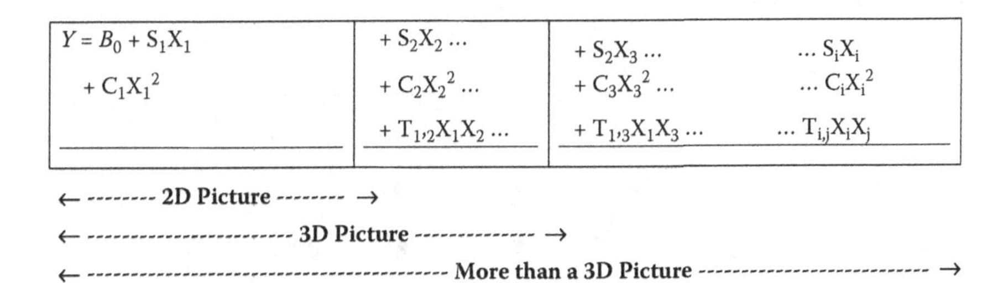

FIGURE 10.24
Description of equation components.

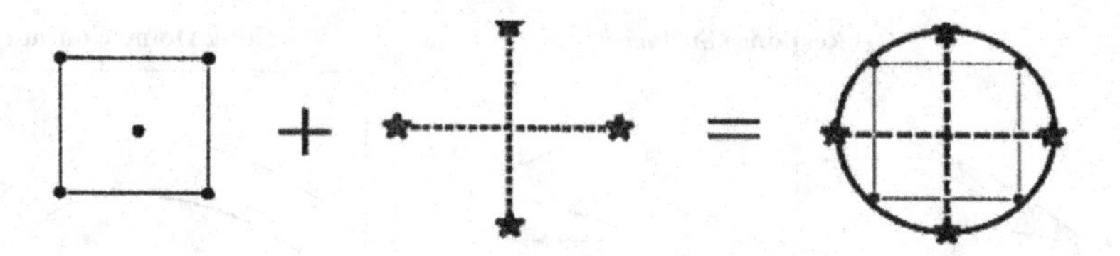

FIGURE 10.25
Generation of a central composite design for two factors.

matrix with center points augmented with a group of "star points" that allow estimation of curvature. If the distance from the center of the design space to a factorial point is ±1 unit for each factor, the distance from the center of the design space to a star point is ± *a*, where l*a*l > '1.

The precise value of (*a*) depends on certain properties desired for the design and the number of factors involved. Similarly, the number of center point runs the design must contain also depends on certain properties required for the design. (See Figure 10.25.)

A central composite design always contains twice as many star points as there are factors in the design. The star points represent new extreme values (low and high) for each design factor. Figure 10.26 illustrates the relationships among three varieties of central composite designs.

In Figure 10.26, note that the CCC explores the largest process space and CCI explores the smallest process space. Both CCC and CCI are rotatable designs, but the CCF is not. In the CCC design, the design points describe a circle circumscribed about the factorial square. For three factors, the CCC design points describe a sphere around the factorial cube. Table 10.17 summarizes the properties of the three varieties of central composite designs.

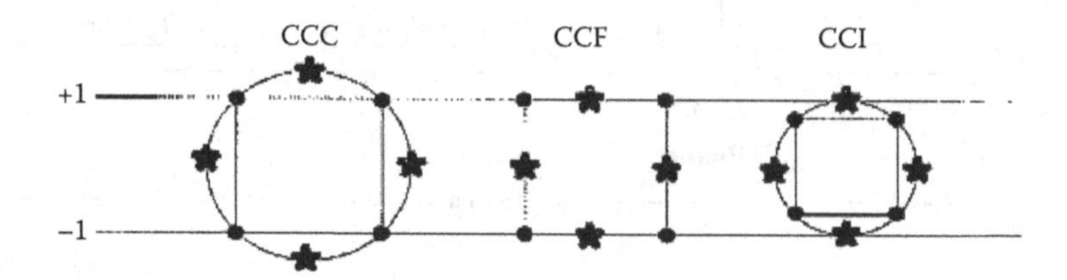

FIGURE 10.26
Comparison of the three types of central composite designs.

TABLE 10.17

Central Composite Designs

Design Type	Comments
Circumscribed CCC	CCC designs are the original form of the central composite design. The star points are at some distance *a* from the center based on the properties desired for the design and the number of factors in the design. The star points establish new extremes for the low and high settings for all factors. Figure 10.27 illustrates a CCC design. These designs have circular, spherical, or hyperspherical symmetry and require five levels for each factor. Augmenting an existing factorial or resolution V fractional factorial design with star points can produce this design.
Inscribed CCI	For those situations where the limits specified for factor settings are truly limits, the CCI design uses the factor settings as the star points and creates a factorial or fractional factorial matrix within those limits (in other words, a CCI design is a scaled-down CCC design with each factor level of the CCC design divided by *a* to generate the CCI design). This design also requires five levels of each factor.
Face-centered CCF	In this design the star points are at the center of each face of the factorial space, so $a = \pm I$. This design requires three levels of each factor. Augmenting an existing factorial or resolution V design with appropriate star points can also produce this design.

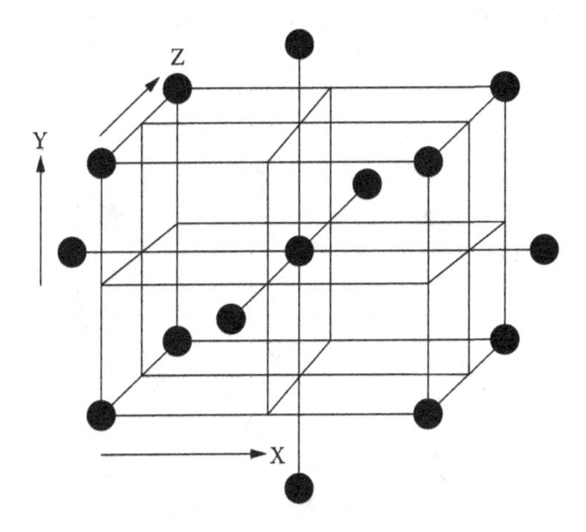

FIGURE 10.27
A CCC design for three factors (15 runs shown).

Determining α in Central Composite Designs

To maintain rotatability, the value of α depends on the number of experimental runs in the factorial portion of the central composite design:

$$[\text{number of factorial runs}]^{1/4}$$

If the factorial is a full factorial, then k = the number of factors.

Box-Behnken Designs

The Box-Behnken design is an independent quadratic design in that it does not contain an embedded factorial or fractional factorial matrix. In this design the treatment combinations are at the midpoints of edges of the process space and at the center. These designs are rotatable (or near rotatable) and require three levels of each factor. These designs have limited capability for orthogonal blocking compared to the central composite designs. Figure 10.28 illustrates a Box-Behnken design for three factors.

The geometry of this design suggests a sphere within the process space, such that the surface of the sphere protrudes through each face with the surface of the sphere tangential to the midpoint of each edge of the space.

Choosing a Response Surface Design

Table 10.18 contrasts the structures of four common quadratic designs one might use when investigating three factors. The table combines CCC and CCI designs because they are structurally identical. For three factors, the

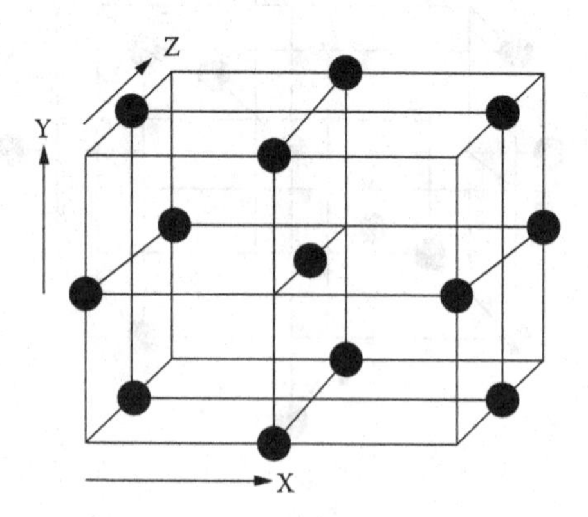

FIGURE 10.28
A Box-Behnken design for three factors (13 runs shown).

TABLE 10.18

Structures of CCC (CCI), CCF, and BB Three-Factor

CCC (CCI)				CCF				Box-Behnken			
Rep.	**X1**	**X2**	**X3**	**Rep.**	**XI**	**X2**	**X3**	**Rep.**	**X1**	**X2**	**X3**
1	−1	−1	−1	1	−1	−1	−1	1	−1	−1	0
1	1	−1	−1	1	1	−1	−1	1	1	−1	0
1	−1	1	−1	1	−1	1	−1	1	−1	1	0
1	1	1	−1	1	1	1	−1	1	1	1	0
1	−1	−1	1	1	−1	−1	1	1	−1	0	−1
1	1	−1	1	1	1	−1	1	1	1	0	−1
1	−1	1	1	1	−1	1	1	1	−1	0	1
1	1	1	1	1	1	1	1	1	1	0	1
1	−1.682	0	0	1	−1	0	0	1	0	−1	−1
1	1.682	0	0	1	1	0	0	1	0	1	−1
1	0	−1.682	0	1	0	0	0	1	0	−1	1
1	0	1.682	0	1	0	0	0	1	0	1	1
1	0	0	−1.682	1	0	−1	−1	3	0	0	0
1	0	0	1.682	1	0	1	1				
6	0	0	0	6	0	0	0				
Total 20 runs (15 minimum)				Total 20 runs (15 minimum)				Total 15 runs (13 minimum)			

Box-Behnken design offers some advantage in requiring a fewer number of runs. Central composite designs require fewer treatments than the Box-Behnken when the factor count is 5 or higher.

Table 10.19 summarizes properties of the classical quadratic designs. Use this table as a guideline when attempting to choose among the four available designs. Table 10.20 compares the number of runs required for CC and BB designs.

Evolutionary Operations (EVOP)

Prior coverage of steepest ascent/descent designs emphasized a bold strategy for improvement. In contrast, EVOP emphasizes a conservative experimental strategy for continuous process improvement. Refer to Figure 10.29.

Tests are carried out in phase A until a response pattern is established. Then phase B is centered on the best conditions from phase A. This procedure is repeated until the best result is determined. When nearing a peak, switch to smaller step sizes or examine different variables. EVOP

TABLE 10.19

Summary of Properties of Classical Response Surface Designs

Design Type	Comment
CCC	CCC designs provide high-quality predictions over the entire design space, but require using settings outside the range originally specified for the factorial factors. When the possibility of running a CCC design is recognized before starting a factorial experiment, factor spacings can be reduced to ensure that plus or minus α for each coded factor corresponds to feasible (reasonable) levels. Five levels are required for each factor.
CCI	CCI designs use only points within the factor ranges originally specified, but do not provide the same high-quality prediction over the entire space compared to the CCC. Five levels are required for each factor.
CCF	CCF designs provide relatively high-quality predictions over the entire design space and do not require using points outside the original factor range. However, they give poor precision for estimating pure quadratic coefficients. They require three levels for each factor.
Box-Behnken	These designs require fewer treatment combinations than a central composite design in cases involving three or four factors. The Box-Behnken design is rotatable (or nearly so), but it contains regions of poor prediction quality like the CCI. Its "missing corners" may be useful when the experiment should avoid combined factor extremes. This property prevents a potential loss of data in those cases. Requires three levels for each factor.

can entail small incremental changes so that little or no process scrap is generated. Large sample sizes may be required to determine the appropriate direction of improvement. The method can be extended to more than two variables, using simple main effects experiment designs.

The EVOP approach does not recommend the formal calculation of the steepest ascent path, and a more informal experimental procedure seems appropriate. The experimenter naturally tends to change variables in the direction of expected improvement and thus follows an ascent path. In EVOP experimentation there are fewer considerations to be taken into

TABLE 10.20

Number of Runs Required by CC and BB Designs

Factors	CC	BB
2	13 (5 center points)	—
3	20 (6 center point runs)	15
4	30 (6 center point runs)	27
5	33 fractional or 52 full	46
6	54 fractional or 91 full	54

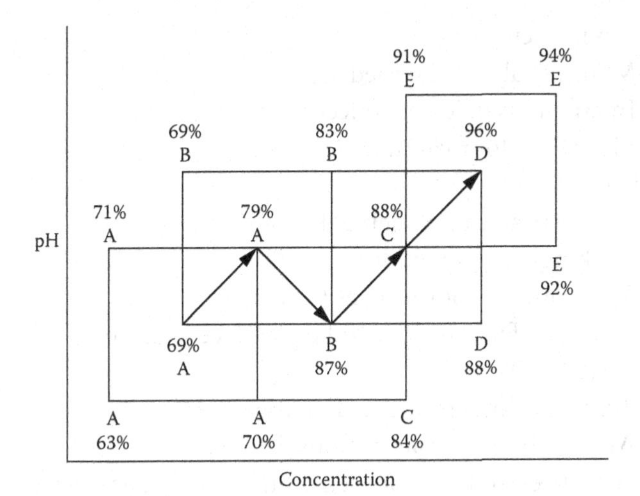

FIGURE 10.29
Illustration of EVOP experimentation.

account since only two or three variables are involved. The formal calculation of the direction of steepest ascent is not particularly helpful.

When to Use Which Design

1. Sandbox
   a. A research tool for brainstorming
2. Simple comparison (paired comparison)
   a. Only one or two variables and no interactions
3. Random strategy
   a. A starting point
   b. Usually involves a large number of variables
   c. Little is known about the process variables and their interactions
   d. Process complete, but is it capable over the process specification range?
   e. A starting point for developing a process
   f. An ending point to snapshot the process stability
4. Plackett-Burman
   a. A lot of variables
   b. Little known about the variables
   c. Need to reduce the number variables by identifying the most important
   d. A scanning procedure identifying the most probable significant variables and leading to further experimentation

5. Fractional factorials
   a. Main variables identified
   b. Interactions of lesser concern
   c. More than four variables involved
6. Full factorials
   a. Main variables well defined
   b. Looking for interactions
   c. Small number of variables (usually < 5)
7. Response surface—Box-Behnken and central composite rotatable design (CCRD)
8. Mixture and simplex-lattice designs
   a. All variables and interactions defined
   b. Want to establish response over the combination of variable limits
   c. Can be used at the end or beginning of a process to give a response surface for the process
   d. Gives an equation that is able to predict the response of the process given the independent variable (factor) settings
9. Regression analysis
   a. Least desirable in terms of hard knowledge about the cause-and-effect relationships between variables
   b. Gives trends versus variables
   c. Can be used to analyze:
      i. Incomplete data
      ii. Random strategy data
10. Process capability study
   a. Establishes the observable variation in response when all independent variables are controlled or held constant

Note: You can use X (X bar) and R-charts to plot experimental responses. The result is a visual presentation of what the statistical analysis numbers will tell you.

Project Strategies

Experimental designs are the building blocks of process definition, development, and optimization. The *project strategy* is a system of networking experimental designs together in order to:

1. Achieve the goal
2. Know when the goal can't be achieved (need to change the process)

The elements in choosing the appropriate experimental design are:

1. What do you want to know?
2. What do you know about the process?
   a. How many dependent variables (responses) are there?
   b. How many independent variables (factors) are there?
   c. How many variables are to be held constant?
   d. How many interactions are there (known and potential)?
   e. Is there a process capability study?
   f. What stage of development is the process in?
      i. Fresh from research (feasibility, sandbox)
      ii. First given to development
      iii. Refining the development process
      iv. Making the development process manufacturable
      v. The initial manufacturing process
      vi. The manufacturing process capability
         a. Establish its initial limits
         b. Identify the significant variables
      vii. Refining the manufacturing process
      viii. Stabilizing the manufacturing process
   g. What measurement tools are you going to use?
      i. How do they relate to the function?
      ii. How do they relate to what you want to know?
      iii. How do they relate to the specifications?
      iv. How good is the measurement?
         a. Precision
         b. Accuracy
         c. Repeatability
3. Based on 1 and 2 above, what is an appropriate experimental design to start with?
4. Where are you in the network?
   a. Plotting your path:
      i. What possible paths can you follow?
      ii. Establish alternative paths.
      iii. What criteria for decisions do you need? (success and failure)
   b. The process of maximum slope:
      i. How to use the results from one experiment to build the set of levels of the independent variables to be used in the next experiment.

c. Documentation
   i. Before the experiment
   ii. Actual versus planned levels of the independent variables (factors)
   iii. Data collection
   iv. Data summary
   v. Data presentation
   vi. The final report

A comparison of the number of trials required for the basic experimental designs is shown in Table 10.21.

Data Analysis

Graphical
- Control charts
- Scatter plots
- Regression
- Surface response

Numerical
- Exploratory data analysis
- Analysis of means
- Analysis of variance
- Student's *t*-test
- Paired comparison
- Chi-squared

Signal to noise

Experimental Designs

Full factorial:	All main effects and interactions.
Half fractional factorial:	All main effects and interactions (no estimate of variance).
Quarter fractional factorial:	All main effects (second-order effects confounded).

TABLE 10.21

Number of Trials Required

No. of Variables	Simple Comparison	Full Factorial 2^n	Full Factorial 3^n	Full Factorial 5^n	1/2 Fractional Factorial	1/4 Fractional Factorial	CCRD	Plackett-Burman	Random Strategy
1	2	2	3	5	—	—	—	—	≤20
2	4	4	9	25	—	—	13	—	≤20
3	6	8	27	125	4	—	20	—	≤20
4	8	16	81	625	8	—	31	—	≤20
5	10	32	243	3,125	16	—	32	—	≤20
6	12	64	728	15,625	32	16	53	—	≤20
7	14	128	2,187	78,125	64	16–32	92	8	≤20
8	16	256	6,552	390,625	128	16–64	164	16	≤20

Response Surface Designs

CCRD and Box-Behnken: All effects and interactions and a mathematical equation of the response surface.
Plackett-Burman (PB): Relative main effect magnitude.
Random strategy: A starting point.

Project Strategy Decision Table (Figure 10.30)

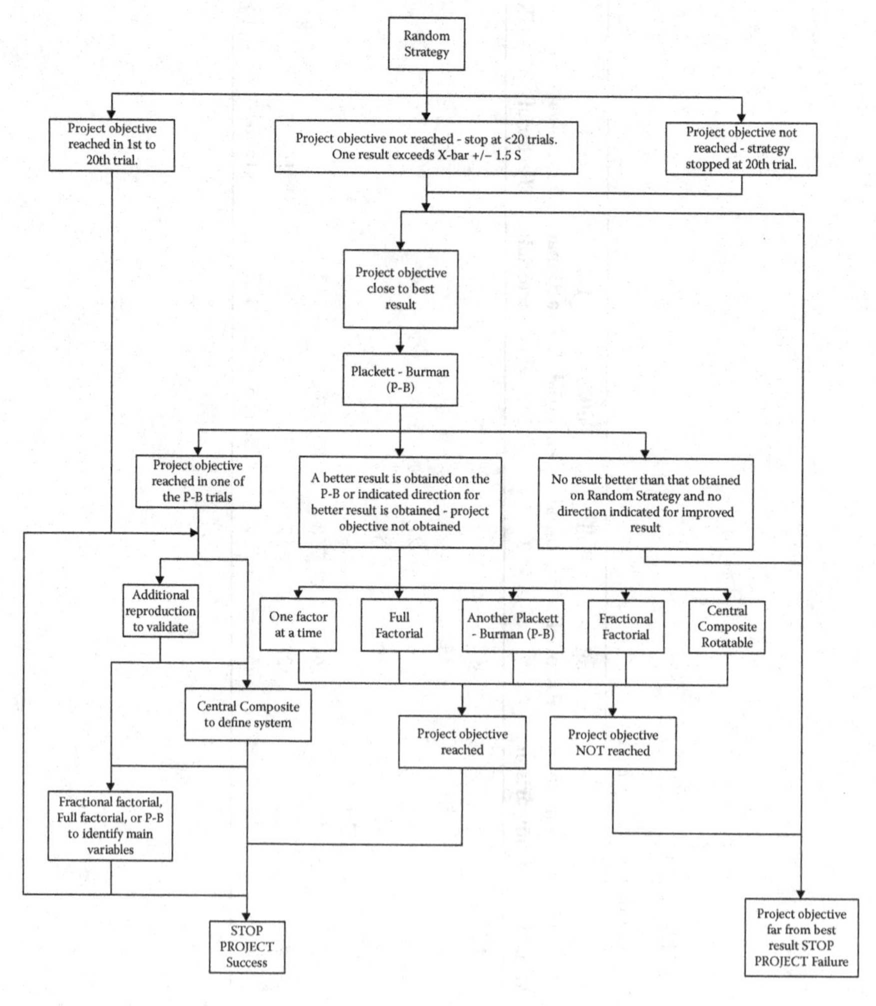

FIGURE 10.30
Project Strategy Decision Table.

Bibliography

1. Box, G. E. P., and K. B. Wilson. On the Experimental Attainment of Optimum Conditions. *Journal of the Royal Statistical Society, Series B*, 1951, 13, 1.
2. Byrne, D. M., and S. Taguchi. *The Taguchi Approach to Parametric Design* (Milwaukee, WI: ASQ Quality Press, 1987).
3. Cornell, J. A. *Experiments with Mixtures: Design, Models and the Analysis of Mixture Data*, 2nd ed. (New York: John Wiley & Sons, 1990).
4. Hahn, G. J., and S. S. Shapiro. *A Catalog and Computer Program for the Design and Analysis of Orthogonal Symmetric and Asymmetric Fractional Factorial Experiments* (Schenectady, NY: General Electric, Research and Development Center, 1966).
5. NIST, Information Technology Laboratory, Statistical Engineering Division. *Engineering Statistics Handbook* (2001). Retrieved September 21, 2001, from http://www.itLnist.gov/div898/handbook/.
6. Plackett, R. L., and J. P. Burman. The Design of Optimal Multifactorial Experiments. *Biometrika*, 1946, 33.
7. Wheeler, D. J. *Understanding Industrial Experimentation*, 2nd ed. (Knoxville, TN: SPC Press, 1989).
8. Wortman, B. L., and D. R. Carlson. *CQE Primer* (Terre Haute, IN: Quality Council of Indiana, 1999).

Background References

American National Standards Institute. *ASQC Standard B3/ANSI Zi.3: Control Chart Method of Controlling Quality During Production* (Milwaukee, WI: American Society for Quality Control, 1958, revised 1975).

American National Standards Institute. *ASQC Standards BI and B2/ANSI Zi.I and Zi.2: Guide for Quality Control and Control Chart Method of Analyzing Data* (Milwaukee, WI: American Society for Quality Control, 1958, revised 1975).

American Society for Testing and Materials. *Manual on Presentation of Data and Control Chart Analysis (STP-15D)* (Philadelphia: ASTM, 1976.)

Box, George E. P., William G. Hunter, and Stuart J. Hunter. *Statistics for Experimenters* (New York: John Wiley & Sons, 1978).

Charbonneau, Harvey C., and Gordon L. Webster. *Industrial Quality Control* (Englewood Cliffs, NJ: Prentice-Hall, 1978).

Davies, Owen L. *Design and Analysis of Industrial Experiments* (London: Oliver and Boyd, 1954).

Deming, W. Edwards. *Quality, Productivity and Competitive Position* (Cambridge: Massachusetts Institute of Technology, Center for Advanced Engineering Study, 1982).

Diamond, William J. *Practical Experiment Designs for Engineers and Scientists* (Belmont, CA: Lifetime Learning Publications, 1978).

Duncan, Acheson J. *Quality Control and Industrial Statistics*, 4th ed. (Homewood, IL: Richard D. Irwin, 1974).

Farago, Francis T. *Handbook of Dimensional Measurement* (New York: Industrial Press, 1982).

Grant, Eugene L., and Richard S. Leavenworth. *Statistical Quality Control*, 5th ed. (New York: McGraw-Hill, 1980).

Hahn, Gerald J., and Samuel S. Shapiro. *Statistical Models in Engineering* (New York: John Wiley & Sons, 1968).

Harris, Forest K. *Electrical Measurements* (New York: John Wiley & Sons, 1962).

Hicks, Charles R. Fundamentals of Analysis of Variance, Parts I, II and III. *Industrial Quality Control*, August–October 1962.
Ishikawa, Kaoru. *Guide to Quality Control*, rev. ed. (Tokyo: Asian Productivity Organization, 1976).
Juran, J.-M., and Frank M. Gryna Jr. *Quality Planning and Analysis* (New York: McGraw-Hill, 1980).
Juran, J. M., Frank M. Gryna Jr., and R. S. Bingham Jr. *Quality Control Handbook*, 3rd ed. (New York: McGraw-Hill, 1979).
Military Specification. *MIL-C-45662A: Calibration System Requirements* (U.S. Government Printing Office, 1962).
Montgomery, Douglas C. *Design and Analysis of Experiments* (New York: John Wiley & Sons, 1976).
Natrelia, Mary G. *Experimental Statistics*, AIBS Handbook 91 (U.S. Government Printing Office, 1966).
Ott, Ellis R. *Process Quality Control* (New York: McGraw-Hill, 1975).
Plackett, R. L., and J. P. Burmen. The Design of Multifactorial Experiments. *Biometrika*, 1946, 305–325.
Statistical Engineering Laboratory. *Fractional Factorial Experiment Designs for Factors at Two Levels*, NBS Applied Mathematics Series 48 (1957).
Western Electric Co. *Statistical Quality Control Handbook* (Indianapolis: Western Electric Co., 1956).

MEASUREMENT SYSTEMS ANALYSIS (MSA)

Definition

Measurement systems analysis (MSA) is a mathematical procedure to quantify variation introduced to a process or product by the act of measuring. The item to be measured can be a physical part, document, or scenario for customer service.

Just the Facts

How well can we trust the data that we have? Six Sigma is data dependent and decisions are made based on that data. What if the data are wrong from the start? Remember the old adage "garbage in, garbage out." As a member of a Six Sigma Team, you need to assure yourself and those depending on you to make the right decisions that the data you are using are "good data."

Operator can refer to a person or can be different instruments measuring the same products.
Reference is a standard that is used to calibrate the equipment.

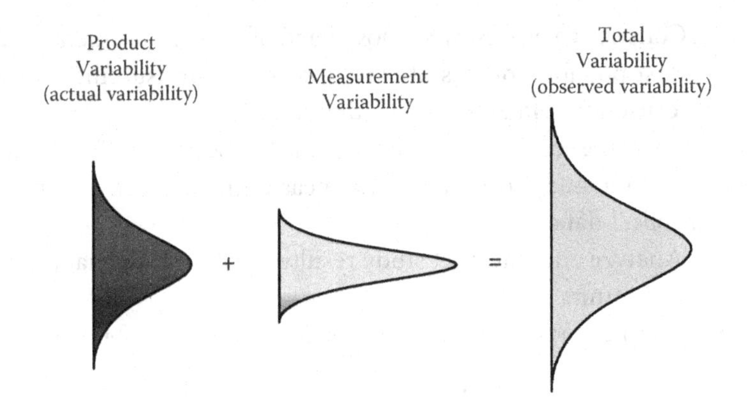

FIGURE 10.31
Total product, process, and measurement error.

Procedure is the method used to perform the test.
Equipment is the device used to measure the product.
Environment is the surroundings where the measures are performed.

Is the variation (spread) of my measurement system too large to study the current level of the process variation in Figure 10.31?

Important questions to ask include:

- How big is the measurement error (i.e., measurement variation)?
- What are the sources of measurement error/variation?
- Is the measurement system stable over time?
- Are the measurements being made with measurement units that are small enough to properly reflect the variation present?
- Are the measurements accurate?
- Can we tell when and if the process changes?
- How do we improve the measurement system?

Example

The following is a step-by-step approach to MSA:

Step 1: Identify the measurement system being used to judge the characteristic of interest. Is it appropriate?
Step 2: Identify the individual(s) responsible for ensuring quality of the measurement system.

Step 3: Construct a process map (flowchart) of the measurement process.
Step 4: List possible sources of variation on the measurement process on a cause-and-effect (fishbone) diagram.
Step 5: Assess the accuracy of the measurement process—verify calibration.
Step 6: Plan a study to evaluate the measurement process, run trials, and collect data.
Step 7: Analyze and interpret study results, verifying adequacy of measurement units.
Step 8: Continue to monitor the measurement process for stability over time.

Approaches to Attribute MSA

Expert reevaluation (truth known): Investigates accuracy only and reports back the percent of correct decisions.
Round-robin study (truth known): Explores accuracy and operator-to-operator agreement (precision).
Inspection concurrence (truth known): Assesses repeatability within operators and reproducibility between operators when the truth is known.

When evaluating attribute processes, we want the percentage of correct decisions and the degree of agreement to be at least 90%.

METHOD OF LEAST SQUARES

Definition

The method of least squares is based on the concept that the best-fit equation for a set of data is obtained when the sum of the squared difference between the actual data points and the points calculated from the equation of the fitted line is a minimum.

Just the Facts

The best-fitting curve to a given set of points is obtained by minimizing the sum of the squares of the offsets (the residuals) of the points from the curve. The residuals are treated as a continuous differentiable quantity; however, because squares of the offsets are used, outlying points can have

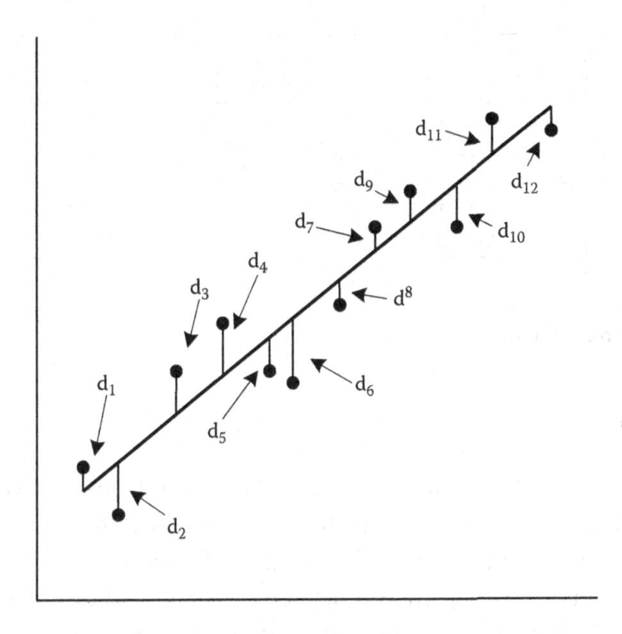

$$\text{II} = d_1^2 + d_2^2 + \cdots + d_n^2 = \sum_{i=1}^{n} d_i^2 = \sum_{i=1}^{n} [y_i - f(x_i)]^2 = \text{a minimum}$$

FIGURE 10.32
Least squares.

a disproportionate effect on the fit, a property that may or may not be desirable depending on the problem at hand.

Example

When the means of all of the squared errors between the fitted line and actual data are minimized, you have achieved the best fit for the given equation being fitted (Figure 10.32).

MULTIVARI CHARTS

Definition

Multivari charts illustrate how multiple input variables impact the output variable or response.

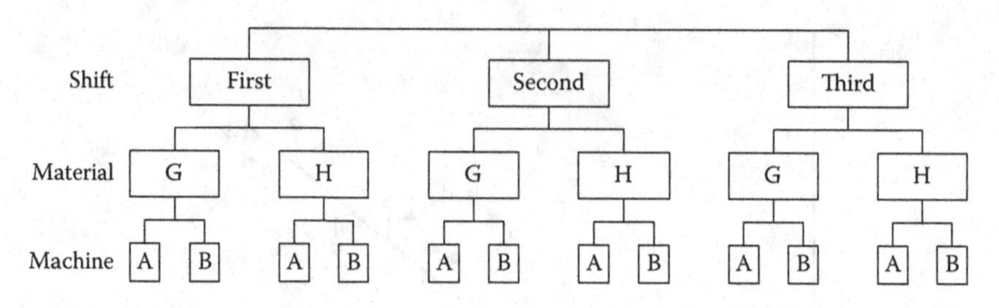

FIGURE 10.33
Nested and crossed design.

Just the Facts

Multivari charts are useful in identifying which independent variable (x) has the most significant effect on the response of a system. Other uses of multivari charts are to do initial screening of complex systems, and the impacts of variations within and between variables. They are also useful in identifying the impact of discrete (categorical) variables on a system, such as different materials. They are also good at graphically representing results from nested and crossed designs.

Example

Consider the desire to examine the effect of shift, material, and machine on the product response. A set of samples are collected from each machine on each shift for material G. It was repeated for material H. The relationship is illustrated in the chart in Figure 10.33.

The results of the study need to separate the effect of shift, material, and machine. The resultant multivari chart (Figure 10.34) clearly separates each independent variable and illustrates its effect on the response.

NONPARAMETRIC STATISTICAL TESTS

Definition

Data that do not fit into a continuous distribution are called nonparametric and are more difficult to make reject/not reject decisions for.

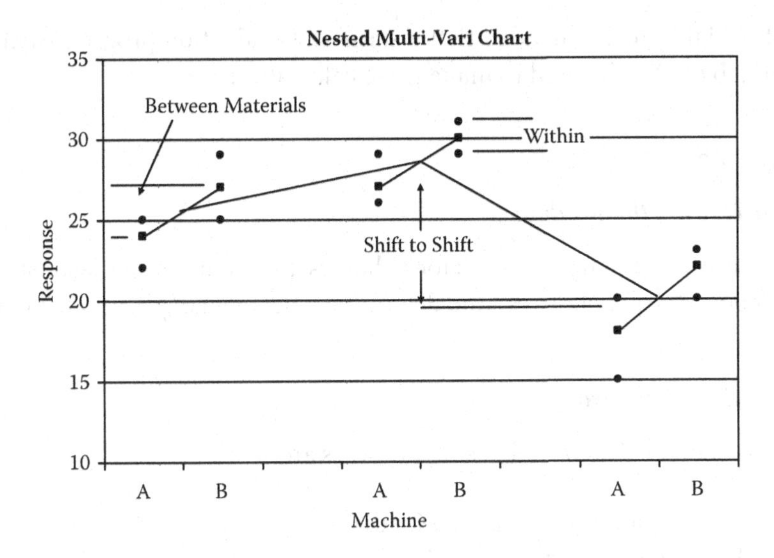

FIGURE 10.34
Nested multivari chart.

Just the Facts

There are several nonparametric statistical tests that have been developed depending on what is being investigated. The following is a list of the most common nonparametric tests.

- One-sample sign test: Estimates the point value of the median and confidence interval.
- One-sided Wilcoxon test: Performs a signed rank test of the median similar to the one-sample sign test.
- Two-sample Mann Whitney: Performs a hypothesis test of the equality of two population medians and calculates the corresponding point estimate and confidence interval.
- Kruskal-Wallis: Performs a hypothesis test of the equality of population medians for a one-way design. This test is a generalization of the procedure used by the Mann-Whitney test.
- Mood's median test: Similar to the Mann-Whitney and Kruskal-Wallis tests. It is more robust against outliers but produces larger confidence intervals.
- Friedman test for a randomized block design: Performs a nonparametric analysis of a randomized block experiment. This test requires exactly one observation per treatment-block combination.

It is highly recommended that a statistical evaluation program such as Minitab or JMP be used to make these calculations.

Example

Sign Test for the Median

Evaluate the median price index for 30 homes in a suburban area against similar homes the previous year and determine if the median price has changed.

One-Sided Wilcoxon Test

The Wilcoxon signed rank test hypotheses are:

- H_0: median = hypothesized median
- Ha: median ≠ hypothesized median

The assumption is that the data are a random sample from a continuous, symmetric distribution.

Two-Sample Mann-Whitney

This is a nonparametric hypothesis test to determine whether two populations have the same population median (*h*). It tests the null hypothesis that the two population medians are equal (H_0: $h1 = h2$). The alternative hypothesis can be left-tailed ($h1 < h2$), right-tailed ($h1 > h2$), or two-tailed ($h1 \neq h2$). The Mann-Whitney test does not require the data to come from normally distributed populations, but it does make the following assumptions:

- The populations of interest have the same shape.
- The populations are independent.

Kruskal-Wallis

Similar to the one-way ANOVA test. It is one of the most versatile tests. When the *p* value is less than 0.05, conclude that at least one mean is different:

H_0: $\mu_1 = \mu_2 = \mu_3 = \ldots = \mu_n$
Ha: At least one mean is different.

Mood's Median Test

Mood's median test can be used to test the equality of medians from two or more populations and, like the Kruskal-Wallis test, provides a nonparametric alternative to the one-way analysis of variance. Mood's median test is sometimes called a sign scores test. Mood's median test tests:

H_0: The population medians are all equal.
H_1: The medians are not all equal.

An assumption of Mood's median test is that the data from each population are independent random samples and the population distributions have the same shape. Mood's median test is robust against outliers and errors in data and is particularly appropriate in the preliminary stages of analysis. Mood's median test is more robust than the Kruskal-Wallis test against outliers, but is less powerful for data from many distributions, including the normal.

Friedman Test for a Randomized Block Design

The analysis of variance tests are quite robust with respect to the violation of their assumptions, providing that the k groups are all of the same size. There are certain kinds of correlated-samples situations where the violation of one or more assumptions might be so extreme that their use may cast doubt on any result produced by an analysis of variance. In cases of this sort, a useful nonparametric alternative can be found in a rank-based procedure known as the Friedman test.

In both of the situations the assumption of an equal-interval scale (like a ruler) of measurement is clearly not met. There is a good chance that the assumption of a normal distribution of the source population(s) would also not be met. Other cases where the equal-interval assumption will be thoroughly violated include those in which the scale of measurement is intrinsically nonlinear, for example, the decibel scale of sound intensity, the Richter scale of earthquake intensity, or any logarithmic scale.

Use this test when there are two kinds of correlated-samples situations where the use of the nonparametric alternative of Friedman would be advisable:

- The first would be the case where the k measures for each subject start out as mere rank orderings. Example: to assess the likely results of an upcoming election, the 30 members of a representative "focus group" of eligible voters are each asked to rank the three candidates, A, B, and C, in the order of their preference (1 = most preferred, 3 = least preferred).

- A second would be the case where measures start out as mere ratings. Example: The members of the focus group are instead asked to rate candidates on a 10-point scale (1 = lowest rating, 10 = highest).

POPULATIONS AND SAMPLES

Definition

A population is the entire collection of units whose characteristics are of interest. A sample is a portion or subset of units taken from the population whose characteristics are actually measured.

Just the Facts

We would like to base our decisions on the "true" characteristics of the process as a whole—the population. However, in practice, we usually work with a limited amount of data that we observe and record as units are produced—a sample. (See Figures 10.35 and 10.36.)

Definitions

- A *population* is the entire collection of units whose characteristics are of interest.
- A *parameter* describes the "true" value of a characteristic. A parameter's value is fixed but usually unknown.
- A *sample* is a portion or subset of units taken from the population whose characteristics are actually measured.
- A *statistic*, any number calculated from sample data, describes a sample characteristic. A statistic's value is known for a specific sample, but usually changes from sample to sample.

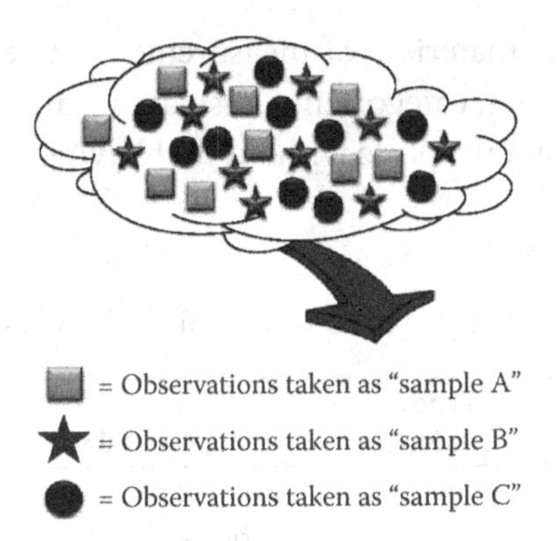

FIGURE 10.35
Representative elements of a population.

Uncertainty in the Mean—Conclusions

How well a sample represents the population depends on:

- The sampling method
- The amount of variation from sample to sample
- The number of units in the sample (sample size)

The larger the sample, the more confidence we have in the estimate.

Data Defined

Raw Facts

- Qualitative or quantitative
- Obtained by observing a population, product, process, or service

Population		**Sample**
Parameter	**Nomenclature**	**Statistic**
Greek letter	Symbol	Arabic letter
μ (mu)	Mean	**X**
σ (sigma)	Standard Deviation	**s**

FIGURE 10.36
Population versus sample notation.

Data are the "raw material" of measurement. Rarely can we use data without first compiling, categorizing, displaying, and/or analyzing it. But no measurement can take place without first having some data.

Summary of Data Types

Process performance is measured in one of three ways:

When you:	Type of Problem:	Type of data:	Type of distribution:
Classify	Defectives	Attribute	Binomial
Count	Defects	Attribute	Poisson
Measure	Continuous	Continuous Variable	Normal

The type of data you collect will determine which LSS tools may be applied to your project.

Process Measurements Summary

For any process, regardless of data type:

- Decisions driven by data are superior to those made without data.
- Sample statistics give us information, which allows us to find and measure process improvements.

To increase customer satisfaction, the goal of Lean Six Sigma is to reduce variation such that:

- For defectives data, units are acceptable as initially produced correct the first time.
- For defects data, units are produced without defects.
- For continuous data, units show minimal deviation from the specified target.

Your choice of a unit and a defect will determine how to measure your project's success.

Example

You take a sample of 20 employees and measure the height of each of them and find that the heights are distributed with the shortest person being 5 feet 3 inches, the tallest being 6 feet 2 inches, and a mean height is 5 feet 11¾ inches tall. If this is a representative sample of the 200 employees in the company, then you would expect to find the population distribution of heights to give a similar result. Since this is only a sample, we would actually expect the population sample to be broader, and the mean will most probably vary from the sample mean.

REGRESSION ANALYSIS

You can't vary everything at the same time and understand what the problem is.

Definition

Regression analysis is a mathematical method of modeling the relationships among three or more variables. It is used to predict the value of one variable given the values of the others. For example, a model might estimate sales based on age and gender. A regression analysis yields an equation that expresses the relationship. (*Computer Desktop Encyclopedia* © 2011)

A statistical technique for analyzing the relationship between two or more variables, and which may be used to predict the value of one variable from the other or others. (*The Oxford English Dictionary*)

Just the Facts

The relationship between independent variables and the dependent variable can be represented by a mathematical model called a *regression equation*. The regression model is fit to a set of sample data. The sample data should come from a well-designed experiment in order to guarantee that the full range of possible combinations of the independent variables is represented and that the design is balanced. In most cases, the true relationship between the X's and Y's is unknown and the experimenter chooses a polynomial model to approximate these relationships.

$$Y = b_0 + b_1X_1 + b_2X_2 + b_{12}X_1X_2 + b_{11}X_1^2 + b_{22}X_2^2$$

Similar analysis can be applied to other model equations.

Regression methods are often used to analyze data from raw production data such as historical records. While this is useful as an approximation, it can also lead to erroneous relationships, especially if there are interactions between two or more of the independent variables. Applying a regression analysis to a statistically designed experiment, especially one specifically designed to fit an equation, is more likely to identify which independent variables are more important and which independent variables interact with each other to significantly affect the dependent variable. The regression builds a quantitative model that identifies these relationships.

Simple Linear Regression

Assuming that there is only one independent variable (X) and the dependent response variable (Y), we can determine the relationship between X and Y. In order for this relationship to accurately represent this relationship, the X variables should be continuous over the range of interest; i.e., there must be a continuous response over the range of X's used. Designing an experiment, the X values are carefully chosen by the experimenter and the resulting responses observed. If we assume that the relationship between X and Y is a straight line, then the expected value of Y for each X is:

$$\hat{Y} = \beta_0 + \beta_1X$$

where $\hat{Y}$ is the expected value for the Y response and the parameters of the straight line β_0 and β_1 are unknown constants. We assume that each observation Y can be described by the model

$$\hat{Y} = \beta_0 + \beta_1X + \varepsilon$$

where ε is a random error with mean zero and variance σ^2. The $\{\varepsilon\}$ are also assumed to be uncorrelated random variables. The regression model involving only a single independent variable x is often called the *simple linear regression model.*

Multiple Linear Regression

If there is more than one independent variable involved, then the equation is expanded. If we assume that there are three *X* variables involved, then the regression model for a linear relationship would be:

$$\hat{Y} = \beta_0 + \beta_1 X_1 + \beta_2 X_2 + \beta_3 X_3 + \varepsilon$$

Curvilinear Regression

These fit equations such as polynomial, exponential, and lognormal equations to a set of data where linearity is not indicated.

Other Linear Regression Models

The linear model $y = X\beta + \varepsilon$ is a general model. It can be used to fit any relationship that is *linear* in the unknown parameters β. An example would be the (kth) degree polynomial in one variable:

$$y = \beta_0 + \beta_1 x + \beta_2 x^2 + \cdots + \beta_k x^k + \varepsilon$$

Other examples include the second-degree polynomial in two variables:

$$y = \beta_0 + \beta_1 X_1 + \beta_2 X_2 + \beta_{11} X_1^2 + \beta_{22} X_2^2 + \beta_{12} X_1 X_2 + \varepsilon$$

and the trigonometric model:

$$y = \beta_0 + \beta_1 \sin x + \beta_2 \cos x + \varepsilon$$

Use the *general linear model* $y = X\beta + \varepsilon$ to fit models.

Caution

Regression analysis is often misused. Care must be taken when selecting the independent variables used to generate the regression model. If the data are not carefully chosen, it is possible to create a model that is meaningless.

It is imperative that the regression equation only be applied to the range of values of the independent variables used in generating the equation. Extrapolating beyond this range can lead to significant errors. If we assume a linear relationship and the regression equation fits with a minimum error, there is no guarantee that the relationship will not become nonlinear outside of the range of X's used in the experimental data. Regression models should never be used for extrapolation.

There are several commercially available programs for designing and evaluating experiments. While these programs make performing the calculations easier, they should be used with caution and the raw data should be examined to determine if there are any outliers that could significantly affect the resultant regression equation. Minitab and JMP are among the leading programs for statistical designing and evaluation of experimental data.

Example

The sum of squares is calculated by the following equation:

$$SS_E = \sum_{i=1}^{n} y_i^2 - n\bar{y}^2 - \hat{\beta}_1 S_{xy}$$

which is just the corrected sum of squares of the y's, so we may write SS_E as

$$SS_E = S_{yy} - \hat{B}_1 S_{xy}$$

By taking expectation of SS_E, we may show that $E(SS_E) = (n - 2)\sigma^2$. Therefore,

$$\sigma = \frac{SS_E}{n-2} \equiv MS_E$$

is an unbiased estimator of σ^2. Note that MS_E is the *error* or *residual mean square.*

ROLLED-THROUGHPUT YIELD

One percent loss at an operation is bad, but if you have it at many operations in the process, it is a disaster.

Definition

Rolled-throughput yield (RTY) is calculated by dividing the total units reworked and scrapped divided by the total units started.

Just the Facts

The use of first-pass yield (FPY) and rolled-throughput yield (RTY) calculations provides a good measure of quality performance at the task and process levels. Only units that meet requirements at each step count as "done right." (See Figure 10.37.)

From which measure of process performance should we calculate process sigma? Rolled-throughput and first-pass yield performance is often used to calculate process sigma. Why?

- Defects/errors, once produced, add waste and cost (some costs are easy to quantify and some not).
- Even the best inspection processes cannot catch all defects/errors.
- The payback is generally bigger when keeping defects/errors from occurring.

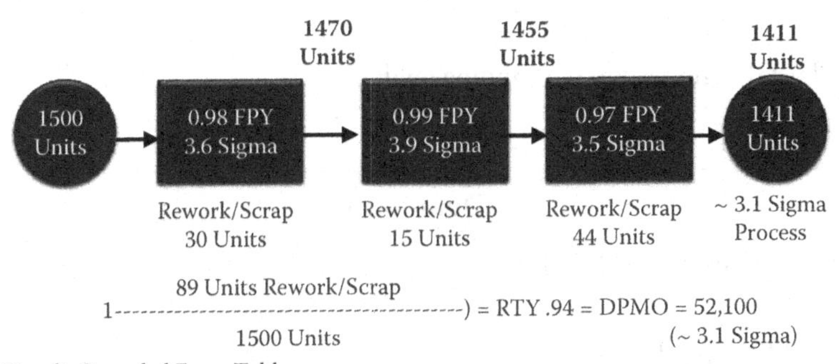

FIGURE 10.37
Rolled-throughput yield example.

Example

Calculating Process Sigma

1. Number of units processed $N = 500$
2. Total number of defects made (include defects made and later fixed) $D = 57$
3. Number of defect opportunities per unit $O = 3$
4. Solve for defects per million opportunities

$$DPMO = 1{,}000{,}000 \times \frac{D}{N \times O}$$

When using the sigma conversion table round sigma down $= 1{,}000{,}000 \times \frac{57}{(500)\,(3)} = 38{,}000$

Look up process sigma in sigma conversion table = 3.275

(with 1.5 sigma shift)..Sigma = ~3.3

TAGUCHI METHOD

Robust designs lead to outstanding quality.

Definition

The Taguchi method defines:

- Ideal quality
- Quality loss function
- His philosophy of a robust design

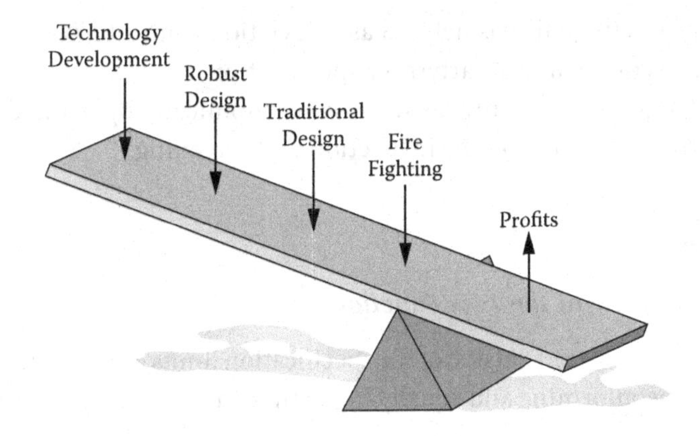

FIGURE 10.38
The Taguchi method.

Just the Facts

To understand the impact of variation in terms of cost, you must understand the Taguchi method (Figure 10.38).

Taguchi Quality Definitions

Ideal Quality

- Reference point (target value) for determining the quality level of a product or service
- Delivery of a product or a service that reliably performs its intended use throughout its life cycle under reasonable conditions

Robust Design

- An Engineering methodology for improving productivity during research and development so that high-quality products can be produced quickly and at low cost.
- Products and services designed defect-free and of high quality
- A design with minimum sensitivity to variations in uncontrollable factors

Quality Loss Function Fundamental Concepts

- Economic and societal penalties incurred as a result of purchasing a nonconforming product.

- Any loss in quality is defined as a deviation from a target, not a failure to conform to an arbitrary specification.
- High quality can only be achieved economically by being designed in from the start, not by inspection and screening.

Example

Traditional View of the Loss Function

Only parts considered outside of the specification limits are considered to be equally nonconforming and equally impacting on the customer and society.

Taguchi Approach

In the Taguchi approach, loss relative to the specification limit is not assumed to be a step function. For example, is it realistic to believe that there is no exposure (loss) when a unit is barely within the specification limit and the maximum loss is assigned when the part is just outside the specification limit? Taguchi rejects the concept of loss assignment only at the step function (Figure 10.39). Instead, it assigns a parabola curve starting at the nominal target. Most people now agree with the Taguchi approach to loss function.

Specify a Target

This is a standard quality graph. Its *X* axis represents performance of characteristics, and the *Y* axis represents a loss to society. Taguchi disagreed with the use of LSL/USL exclusively and contended that any movement from the target would eventually negatively impact society. It was not good enough to prove process capability (CPK), but that any movement from the target is bad. (See Figure 10.40.)

The Quadratic Loss Function (QFL)

The added curve line represents the quadratic loss function, and how losses to society rise quickly as performance moves away from the nominal target. Taguchi used a quadratic loss function to show that any variation from a specified target causes a loss to the customer and society. Minimizing variation will prevent both losses. (See Figure 10.41.)

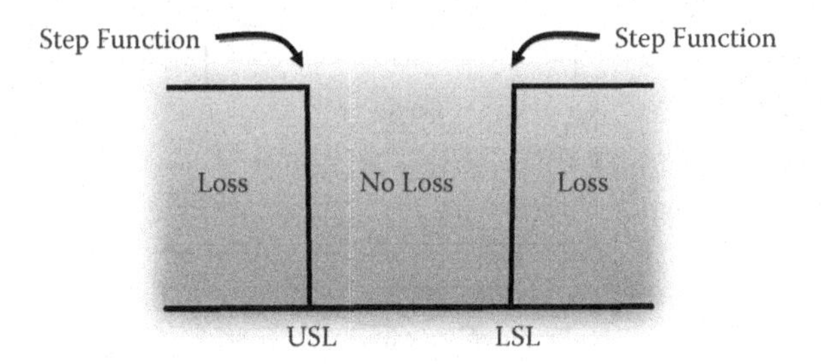

FIGURE 10.39
The Taguchi cost loss factor.

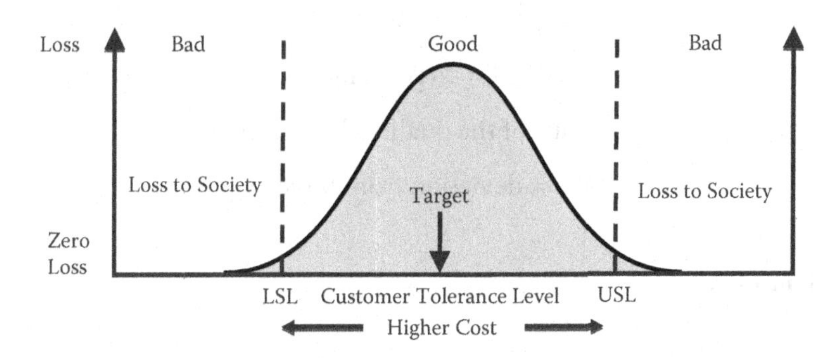

FIGURE 10.40
Product quality.

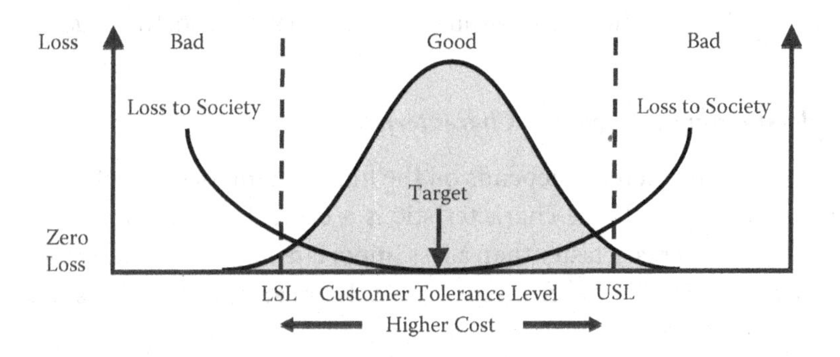

FIGURE 10.41
Cost loss.

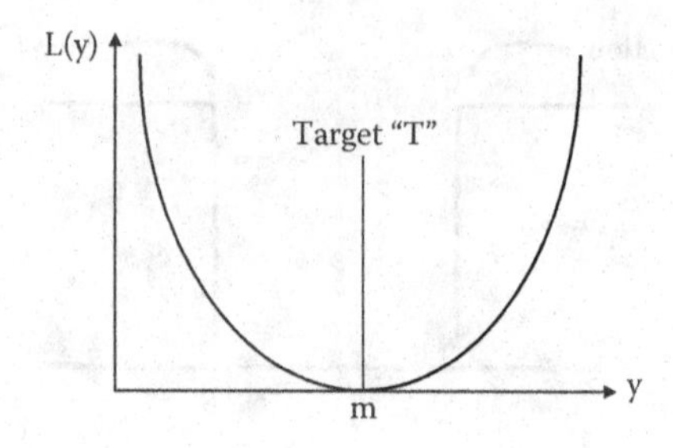

FIGURE 10.42
The QLF model.

QLF is a quadratic function and an approximation of variation around the target value, *m*. (See Figure 10.42.)

$$L = \text{loss}$$

$$M = \text{target value}$$

$$y = \text{value of the quality characteristic}$$

$$y - m = \text{deviation from target at } y$$

QLF model:

$$L(y) = k(y - m)2$$

1. K is a constant (the cost coefficient).
2. $y - m$ is deviation from the target.
3. Loss is proportional to the square of the deviation from the target value.

Understanding the Quality Characteristic

The rate of loss increase depends on the financial importance of the quality characteristic. If the characteristic is a critical dimension for safety, the loss will increase faster than a less important characteristic as the performance moves farther from the target. There are two questions to ask:

1. What is the financial importance of the characteristic?
2. What is the rate of acceleration as the characteristic moves away from the target?

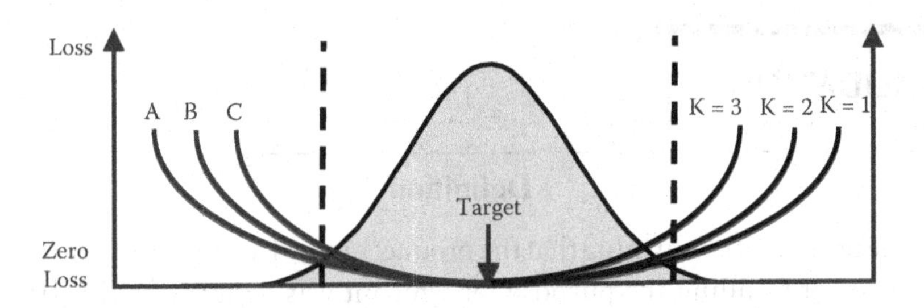

FIGURE 10.43
The QLF model.

Mathematically the question is: What is the steepness of the slope of the parabola?

Observing the Slope

Loss accelerates more quickly with characteristic C than A. The functional loss is the maximum permissible deviation from target, not necessarily the specification. The customer's loss is the average loss generated when deviation has a cost impact on the customer. (See Figure 10.43.)

Determining Customer Impact

The average customer loss due to poor quality usually calculates costs such as:

- Cost to repair
- Cost to replace
- Time considerations to deal with the issue

The Cost of Not Being on Target

- Specifications allow variance—a target is an attempt to have no variance—which means what?
    - There are costs for being above or below the target value.
- As we get further from the target value, the loss increases:
    - Tolerance stack-up
    - Increased opportunity for scrap or rework
    - Lower performance
    - Wasted material

VALIDATION

Definition

The process of confirming that the product or service meets the requirements of fulfilling the purpose for which it was designed. Validation tests inputs and outputs and differs from verification, which tests the product or service in the actual environment that it is to be used. There can be a disconnect between validation and verification.

Just the Facts

Standard validation tools include periodic process reviews, audits, development history record generation, and review and objective testing. Validation, therefore, involves several tools, starting with a robust risk analysis. Before we can test the failure of a product or process, we need to identify the modes in which the product or service can fail and put into place processes to mitigate and eliminate them.

Validation, therefore, evaluates the effectiveness of the change or innovation that has been implemented. If validation is not performed, then there is no assurance that the change is effective in achieving the performance that the change or innovation was designed to achieve.

The validation protocol confirms or denies the effectiveness of a Six Sigma project on achieving the desired results.

The Failure Modalities

Failures may be:

- Gradual wear over time of use
- Catastrophic, failing all at once
- A combination of the two

A light bulb is an example of a catastrophic failure. Automobile brake pads are an example of gradual wear to failure. Automobile tires are a hybrid failure mode, as catastrophic failure may occur from road hazards.

Accelerated failure testing techniques have been developed and are used to estimate the service time for a product. An example is automobile tire wear. The tire companies test the wear properties of their tires on special equipment that is designed to represent the kind of environment that the tire will experience on a vehicle. To the degree that this test accurately represents the actual environment that the product will experience is the degree to which the outcome will match the actual performance of the tire.

Some Risk Assessment Tools

Besides risk analysis, other tools that are useful in evaluating and validating the performance of a product include, but are not limited to, the following:

- Failure modes and effects analysis (FMEA)
- Process failure modes and effects analysis (PFMEA)
- Accelerated life testing
- Highly accelerated life test (HALT)
- Highly accelerated stress screening (HASS)
- Burn-in
- Mean time between failures (MTBF)
- Mean time to failure (MTTF)
- Risk analysis programs

While it is not our intention to cover all of the tools, we will give an overview of some of them here. It is recommended that the LSSBB study and become familiar with these and other accelerated tests so that he or she will be able to choose the appropriate accelerated test to use for the product or service.

- HALT testing is commonly used on electronic components and finished products and involves temperature, random vibration, marginal power, and power cycling testing over time and identifying all modes of failure as either recoverable (the circuit recovers its function after testing) or permanent (the circuit is no longer functional in one or more modes). This test is used mainly to determine failure modes and not an estimate of expected life of a completed product.
- HASS is usually applied to the process of estimating expected life by accelerating the failure rate by testing under elevated temperatures. It is usually applied to production products.
- MTBF and MTTF are used to estimate the expected service life of a product. To use these, we need to establish the kind of failure that we expect. Failures may be gradual wear over time of use, catastrophic, failing all at once, or a combination of the two.

A robust risk analysis program will help to identify and mitigate potential failure modes.

Guidelines and Tips

A risk analysis program may follow the following sequence:

- FMEA and PFMEA
  - Identify potential failure modes
  - Identify mitigation and elimination design activities
- Identifying the failure modes
  - Catastrophic
  - Gradual over time
  - Combinations
    - Initial catastrophic failure
    - Gradual failure to a catastrophic failure
    - Gradual failure to end of life
    - Some initial failure and then catastrophic end-of-life failure
- Selecting test methods to test the failure modes
- Select accelerated life tests to:
  - Confirm the failure mode
  - Establish the serviceable life
  - Give feedback to design for improvements

Example

Several electronic units are tested at elevated temperatures and their performance monitored over time. The elevated temperature is believed to be the main function in accelerating the failure of the electronic circuit. When a failure occurs, its failure mode is evaluated, recoverable or nonrecoverable. From the sample performance, the estimated serviceable life can be calculated.

Appendix A: Glossary

5W's and 2H's: A rigid, structured approach that probes into and defines a problem by asking a specific set of questions related to a previously defined opportunity or problem statement. The 5W's and 2H's stands for:

- W1: What?
- W2: Why?
- W3: Where?
- W4: Who?
- W5: When?
- H1: How did it happen?
- H2: How much did it cost?

Acceptable quality level (AQL): The percentage or proportion of defects or defectives that is considered satisfactory quality performance for a process or product.

Acceptance decisions: The process of making the choice to accept or reject an output based on the risk related to accepting that output and/or your evaluation of the output that is provided. Acceptance decision is the highest number of nonconforming units or defects found in a sample that permits the acceptance of the lot.

Accumulative distribution function: The area beneath the probability density function to the left of X. Mathematically, the accumulative distribution function is equal to the integral of the probability density function to the left of X.

Activity-based costing (ABC): A technique for accumulating product cost by determining all costs associated with the activities required to produce the output.

Activity plan: A simple chart that shows a list of implementation activities listed in sequence. It identifies the individual responsible for a particular activity and the projected timing of that activity.

Adaptability: The flexibility of a process to handle future, changing customer expectations and today's individual, special customer requirements. It is managing the process to meet today's special needs and future requirements. Adaptability is an area largely ignored but is critical for gaining a competitive advantage in the marketplace. Customers always remember how you handle or don't handle their special needs.

Advantage/disadvantage technique: Lists of advantages and disadvantages of each proposed solution are made. The solution of the most favorable ratio of advantages to disadvantages is assumed to be the best solution.

Advocate: An individual/group that wants to achieve change but does not have sufficient sponsorship.

Affinity diagrams: A technique for organizing a variety of subjective data (such as options) into categories based on the intuitive relationships among individual pieces of information. Often used to find commonalties among concerns and ideas.

Appraisal costs: The costs that result from evaluating already completed output and auditing the process to measure compliance to established criteria and procedures. To say it another way, appraisal costs are all the costs expended to determine if an activity was done right every time.

Area activity analysis (AAA): A proven approach used by each natural work team (area) to establish efficiency and effectiveness measurement systems, performance standards, improvement goals, and feedback systems that are aligned with the organization's objectives and understood by the employees involved.

Area graphs: Convenient methods of showing how 100% of something is apportioned. The most commonly used area graph is the pie chart.

Arrow diagrams: A way to define the most effective sequence of events and control the activity in order to meet a specific objective in a minimum amount of time. It is an adaptation of program evaluation and review technique (PERT) or the critical path method (CPM).

Ask "why" 5 times: A systematic technique used to search for and identify the root cause of a problem.

Assumption evaluation: Provides a way of redefining problem statements, analyzing solutions, and generating new ideas.

Attribute control chart: A plot of attributes data of some parameter of a process's performance, usually determined by regular sampling of the product, service, or process as a function (usually) of time or unit number or other chronological variable. This is a frequency distribution plotted continuously over time, which gives immediate feedback about the behavior of a process. A control chart will have the following elements:

- Center line (CL)
- Upper control limit (UCL)
- Lower control limit (LCL)

Attributes data: Counted data that can be classified as either yes/no, accept/reject, black/white, or go/no-go. These data are usually easy to collect because they require only counting and are not measuring the process, but they often require large samples.

Automation: The use of robots, machinery, or software to eliminate repetitive and boring jobs previously done by people. Automation is the automatic operation and control of machinery or processes by devices such as robots that can make and execute decisions without human intervention.

Average incoming quality (AIQ): The average quality level going into an inspection point.

Average outgoing quality (AOQ): The average quality level leaving an inspection point when the rejected parts have been removed from the line.

Axiomatic design: This approach provides a framework of principles that guide the design engineers of products, services, or processes. The approach reduces the complexity of the design process. Its purpose is to make the human designer more creative by reducing the random search process, thereby minimizing the trials and errors that are made during the design process.

Bar graph: Have bands positioned horizontally (bars) or vertically (columns) that, by their height or length, show variations in the magnitude of several measurements. The bars and columns may be multiple to show two or more related measurements in several situations.

Bathtub curve: A picture of an item's failure rate versus time. It shows how the failure rate decreases during the item's early life to its intrinsic failure rate level and remains at the level until the item starts to wear out and its end-of-life rate begins to increase.

Bell-shaped curve: The shape of a normal distribution curve.

Benchmark: A reference point where other items can be compared. It can be a location, a process, a measurement, or a result.

Benchmarking (BMKG): A systematic way to identify, understand, and creatively evolve superior products, services, design, equipment, processes, and practices to improve your organization's real performance.

Best practice: A process or a method that is superior to all other known methods.

Best-value future state solution: A solution that results in the most beneficial new item as viewed by the item's stakeholders. It is the best combination of implementation cost, implementation cycle time, risk, and performance results (examples: return on investment, customer satisfaction, market share, risk, value-added per employee, time to implement, cost to implement, etc.).

Bivariate distribution: A three-dimensional plot where the *X* and *Y* axes represent the independent variables, and the Z axis represents the frequency for discrete data or the probability of continuous data.

Black Belts: Highly trained team leaders responsible for implementing process improvement projects within an organization. They have a deep understanding of statistical methods and have a detailed understanding of how to use DMAIC and DMADV. They are normally full-time assignments, for each Black Belt is expected to save the organization a minimum of $1 million per year. Black Belts focus on customer/business alignment, and that includes both hard and soft dollar savings. They have been trained to manage the projects by fact, process, and project management methodology, not by gut feel. Black Belts coach Green Belts and receive coaching support from Master Black Belts.

Block: A part of the experimental material that is likely to be more homogeneous than the whole.

Block diagrams: A pictorial method of showing activity flow through a process, using rectangles connected by a line with an arrow at the end of the line indicating direction of flow. A short phrase describing the activity recorded in each rectangle.

Brainstorming: A technique used by a group to quickly generate large lists of ideas, problems, or issues. The emphasis is on quantity of ideas, not quality.

Budget: Provides the resources required to implement the tactics.

Bureaucracy elimination method: An approach to identify and eliminate checks and balances activities that are not cost justified.

Business case development: An evaluation of the potential impact a problem has on the organization to determine if it is worthwhile investing the resources to correct the problem or take advantage of the opportunity.

Business objective: Defines what the organization wishes to accomplish over the next 5 to 10 years.

Business plan: A communication, planning, and business system that reaches and involves every employee in support of common goals and objectives. It is a three-way interactive process that provides direction, expectations, and funding. It also defines the activities required to meet the agreed upon expectations. It includes the following 11 outputs:

- Direction
  1. Visions
  2. Mission
  3. Values
  4. Strategic focus
  5. Critical success factors
- Expectation
  6. Business objective
  7. Performance goals
- Action
  8. Strategies
  9. Tactics
  10. Performance plans
  11. Budget

Business process improvement (BPI): A breakthrough methodology that includes process redesign, process re-engineering, process benchmarking, and fast-action solution teams.

Cp: *See* process capability index.

Cpk: *See* time-related process capability index.

Calibration: Comparing an instrument or measurement equipment performance to a standard of known accuracy. Normally the standards are traced back to the National Bureau of Standards.

Cause-and-effect diagram: A visual presentation of possible causes of a specific problem or condition. The effect is listed on the right-hand side and the causes take the shape of fish bones. This is the reason it is sometimes called a fishbone diagram. It is also called an Ishikawa diagram.

Cause-and-effect matrix: A tool used to evaluate the net impact of potential *X*'s versus various *Y*'s or goals in order to make a first pass at setting aside potential *X*'s that are not likely to impact the *Y*'s, thereby eliminating elements that do not have to be statistically evaluated. It is an excellent tool to align significant process inputs with customer requirements.

***c*-Charts:** Plot the number of defects per sample, with a constant sample size.

Cellular manufacturing: A tool used to produce your product in the least amount of time using the least amount of resources. When applying the cellular manufacturing tool, you group products by value-adding process steps, assess the customer demand rate (Takt time), then configure the cell using Lean Six Sigma concepts and tools. This is a powerful tool to allow the use of many Lean concepts and tools together to achieve dramatic process improvements.

Central composite design: This design contains an embedded factorial or fractional factorial matrix with a center point augmented with a group of star points that allows the curvature to be estimated. This design always contains twice as many star points as there are factors in the design, and the star points represent the extremes, high and low, of each design factor.

Central tendency: A measure of the center of the distribution.

Certification: Applies to a single operation or piece of equipment. When an acceptable level of confidence has been developed that proves that the operation and/or equipment is producing products to specification when the documentation is followed, that item is then certified. Typically, a Cpk (process capability index) of 1.4 is required to be certified.

Change agent: Individual/group who is responsible for implementing the change.

Changee: Individual/group who must actually change. A changee is also called a change target.

Check sheet: A simple form on which data are recorded in a uniform manner. The forms are used to minimize the risk of errors and to facilitate the organized collection and analysis of data.

Collecting data: A systematic way of acquiring information about a specific point of interest.

Common cause: A source of errors that is always present because it is part of the random variation in the process itself. These types of failures are normally traced back to the process, which only management can correct.

Communication techniques: The many processes that are available to deliver and send messages through an organization by various channels, such as e-mail, meetings, gossip, newsletters, etc.

Comparative analysis: A systematic way of comparing an item to another item to identify improvement opportunities and/or gaps. (It is the first three phases in the benchmarking process.)

Comparative experiments: An experiment whose objective is to compare the treatments rather than determine absolute values.

Competitive benchmarking: A form of external benchmarking that requires investigating a competitor's products, services, and processes. The most common way to do this is to purchase competitive products and services and analyze them to identify competitive advantages.

Confidence limits: A calculated measure of the accuracy of the results obtained from pulling a sample of a complete population. For example: Your confidence level may be 90% that the cycle time is 3 hours plus or minus 2%.

Conflict resolution: An approach to find a win-win solution when two or more parties are in disagreement with each other. Often, conflict resolution ends up with a compromise on the position each party took in the original conditions.

Consensus: An interactive process, involving all group members, where ideas are openly exchanged and discussed until all group members accept and support a decision, even though some of the group's members may not completely agree with it. To reach a consensus is time-consuming and often involves individual compromising.

Constants: Independent variables that are deliberately held constant during the experiment.

Continuous flow manufacturing (CFM): A manufacturing system that is set up where there is no buffer between individual activities. The product is continuously moving without going into a storage area.

Control chart: A graphic representation that monitors changes that occur within a process by detecting variation that is inherent in the process and separating it from variation that is changing the process (special causes).

Controllable poor-quality costs: The costs that management has direct control over to ensure that only acceptable products and services are delivered to the customer. They are divided into two subcategories: prevention costs and no-value-added costs.

Corrective action: Action that is taken to prevent re-occurrence of a problem. It is usually taken when an error/nonconformity is detected that warrants expending effort and money to prevent it from re-occurring.

Correlation coefficient (*r*): Used to quantify the degree of linear association between two variables. Its value can range all the way from –1 to +1. In a formula it is usually represented by a small *r*.

Cost driver: Any factor that causes a change in cost of an activity.

Cost of quality: A process developed by Val Feigenbaum when he was quality director at a General Electric Division in the 1950s and put all the quality-related activities into a single cost base that could be added together. It is made up of four parts: prevention costs, internal defect costs, external defect costs, and appraisal costs.

Cost, quality, features, and availability (CQFA): Customers evaluate and select suppliers based upon the four factors of cost, quality, features, and availability. An organization must excel in one of these to stay on the market. The more of these four factors that an organization excels in, the greater value they provide. Organizations survive based upon the value provided in the CQFA grid.

Creative thinking: A methodology designed to stimulate and encourage creativity and innovation within an organization and individuals.

Creativity: Developing new or different ideas.

Critical path methodology: Normally used with a project work breakdown structure where there is one path through the complex process that determines when the process is completed. By identifying this path the project manager can focus on ensuring that cycle time and cost are optimized, thereby minimizing the risk of not completing the budget on time and on schedule.

Critical to quality (CTQ): Key measurable characteristics of a product or a process that are set to ensure customer satisfaction. They help ensure that the improvement activities are in line with the customer requirements. These customers can be either internal or external customers.

Cultural roadblocks: Each organization has its own set of acceptable and unacceptable behavioral patterns. Cultural roadblocks are those unacceptable behavioral patterns that will have a negative impact upon completing a project.

Current state maps: Flow diagrams of the present process as it is operating prior to implementing a change.

Customer-dissatisfaction poor-quality costs: The lost profits because customers buy competitive products because they perceive that the competitor's product is better quality or because the customer has had or knows someone that has had an unsatisfactory experience with the organization.

Customer-incurred poor-quality costs: The costs that the customer incurs when a product or service fails to perform to the customer's expectations. Example: Loss of productivity while equipment is down, or travel costs and time spent to return defective merchandise, and the repair cost after the warranty period.

Customer requirements: Stated or implied terms that the customer requires to be provided with in order for him or her not to be dissatisfied.

Customer surveys: Obtaining customers' opinions related to the service or products supplied. This can be done in many ways, including phone calls, written surveys, focus groups, one-on-one meetings, etc.

CuSum control charts (cumulative sum): An alternative to the Shewhart concept of control charts. It was primarily developed in the late 1990s to create a CuSum chart to collect *m* sample groups, each the size of an *m*, and compute the *X* bar sub *i* of each sample. Determine Ssubm or *S* prime subm using the appropriate formulas. CuSum control charts are very effective at discerning shifts in the process mean that are less than two sigma.

Cycle time: The actual time from the point when all of the input has been received by the task until an output has been delivered to the next task.

Cycle time analysis: An approach to reduce the time that it takes to move an item through a process.

Decision-making matrix: The team defines the desired results. Then it makes a list of the criteria that are givens (must have) and wants (would like to have). The alternative solutions are compared to the givens and wants list, and a risk analysis is made.

Defects per million opportunities (DPMO): Average number of errors that would occur in a million opportunities to make an error. It is not the number of defects in a million items. For example, if an item had 10 opportunities for being defective and there were 15 errors in the 1 million units, that would be 1.5 errors per million opportunities. We use errors in place of defects because Six Sigma is now being applied to service areas where defects are not normally seen, but there are many opportunities to make errors.

Delphi narrowing technique: A technique where team members' priorities are used to reduce a list of alternatives to a few of the most important alternatives.

Dependent variable: A variable that we measure as a result of changes in the independent variables.

Design for maintainability and availability: A methodology and tool set that is directed at analyzing the maintenance of a product to minimize the time to repair it and to maximize its total reliability. The object is to minimize downtime. Often it involves modular replacement rather than individual component replacement.

Design for manufacturing and assembly (DFMA): A methodology that is used to determine how to design a product for ease of manufacturing. It is usually done by performing concurrent engineering, where manufacturing engineering develops the manufacturing process along with the design.

Design for Six Sigma (DFSS): *See* DMADV.

Design for *X* (DFX): An approach when the design team develops a product or service with as many desirable characteristics as possible, when viewed from the consumers' standpoint. This approach includes characteristics like safety, friendliness, serviceability, reliability, quality, maintainability, cost, and features. It is based principally on work done by Watson and Radcliffe in 1988. It includes factors like safety, quality, reliability, testability, manufacturability, design for assembly, environmental, serviceability, maintainability, repairability, user-friendliness, ergonomics appearance, packaging, features, and time-to-market.

Design of experiments (DOE): A structure-organized method of determining the relationship between factors affecting a process and the output of that process. It is a structured evaluation designed to yield a maximum amount of information at a defined confidence level at the least expense. DOE is a set of principles and formulas for creating experiments to define regions of variable value that support customer satisfaction or to define relations between variables for having more accurate models of phenomena.

Direct poor-quality costs: Costs that can be identified in the organization's ledger.

Discrete data: Based on count. It cannot be broken down into subdivisions. For example, it is the number of customer complaints that are received per week. It is also referred to as *qualitative data.*

DMADV: Define, Measure, Analyze, Design, and Verify. It is Six Sigma's approach to using data for designing products and processes that are capable of performing at the Six Sigma level.

DMAIC: Define, Measure, Analyze, Improve, and Control. It is Six Sigma's version of Shewhart's Plan-Do-Check-Act problem analysis technique. Each step in this cycle is designed to ensure the best possible results.

DMEDI: Define, Measure, Explore, Develop, and Implement. It is equivalent to the design for Six Sigma approach under a different set of titles.

Effect of a factor: The change in response produced by a change in the level of the factor. (Applicable only for factors at two levels each.)

Effectiveness: The extent to which an output of a process or sub-process meets the needs and expectations of its customers. A synonym for effectiveness is quality. Effectiveness is having the right output at the right place at the right time at the right price. Effectiveness impacts the customer.

Efficiency: The extent to which resources are minimized and waste is eliminated in the pursuit of effectiveness. Productivity is a measure of efficiency.

Equipment certification: An evaluation of each piece of equipment to define its accuracy, repeatability, drift, and capabilities so that it can be matched to the product specifications.

Equipment poor-quality costs: The cost invested in equipment used to measure, accept, or control the products or services, plus the cost of the space the equipment occupies and its maintenance costs. This category also includes any costs related to preparing software to control and operate the equipment.

Error proofing: Designing processes and products so that it is difficult or impossible for errors to occur during creation and delivery to your customers.

Establish the burning platform: Define why the as-is process needs to be changed and prepare a vision that defines how the as-is pain will be lessened by the future state solution.

Executive error rate reduction (E^2R^2): A way to establish acceptable executive behavior standards and measure compliance to them.

Experiment: A sequence of trials consisting of independent variables set at predesigned levels that lead to measurements and observations of the dependent variables. A planned set of operations that lead to a corresponding set of observations.

Experimental design: The building blocks of process definition, development, and optimization.

Experimental unit: An experimental unit is one item to which a single treatment is applied in one replication of the basic experiment.

Exponential distribution: Used to model items that consist of failure rates, usually electronic items. This exponential distribution is closely related to the Poisson distribution. It is usually used to model the mean time between occurrences, such as arrivals or failures. It usually measures probability of occurrence per time interval.

External and internal customers: All organizations have internal and external customers. The output from any activity within an organization that goes to another individual within the organization has created an internal customer-supplier relationship. The person who receives the input is the internal customer. External customers are individuals or organizations that are not part of the organization that is producing the product. They typically buy the product for themselves or for distribution.

***F*-test:** An evaluation of two samples taken from different populations to determine if they have the same standard deviation at a specific confidence level.

Facilitor of teams: An individual who is assigned to work with the team to make the meetings run more effectively. They work to ensure that the team functions correctly, not to participate in solving the problem.

Factor-independent variable: A feature of the experimental conditions that may be varied from one observation to another. These may be qualitative, fixed, or random. Qualitative factors would be items such as good or bad agitation, which could either be set (like a power switch) or naturally occurring settings. Represented by *X*'s.

Failure mode and effects analysis: Identifies potential failures or causes of failures that may occur as a result of process design weaknesses.

Fast-action solution technique (FAST): A breakthrough approach that focuses a group's attention on a single process for a 1- or 2-day meeting to define how the group can improve the process over the next 90 days. Before the end of the meeting, management approves or rejects the proposed improvements.

First-time yield (FTY): The number of good parts that go into an operation divided by the number of acceptable parts going out of the operation without any rework. First-time yield for a total process is calculated by multiplying the first-time yield at each activity times the first-time yield at each activity in the process. It represents the number of parts that go through the process without being reworked or scrapped. It is also called roll-through yield (RTY).

Five S's or five pillars: A system designed to bring organization to the workplace. A translation of the original 5S terms from Japanese to English went like this:

- Seiri—Organization
- Seiton—Orderliness
- Seiso—Cleanliness
- Seiketsu—Standardized cleanup
- Shitsuke—Discipline

In order to assist users of this tool to remember the elements the original terminology has been retranslated to the following 5S's.

- Sort
- Set-in-order
- Shine
- Standardize
- Sustain

Five whys (5W's): A technique to get to the root cause of the problem. It is the practice of asking five times or more why the failure has occurred in order to get to the root cause. Each time an answer is given, you ask why that particular condition occurred.

Flowchart: A method of graphically describing an existing process or a proposed new process by using simple symbols, lines, and words to pictorially display the sequence of activities in the process.

FOCUS: An acronym for:

- Find a process to improve
- Organize an effort to work on improvement
- Clarify current knowledge of the process
- Understand process variation and capabilities
- Select a strategy for continuous improvement

This was developed by W. Edwards Deming and provides his model for improving processes. It was based upon Shewhart's Plan-Do-Check-Act approach.

Focus groups: A group of people who have a common experience or interest is brought together where a discussion related to the item being analyzed takes place to define the group's opinion/suggestions related to the item being discussed.

Force field analysis: A visual aid for pinpointing and analyzing elements that resist change (restraining forces) or push for change (driving forces). This technique helps drive improvement by developing plans to overcome the restrainers and make maximum use of the driving forces.

Fractional factorial: A type of design of experiment (DOE) where selected combinations of factors and levels are analyzed. It is useful when a number of potential factors are involved in causing an error to occur because it reduces the total number of runs required to define the high potential root causes.

Full factorial: A design of experiment that measures the response of every possible combination of factors and levels. It provides information about every main effect and each interacting effect. It is normally not used when there are more than five factors involved in the evaluation.

Full factorial design: A design in which every setting of every factor appears with every setting of every other factor is a full factorial design. This is not recommended for five or more factors.

Function diagrams: A systematic way of graphically displaying detailed tasks related to broader objectives or detailed issues related to broader issues.

Future state mapping: This usually takes the form of a flow diagram or a simulation model where a proposed change is drawn out pictorially to better understand the process. In the case where a simulation model is developed, the process can be operated over a period of time based on the assumptions made in the simulation model to determine how effectively it will operate.

Gantt chart: A Gantt chart is a bar chart laid on its side. It is typically used for conveying a project schedule. It is an effective way of identifying interrelationships between tasks and helping to define critical paths through a process or project.

Gap analysis: A gap analysis is used to compare a present item to a proposed item. It typically will compare efficiency and effectiveness measurements between one product and a competitor's product or one process and another process. It reveals the amount of improvement necessary to bring it in line with the process or product it is being compared to.

Graeco-Latin design: This experimental design is often useful in eliminating more than two sources of variability in an experiment. This is an extension of the Latin square design with one additional blocking value, resulting in a total of three blocking variables.

Graphs: Visual displays of quantitative data. They visually summarize a set of numbers or statistics.

Green Belt: An individual who has been trained in the improvement methodologies of Six Sigma and will be able to lead a Six Sigma process improvement team or work on a process improvement team that is led by a Black Belt or Master Black Belt. This is a part-time job and Green Belts maintain their full-time job while performing this activity. They work under the guidance of a Black Belt.

Group: Individuals who are gathered together for administrative purposes only. Individuals work independently, sometimes at cross-purposes with others in the group.

Hard consensus: When all members of the team absolutely agree with the outcome or solution.

High-impact team (HIT): A methodology that designs and implements a drastic process change in a dozen days.

Histograms: A visual representation of the spread or distribution. It is represented by a series of rectangles or bars of equal class sizes or width. The height of the bars indicates the relative number of data points in each class.

Hoshin kanri: This in an annual planning process that is used to develop the hoshin plan or policy development. It is used to set the direction of the improvement activities within the organization. Hoshin is made up of two Chinese words: *ho*, which means "method or form," and *shin*, which means "shiny needle or compass." *Kanri* means "control or management." It is a very systematic, step-by-step planning process that breaks down strategic objectives against daily management tasks and activities.

House of quality: A matrix format used to organize various data elements, so named for its shape. It is the principle tool of QFD.

Hypothesis testing: Hypothesis testing refers to the process of using statistical analysis to determine if the observations that differ between two or more samples are caused by random chance or by true differences in the sample. A null hypothesis (Ho) is a stated assumption that there is no difference in the parameters of two or more populations. The alternate hypothesis (Ha) is a statement that the observed differences or relationships between the populations are real and are not the results of chance or an error in the sampling approach.

Independent variable: An independent variable is an input or process variable that can be set directly to achieve a desired result. A variable that we control during an experiment.

Indirect cost: The costs that are imposed on an output that is not directly related to the cost of the incoming materials or the activities that transform it into an output. It is all the support costs that are needed to run the business that are applied against the product in order to make a profit, for example, the cost of accounting, personnel, ground maintenance, etc.

Indirect poor-quality costs: Costs that are incurred by the customer or costs that result from the negative impact poor quality has on future business, or lost opportunity costs.

Inherent process capability: The range of variation that will occur from the predictable pattern of a stable process.

Initiating sponsor: Individual/group who has the power to initiate and legitimize the change for all of the affected individuals.

Innovation: Converting ideas into tangible products, services, or processes.

Intangible benefits: These benefits are gains attributed to an improvement project that are not documented in the formal accounting process.

They are often called soft benefits. Frequently they are savings that result from preventive action that stops errors from occurring.

Interaction: If the effect of one factor is different at different levels of another factor, the two factors are said to interact or to have interaction.

Internal error costs: The costs incurred by the organization as a result of errors detected before the organization's customer accepts the output. In other words, these are the costs the organization incurs before a product or service is accepted by the customer because someone did not do the job right the first time.

Interrelationship diagrams: A way to graphically map out the cause-and-effect links among related items.

Interviewing: A structured discussion with one or more other people to collect information related to a specific subject.

ISO 9000 Series: A group of standards released by the International Organization for Standardization, Zurich, Switzerland, that defines the fundamental building blocks for a quality management system and the associated accreditation and registration of QMS.

IT applications: All of the IT tools that are used to bring about performance improvement in the organization. They are usually used to eliminate the tedious jobs that employees continuously do and to reduce the potential for employees making errors.

Just-in-time: A major strategy that allows an organization to produce only what is needed—when it's needed—to satisfy immediate customer requirements. Implemented effectively, the just-in-time concept will almost eliminate in-process stock.

Kaikaku: A transformation of thinking. This is a revolutionary type of activity, while Kaizen is evolutionary. Kaikaku is similar to process re-engineering or redesign. Kaikaku can also be described as a series of Kaizen activities completed in unison, forming and exhibiting the presence of a Lean mindset.

Kaizen: A Japanese term that means continuous improvement. *Kai* means "change" and *zen* means "good or for the better."

Kaizen blitz: This means a sudden, overpowering effort to take a process, system, product, or service apart and put it back together in a better way.

Kakushin: Kakushin is innovation. The creation of something completely new. It is essential for the growth of all companies.

Kanban: Usually a printed card that contains specific information related to parts, such as the part name, number, quantity needed, etc. It is the primary communication used in just-in-time manufacturing. It is used to maintain effective flow of materials through an entire manufacturing system while minimizing inventory and work-in-process. It is used in place of complex production control computer systems.

Kano model: A model that was created by Prof. Noriaki Kano that classifies customer preferences into five categories: attractive, one-dimensional, must-be, indifferent, and reverse. The Kano model of customer satisfaction classifies product attributes based on how they are perceived by the customer and their effect on customer satisfaction. This model is useful in guiding the organization in determining when good is good enough and more is better.

Key performance indicator (KPI): These measurements indicate the key performance parameters related to a process, organization, or output. They are the key ways by which that item is measured and are usually used to set performance standards and continuous improvement objectives. They are sometimes called critical performance indicators (CPIs).

Knowledge management: A system for capturing the knowledge that is contained within an organization. It groups knowledge into two categories. The first classification is tacit knowledge (soft knowledge). This knowledge is undocumented, intangible factors embodied in an individual's experience. The second classification is explicit knowledge (hard knowledge). This knowledge is documented and quantified.

Kruskal-Wallis one-way analysis: A way to look at differences among the population's medians. It is a hypothesis test of the equality of population medians for a one-way design with two or more populations. It offers a nonparametric alternative to the one-way analysis of variance.

Latin square designs: Essentially a fractional factorial experiment that requires less experimentation to determine the main impacting areas. It is used when it is desirable to allow for two sources of non-homogeneity in the conditions affecting the test. This approach is limited by two conditions:

- There should be no interactions between rows and column factors because these can't be measured.
- The number of rows, columns, and treatments must be the same.

Lean Manufacturing: A focus on eliminating all waste in the manufacturing process. It includes Lean principles like:

- Zero inventory
- Batch to flow, cutting batch size
- Line balancing
- Zero wait time
- Pull instead of push production control systems
- Cutting actual process time

Lean metric: Lean metrics allow companies to measure, evaluate, and respond to their performance in a balanced way, without sacrificing the quality to meet quantity objectives, or increasing inventory levels to achieve machine efficiencies. The type of the Lean metric depends on the organization and can be of the following categories: financial performance, behavioral performance, and core process performance.

Lean thinking: A focus on eliminating all waste within the processes, including customer relations, product design, supplier networks, production management, sales, and marketing. Its objective is to reduce human effort, inventory, cycle time, and space required to produce customer-deliverable outputs.

Levels of a factor: The various values of a factor considered in the experiment are called levels.

Level (of a variable): The point at which an independent variable is set during a trial.

Line graph: The simplest graph to prepare and use is the line graph. It shows the relationship of one measurement to another over a period of time. Often this graph is continually created as measurement occurs. This procedure may allow the line graph to serve as a basis for projecting future relationships of the variables being measured.

Loss function: The mean-square deviation of the object's characteristics from their targeted value. It is used to determine the financial loss that will occur when the quality characteristic deviates from the target value.

Lost-opportunity poor-quality costs: Lost profits caused by poor internal performance. Example: Lost sales because the salesperson did not show up on time or did not do a good job of selling the service. The lost sales that occurred as a result of engineering or manufacturing problems that resulted in the products or services not being available as initially scheduled.

Machine capability index (Cmk): A short-term machine capability index derived from observations from uninterrupted production runs. The preferred Cmk value is greater than 1.67. The long-term machine capability index should be greater than 1.33.

Main effect: The average effect of a factor is called the main effect of the factor.

Management presentations: A special type of formal meeting of work groups and their managers.

Mann-Whitney *U* tests: A hypothesis test that is used as a nonparametric alternative to the two-sample *t*-test. It tests the equality of two population medians and calculates the corresponding point estimates and confidence intervals.

Market segmentation: This occurs when the total market for an individual product or service is sub-divided into smaller groups based upon the individual characteristics of the group. This allows different market strategies to be applied to the segmented market areas.

Master Black Belt: Master Black Belts are experts in the Six Sigma methodology. They are responsible for the strategic implementation of Six Sigma throughout the organization; training and mentoring Black Belts and Green Belts; conducting complex Six Sigma improvement projects; developing, maintaining, and revising the Six Sigma materials; and applying statistical controls to difficult problems that are beyond the Black Belts' knowledge base.

Matrix diagrams: A way to display data to make them easy to visualize and compare.

Mean: The average data point value within a data set. It is calculated by adding all of the individual data points together, then dividing that figure by the total number of data points.

Measure of dispersion: Dispersion within data is calculated by subtracting the high value from the low value (range).

Measurement error: The error that is inherent in every measurement that is taken. No measurement is precise. Measurement error can be caused by many factors, including human error, equipment precision, and equipment calibration.

Measurement systems analysis: An evaluation of the goodness of an individual. There are four characteristics that need to be examined:

- Sensitivity—This should be no greater than one-tenth of the total tolerance in the specification being measured.
- Reproducibility—The ability of the measurement to repeatedly get the same answer.
- Accuracy—How near the true value (the international standard value).
- Precision—The ability to get the same value using the same operator and the same setup.

Measurement tools: Any object that is used to compare another object to a set of defined standards. It can be a ruler, a gauge, an oscilloscope, a scale, etc.

Median control charts: Median charts are used when an odd number of readings are made. This makes their median value more obvious. Another version records the data and plots the median value and range on two separate graphs.

Method of least squares: A statistical procedure to define the best fit's straight line for a series of points plotted on a graph.

Midrange: The midpoint between the highest and the lowest value of a set of data. It is calculated by adding the highest value and the lowest value together and dividing by two.

Milestone graph: Shows the goals or target to be achieved by depicting the projected schedule of the process. A primary purpose is to help organize projects and coordinate activities.

Mind maps: An unstructured cause-and-effect diagram. Also called mind-flow or brain webs.

Mistake proofing: Methods that help operators avoid mistakes in their work caused by choosing the wrong part, leaving out a part, installing a part backward, etc. Also called mistake proofing, poka-yoke (error proofing), and baka-yoke (foolproofing).

Mixture design: In this type of experiment the measured responses are assumed to depend upon the relative proportions of the ingredients or components in the mixture and not upon the amount of the mixture. This is an example when mixture and process examples are treated together. The fact that the portions of the different factors must be summed to 100% complicates the experiment design as well as the analysis of the mixture experiment. There are a number of different mixture design methodologies. The most frequently used one is the simplex-lattice design.

Mood's median test: This tests the equality of medians from two or more populations. It is sometimes just called the median test.

MTBF: Mean time between failures. The average time between mechanical breakdowns.

Muda: Any activity that consumes resources without creating value for the customer. This can consist of activities that cannot be eliminated immediately, or those that can be eliminated quickly through Kaizen.

Multiple linear regressions: An extension of the linear regression approach where only one independent variable is used. By increasing the number of independent variables, a higher proportion of the variation can be analyzed.

Multivariance analysis: Often variation within the output is different from piece-to-piece and time-to-time variation. Variation analysis uses a chart to investigate the stability or consistency of a process. The chart contains a series of vertical lines or other schematics, along a *y* timescale. The length of each line represents the range of values detected in each of the samples. A typical example might be a machined piece of steel that could be measured at a number of different points to determine the variation across a single surface.

Mura: Unevenness in an operation; for example, a gyrating schedule not caused by end-consumer demand but rather by the production system, or an uneven work pace in an operation causing operators to hurry and then wait. Unevenness often can be eliminated by managers through level scheduling and careful attention to the pace of work.

Muri: Overburdening equipment or operators by requiring them to run at a higher or harder pace with more force and effort for a longer period of time than equipment designs and appropriate workforce management allow.

***MX* bar–*MR* charts (moving range charts):** Used in place of an *X* bar and *R* chart when the data are not readily available. These are two separate but related charts, one that plots averages and one that plots range. The same calculations that are used to calculate *X* bar and *R* charts are used for these charts also. Typically the last three parts are added together and averaged to plot the most recent plot.

Negative analysis: A method used to define potential problems before they occur and develop countermeasures.

Nominal group technique (NGT): A special purpose technique, useful for situations where individual judgments must be tapped and combined to arrive at decisions.

Noncomparative experiments: An experiment whose objective is the determination of the properties or characteristics of a population.

No-value-added costs: The costs of doing activities that the customer would not want to pay for because it adds no direct value to him or her. It can be further divided into business-value-added, no-value-added, and bureaucracy costs. It also includes appraisal costs.

Normal distribution: Occurs when frequency distribution is symmetrical about its mean or average.

Normal probability plots: Used to check whether observations follow a normal distribution. $p > 0.05$ = data are normal.

One-piece flow: A concept where a single piece of work moves between workstations instead of a batch process.

On-off technique: A way to direct attention to information on the screen or to the presenter during a presentation. In other words, turn off the projector when you want attention focused on you and not on the screen.

Operational process capability: Determined by the manner in which the process is operated in respect to how this predictable pattern meets specification requirements.

Opportunities: The way Six Sigma error rates are measured. It is anything within the product, process, service, or system that could cause an error that would make the output less than ideal in the customer's eyes. Opportunities are the things that must be right to satisfy the customer. They are not the number of things that could possibly go wrong within the process. For example, in typing a five-letter word, there are five opportunities for making an error that the customer would be dissatisfied with.

Opportunity cycle: A problem-solving cycle that was developed in support of TQM. It consists of five phases: protection, analysis, correction, measurement, and prevention. During the protection phase, action is taken to protect the customer from receiving a defective product. During the analysis phase, data are collected to determine how to correct the problem. During the correction phase, the proposed solution is implemented. During the measurement phase, the results of the corrective action are monitored to ensure it corrected the original problem. If it corrected the problem, the preventive activity is removed from the process. The last phase is prevention, where the information learned during the cycle is applied to any other process, part, or system that could have the same type of problem.

Organizational cultural diagnostics (cultural landscape): One of the organizational change management tools. It is a survey that is conducted to define strengths and weakness related to the organization's culture, based upon the individual's perception of how the organization's culture will impact the implementation of the proposed change. Changes that are in line with the organization's culture are easy to implement. Changes that are in direct conflict with the organization's culture are usually doomed to failure.

Organizational excellence: This methodology is made up of five key elements, called the *five pillars*, which must be managed simultaneously to continuously excel. The five pillars are:

- Process management
- Project management
- Change management
- Knowledge management
- Resource management

Organization change management: A methodology designed to lessen the stress and resistance of employees and management to individual critical changes.

Origin: The point where the two axes on an *X-Y* graph meet. When numbers are used, their value is increased on both axes as they move away from the origin.

Other point of view (OPV): A method that aids in idea generation and evaluation by careful examination of the views of stakeholders involved. It is generally more effective when used early in the process, for idea generation as opposed to idea evaluation. It also tends to be more effective with small groups (two or three people) than with larger ones.

Outcome (response)-dependent variable: The result of a trial with a given treatment combination of *X*'s is called a response (*Y*).

Overall equipment effectiveness (OEE): An effective tool to assess, control, and improve equipment availability, performance, and quality. This is especially important if there is a constraining piece of equipment.

PPM: Parts per million. Typically in Six Sigma, it is used for defects per million opportunities. It is often referred to as DPMO.

Pareto diagrams: A type of chart in which the bars are arranged in descending order from the left to the right. It is a way to highlight the "vital few" in contrast to the "trivial many."

Pattern and trend analysis: Typically, graphic charts are used to analyze changes, both positive and negative, in processes and outputs. Data are usually presented in either summary (static) or time sequence. Analyzing these graphs and/or charts results in detecting:

- Upward trends
- Downward trends
- Unusual variation
- Cycles
- Process shifts
- Increased variability

A typical application would be to plot a first-time yield by week. A prioritization matrix is a means to determine which factors have the biggest impact upon an individual item. The group first defines the criteria that will be used to evaluate the item and weights each criterion from 0 to 1. The sum of all the criteria can be no greater than 1. They then define the factors that will impact the item and prepare a table. The factors are listed along the vertical axis and the criteria along the horizontal axis. Each factor is then evaluated based upon the criteria on a scale of 1 to 10, where 10 is the highest impact. This is multiplied times the weighting factor for specific criteria.

By summarizing the prioritization number for each factor, you will obtain a weighed prioritization for each factor. The higher the number, the more priority that factor should be given.

***p*-Charts:** A type of attribute control chart that shows the percentage of defective units, used when the sample size varies.

Performance goals: Quantifies the results that will be obtained if the business objectives are satisfactorily met.

Performance improvement plan (PIP): A 3-year plan designed to align the environment within an organization with a series of vision statements that drive different aspects of the organization's behaviors.

Performance plan: A contract between management and the employees that defines the employees' roles in accomplishing the tactics, and the budget limitations that the employees have placed upon them.

Performance standard: Defines the acceptable error level of each individual in the organization.

PERT charts: PERT stands for program evaluation review technique. This is a methodology that was developed by the U.S. government in the 1950s. It is a project management tool used to schedule, organize, and coordinate tasks within the project. It provides an effective way of determining interdependencies between activities and timing. It allows for the critical path through the project to be readily defined.

Pictorial graphs: A way to represent data using pictures. Pictograms are a type of pictorial graph in which a symbol is used to represent a specific quantity of the item being plotted. The pictogram is constructed and used like bar and column graphs.

Plackett-Burman design (PBD): This design is best used in screening experiments. It uses a two-level design with a run number where the run number is a multiple of 4. For two-level nongeometric designs, the PBD run numbers would be 12, 20, 24, 28, etc. For example, if the run number was 12, you may be looking at 11 different factors. With run numbers of 20, you would be looking at 19 different factors.

Plan-Do-Check-Act: A structured approach for the improvement of services, products, and/or processes developed by Walter Shewhart.

Plus minus interesting: An idea evaluation weapon that analyzes the idea or concept by making a list of positive (+) and negative (–) things related to the idea or concept. It also uses a third column, called "Interesting," where random thoughts about the item being evaluated are recorded. A technique often used to evaluate a solution that may initially seem like a bad idea.

***pn*-Chart:** A type attribute control chart that shows the number of defect units, used when the sample size is constant.

Point of use storage (POUS): Storing production parts and materials as close as possible to the operations that require them.

Poisson distribution: An approximation of the binomial when p is equal to or less than 0.1 and the sample size is fairly large (p = probability). It is used as distribution of defect counts and can be used as an approximation of the binomial. It is closely related to the exponential distribution. It is used to model rates, like errors per output, inventory turns, or arrivals per hour.

Policy deployment: An approach to planning in which organization-wide long-range objectives are set, taking into account the organization's vision, its long-term plan, the needs of the customers, the competitive and economic situation, and previous results.

Poor-quality cost (PQC): This was an improvement on the quality cost system developed in the 1950s by Val Feigenbaum at General Electric. It extended the concept from direct quality cost to direct and indirect quality cost. It contains the following categories:

I. Direct poor-quality cost
   A. Controllable poor-quality cost
      1. Preventive cost
      2. Appraisal cost
      3. No-value-added cost
   B. Resultant poor-quality cost
      1. Internal error cost
      2. External error cost
   C. Equipment poor-quality cost

II. Indirect poor-quality cost
   A. Customer-incurred cost
   B. Customer dissatisfaction cost
   C. Loss of reputation cost
   D. Lost opportunity cost

It is a methodology that defines and collects costs related to resources that are wasted as a result of the organization's inability to do everything correct every time. It includes both direct and indirect costs.

Portfolio project management: A technique used to manage all of the projects going on within a specific area. In the past when projects were managed independently, resources were not always assigned in the best manner. This technique optimizes the success of the critical projects that have priority within the organization.

Positive correlation: This occurs when both variables increase or decrease together. Negative correlation is when one variable increases while the other one decreases.

Prevention costs: All the costs expended to prevent errors from being made or, to say it another way, all the costs involved in helping the employee do the job right every time.

Preventive action: Action taken that will eliminate the possibility of errors occurring rather than reacting from errors that occurred. It is a long-term, risk-weighted action that prevents problems from occurring based on a detailed understanding of the output and/or the processes that are used to create it. It addresses inadequate conditions that may produce errors.

Primary functions: Those for which the process was designed.

Probability density function: Applied to a histogram. It is used to calculate the probability that a single sample, drawn randomly from the population, will be less than a specific value. In some cases, it is used in just the opposite mode to determine the probability of the single sample being greater than a specified value. It defines the behavior of a random variable, and it is usually used as a shape of a distribution, frequently in a histogram format.

Probability plots: Typical plots where the values of the item being measured are divided into small segments across the horizontal axis, and the number of occurrences within that measurement segment are plotted on the vertical axis. A histogram is a typical example.

Problem tracking log: A systematic way to categorize, monitor, and measure progress of the corrective action process. It is designed to ensure that the correct amount of resources is applied to solving all important problems.

Process: A series of interrelated activities or tasks that take an input and provide an output.

Process benchmarking: A systematic way to identify superior processes and practices that are adopted or adapted to a process in order to reduce cost, decrease cycle time, cut inventory, and provide greater satisfaction to the internal and external customers.

Process capability index (Cp): A measure of the ability of a process to produce consistent results. It is the ratio between the allowable process spread (the width of the specified limits) and the actual process spread at the ±3 sigma level. For example, if the specification was ±6 and the 1 sigma calculated level was 1, the formula would be 6 divided into 12 equals a Cp of 2.

Process capability study: A statistical comparison of a measurement pattern or distribution to specification limits to determine if a process can consistently deliver products within those limits.

Process control: A way the process is designed and executed to maximize the cost-effectiveness of the process. It includes process initiation, selection of the process steps, selection of alternative steps, integration of the individual activities into the total process, and termination of the process. Too frequently, process control and process control charts are used interchangeably, and they should not be.

Process decision program charts (PDPC): A method that maps out the events and contingencies that may occur when moving from an identified problem to one or more possible solutions. It is used to look at various contingencies to steer the project in the required direction, to define countermeasures, or to get the project back on track. It basically uses a tree diagram and extends the charts a couple of levels by identifying risks and countermeasures from the bottom-level tasks in the tree matrix. It is one of the seven tools for management and planning.

Process elements: The sub-units that make up a process. They are normally referred to as activities, and the sub-units to the activities are tasks.

Process flow animation: A process model that shows the movement of transactions within the process and how outside functions impact the process's performance.

Process improvement team: A group of employees assigned to improve a process. It is usually made up of employees from different departments.

Process maturity grid: A six-level grid that sets standards for a process as it matures in its overall performance. The six levels are:

Level 6: Unknown
Level 5: Understood
Level 4: Effective
Level 3: Efficient
Level 2: Error-free
Level 1: World class

Detailed requirements for each level are defined, broken down into eight different categories.

Process owner: The individual responsible for the process design and performance. He or she is responsible for the overall performance from the start of the process to the satisfaction of the customer with the delivered output. It is the responsibility of the process owner to ensure that sub-optimization does not occur throughout the process, as well as setting improvement performance goals for the process.

Process performance analysis: The collection of performance data (efficiency and effectiveness data) at the activities or task level of a flowchart that is used to calculate the performance of the total process.

Process performance matrix: The efficiency, effectiveness, and adaptability measurements related to the process. Particular focus is paid to the effectiveness measurements because they need to reflect customer requirements.

Process qualification: A systematic approach to evaluating a process to determine if it is ready to ship its output to an internal or external customer.

Process redesign: A methodology used to streamline a current process with the objective of reducing cost and cycle time by 30 to 60% while improving output quality from 20 to 200%.

Process re-engineering: A radical methodology that challenges all the paradigms that the organization has imposed on the process. It is usually used when the present process is so obsolete or so bad that you don't want to influence the new process in the design concept. Typically a process re-engineering project takes 6 to 9 months to complete and is used when cost and cycle time need to be reduced by more than 60%.

Process simplification: A methodology that takes complex tasks, activities, and processes and bisects them to define less complex ways of accomplishing the defined results.

Process simulation: A technique that pictorially processes resources, products, and services in a dynamic computer model.

Project champion: The individual who makes sure that the project has the resources and cross-functional support that are needed to be successful. The project champion is the individual that is most accountable to the executive team for the overall results of the project.

Project communications management: A subset of project management that includes the processes required to ensure timely and appropriate generation, collection, dissemination, storage, and ultimate disposition of project information.

Project cost management: A subset of project management that includes the processes required to ensure that the project is completed within the approved budget.

Project decision analysis: The approach that is used in making a decision to start or continue a project. It includes a cost-benefits analysis, an impact analysis on the organization and its support to the strategic plan, an evaluation of the risks associated with the project, and its impact upon the customer. Each risk factor needs to be identified and evaluated. Project risk factors are evaluated by the sum of the probability occurrence times the consequences of the risk.

Project financial benefit analysis: An analysis that is conducted at least at each checkpoint in the process. It evaluates the potential savings compared to the cost of making the change. Early in the project both potential savings and cost are estimated. When the project has been implemented, actual project financial benefit analysis figures can be provided.

Project human resource management: A subset of project management that includes the processes required to make the most effective use of the people involved with the project.

Project integration management: A subset of project management that includes the processes required to ensure that the various elements of the project are properly coordinated.

Project management: The application of knowledge, skill, tools, and techniques to project activities to meet or exceed stakeholders' needs and expectations from the project. It includes the following:

- Project integration management
- Project scope management
- Project time management
- Project financial/cost management
- Project quality management
- Project resource management
- Project communication management
- Project risk management
- Project procurement management
- Project organizational change management
- Project document/configuration management
- Project planning and estimating management

A primary responsibility of the Master Black Belts, Black Belts, and Green Belts.

Project quality management: A subset of project management that includes the processes required to ensure that the project will satisfy the needs for which it was undertaken.

Project risk management: A subset of project management that includes the processes concerned with identifying, analyzing, and responding to project risk.

Project scope: The boundaries within which the project will work; helps prevent project creep.

Project scope management: A subset of project management that includes the processes required to ensure that the project includes all the work required, and only the work required, to complete the project successfully.

Project selection matrix: A matrix that analyzes the various improvement opportunities to define the ones that should be approved or continued. A number of factors need to be considered. Typical factors are:

- Impact on a customer
- In line with the strategic objectives
- Financial returns
- Competitive advantage, etc.

Project time management: A subset of project management that includes the processes required to ensure timely completion of the project.

Pugh concept selection/Pugh matrix: A scoring matrix used to prioritize the selection of improvement opportunities. It is also used to select options in the design phase. The selection is based on the consolidated scores. It is typically done after the voice of the customer has been captured.

Pugh technique: This technique compares the alternatives to the present process. First, a list of key process characteristics is generated by the PIT. Each alternative solution is then compared, characteristic by characteristic, to the present process. If the proposed solution will provide better results than the present process, it is given a plus (+); if it is the same, an *s* is recorded; if it has provided worse results, it is given a minus (–).

Pull system: A production control system that replaces parts and components only when the previous part or component has been consumed. It is designed to eliminate in-process storage and is part of a just-in-time system.

PUSIC: Plan, Understand, Streamline, Implement Continuous Improvement. This technique is the basic ingredient of business process improvement. It is used with re-engineering, redesign, benchmarking, and fast-action solution teams.

Qualification: Acceptable performance of a complete process consisting of many operations that have already been individually certified. For a process to be qualified, each of the operations and all of the equipment used in the process must be certified. In addition, the process must have demonstrated that it can repeatedly produce high-quality products or services that meet specifications.

Qualitative data: Data related to counting the number of items and cannot be broken down into smaller intervals. It is count rather than measurement data. For example: The number of machines shipped in a specific time period.

Quality at the source: Building quality into value-adding processes as they are completed. This is in contrast to trying to "inspect in quality," which only catches mistakes after they have been made. An effective quality @ source campaign can minimize or eliminate much of the expense associated with traditional quality control programs.

Quality function deployment: A structured process for taking the voice of the customer and translating it into measurable customer requirements and measurable counterpart characteristics, and deploying those requirements into every level of the product and manufacturing process design and all customer service processes.

Quality management: All activities of the overall management function that determine the quality policy, objectives, and responsibilities and implement them by means such as quality planning, quality control, quality assurance, and quality improvement within the QMS (ISO 8402).

Quality management system/ISO 9000 (QMS): The organizational structure, procedures, processes, and resources required to determine the quality policy, objectives, planning, control, assurance, and improvement that impact, directly or indirectly, the products or services provided by the organization.

Quality manual: A document stating the quality policy and describing the QMS of an organization (ISO 8402).

Quality plan: A document setting out the specific quality practices, resources, and sequence of activities relevant to a particular product, project, or contract (ISO 8402).

Quality system: The organizational structure, procedures, processes, and resources needed to implement quality management (ISO 8402).

***R*-charts:** A simple range chart plotted in order to control variability of a variable.

Randomized block plans: This analysis approach is used when there are a large number of factors that need to be evaluated and it is desirable to keep all other conditions constant during each individual factor evaluation. With a randomized block plan, factors are grouped into categories where only one condition is varied and the other is held constant, thereby allowing the evaluation team to look at the variable within the block, ignoring the other variables that are occurring. This allows the major influencing factors to be defined for further evaluation.

Regression analysis: A statistical analysis assessing the association between two variables. It evaluates the relationship between the mean value of a random variable and the corresponding value of one or more independent variables.

Reliability analysis: A technique used to estimate the probability that an item will perform its intended purpose for a specific period of time under specific operating conditions.

Reliability management system: Designing, analyzing, and controlling the design and manufacturing processes so that there is a high probability of an item performing its function under stated conditions for a specific period of time.

Resource driver: Describes the basis for assigning cost from an activity cost pool to products or other cost objects.

Response: The numerical result of a trial based on a given treatment combination.

Resultant poor-quality costs: The costs that result from errors. These costs are called resultant costs because they are directly related to management decisions made in the controllable poor-quality costs category. It is divided into two sub-categories: internal error costs and external error costs.

Reverse engineering: The process of purchasing, testing, and disassembling competitors' products in order to understand the competitors' design and manufacturing approach, then using these data to improve the organization's products.

Rewards and recognition: Action taken to reinforce desired behavior patterns or exceptional accomplishments. Categories of rewards and recognition are:

- Financial compensation
- Monetary awards
- Group/team rewards
- Public personal recognition
- Private personal recognition
- Peer rewards
- Customer rewards
- Organizational awards

Risk analysis: An evaluation of the possibility of suffering harm or loss. A measure of uncertainty. An uncertain event or condition that, if it occurred, might have a positive or negative effect on the organization or the project.

Risk assessment: Performing a quantitative analysis of the risks and conditions to prioritize their effects on the project objectives or the organization's performance.

Robustness: The characteristics of a process output or process design that make it insensitive to the variation in inputs.

Robust process: A robust process operates at the Six Sigma level, producing very few defects even when the inputs to the process vary. They have a very high, short-term Z value and a small Z shift value. The critical element in a robust process is the u element.

Roll-through yield (RTY): *See* first-time yield.

Root cause analysis: The process of identifying the various causes affecting a particular problem, process, or issue and determining the real reasons that caused the condition.

Run charts: A graphic display of data, used to assess the stability of a process over time, or over a sequence of events (such as the number of batches produced). The run chart is the simplest form of a control chart.

SCAMPER: A checklist and acronyms made up of the following:

S Substitute
C Combine

A Adapt/adopt
M Modify/magnify/minify
P Put to other uses
E Eliminate
R Reverse/rearrange

This technique is used to generate ideas when each of these questions is asked. It is a technique generated by Michael Michalko.

Scatter diagrams: A graphic tool used to study the relationship between two variables. Scatter diagrams are used to test for possible cause-and-effect relationships. They do not prove that one variable causes the other, but they do show whether a relationship exists and reveal the character of that relationship.

Secondary functions: Those that support the primary functions or are of secondary importance.

Seven basic tools: Seven quality improvement tools that all employees should be familiar with and able to use. They were originally generated by Kaoru Ishikawa, a professor of engineering at Tokyo University and the father of quality circles. The seven tools are:

1. Cause-and-effect diagrams
2. Check sheets
3. Control charts
4. Histograms
5. Pareto charts
6. Scatter diagrams
7. Stratification

Shewhart cycle (PDCA): The same as Plan-Do-Check-Act.

Short-run charts: Short-run charts are used when the universe is so small that it is difficult, if not impossible, to obtain a large enough sample size for a standard control chart. The prerequisite for use of short-run charts is that there is some commonality between the measurements charted. In other words, the processes of the different runs are similar, but different enough that the normal control chart does not work well. When working with variable data, two different plotting methods should be considered: deviation charts and standardized charts. The method used will depend on the variation between processes. Short-run charts were developed to address the problems related to collecting several dozen measurements of a process before control limits can be calculated. It is often difficult to meet these requirements when the universe is very small.

Sigma: A Greek letter statisticians use to refer to the standard deviation of a population. Sigma and standard deviation are interchangeable. They are used as a scaling factor to convert upper and lower specified limits to Z.

Sigma conversion tables: A set of tables used to convert a sigma value into a percent of product that should meet requirements under normal conditions.

Signal-to-noise ratio (S/N ratio): A calculation to quantify the effects of variation in controllable factors resulting from variation in output. Single factors are defined as factors that strongly influence the mean response and usually have very little influence on variation of the output response, as they are controllable. Noise factors influence the variation in the output. They may or may not be controllable. When running an experiment, controllable conditions are varied to get the most desirable signal-to-noise ratio.

Simple language: A way to evaluate the complexity of writing. It indicates the grade level that a person who is reading the document should have reached in order to understand the document. Simple language produces documents that can be read at two grade levels lower than the lowest educational level of the person who will be using the documents.

Simple linear regression: A method that allows you to determine the relationship between a continuous process output (Y) and one factor (X). Its mathematical equation is $Y = b + mX$.

Simplification approaches: A series of techniques that focus on simplifying the way things are done. It could include things like the following:

- Combining similar activities
- Reducing amount of handling
- Eliminating unused data
- Clarifying forms
- Using simple English
- Eliminating no-value-added activities
- Evaluating present IT activities to determine if they are necessary
- Evaluating present activities to determine if IT approaches would simplify the total operations

Simulation modeling: Using computer programs to mimic the item (activity process or system) under study in order to predict how it will perform or to control how it is performing.

Single-minute exchange of dies (SMED) or quick changeover: SMED is an approach to reduce output and quality losses due to changeovers. Quick changeover is a technique to analyze and reduce resources needed for equipment setup, including exchange of tools and dies.

SIPOC: Suppliers, Inputs, Processes, Output, and Customers. It is used to help you ensure that you remember all the factors when mapping a process.

Six Sigma: Six Sigma is a rigid, systematic methodology that utilizes information (managing by fact) and statistical analysis to measure and improve an organization's performance by identifying and preventing errors. It can be thought of in three parts:

1. Metric: 3.4 defects per million opportunities.
2. Methodology: DMAIC/DFSS structured problem-solving tools.
3. Philosophy: Reduce variation in the organization and drive decisions based on knowledge of the customer.

Six Sigma matrix: Divided into four categories:

- Measuring customer opinion
- Determining customer critical to quality factors
- Measuring product outcome
- Correlating product outcomes to critical to quality factors (measure processes with a matrix that correlates to the organization's economics)

Six Sigma program: A program designed to reduce error rates to a maximum of 3.44 errors per million units, developed by Motorola in the late 1980s.

Six-step error prevention cycle: A process to prevent problems from occurring rather than fix them afterwards.

Six-step problem-solving cycle: A basic procedure for understanding a problem, correcting the problem, and analyzing the results.

Six-step solution identification cycle: A procedure for defining how to solve a problem or take advantage of an opportunity.

SMED: Single-minute exchange of dies. It is one of the Lean tools, and it is a key part of just-in-time programs. It is a methodology used to minimize the amount of time of changing a process over to produce another output.

Soft consensus: When some members would prefer a different solution but are willing to support the decision of the team.

Soft savings: Sometimes also referred to as intangible savings. It is the benefit you get from a change that is not directly reflected in the accounting system. It includes things like reduced cycle time, cost avoidance, improved employee morale, lost-profit avoidance, and higher levels of customer satisfaction.

Solution analysis diagrams: Designed to analyze all the possible effects of a proposed solution or cause.

Spearman rank correlation coefficient (rho): Often denoted by the Greek letter ρ (rho), a nonparametric measure of correlation. This is a measure of the association that requires that both variables be measured in at least an ordinal scale so that the sample or individuals to be analyzed can be ranked in two orderly series.

Spider diagrams/radar charts: Used to show or compare one or more sets of data to each other. Often used to indicate the status quo (current state) against the vision (future state).

Stakeholder analysis plan: A system to identify key stakeholders or individuals that have a stake in the overall success/failure of the process.

Standard deviation: An estimate of the spread (dispersion) of the total population based upon a sample of the population. Sigma (σ) is the Greek letter used to designate the estimated standard deviation.

Standard work: Standard work is a systematic way to complete value-added activities. Having standard work activities is a fundamental requirement of Lean Six Sigma organizations.

Statistical process control (SPC): Using data for controlling processes, making outputs of products or services predictable. A mathematical approach to understanding and managing activities. It includes three of the basic statistical quality tools: design of experiments, control charts, and characterization.

Statistical thinking: Having a complete situational understanding of a wide range of data where several control factors may be interacting at once to influence an outcome.

Statistics: Common sense put to numbers.

Storyboard: A series of pictures and accompanying narrative that is used to define how something is done or what is going on related to a problem or situation.

Strategy: The approach that will be used to meet the performance goals.

Stratification: A technique used to analyze data where the universal population is divided into homogeneous sub-groups that are analyzed independently. For example, the market may be stratified into individual market segments.

Structural roadblocks: Obstacles that must be overcome for a process or an organization to transform from one state into another.

Student's *t*-distribution: A combination of the standard normal random variables and the chi-square random variables analysis. It is calculated by dividing the standard, normal random variable by the square root of the chi-squared random variable by the degrees of freedom.

Supplier controls: The preventive measures that are put into place to minimize the possibility of suppliers providing an unacceptable product. They include things like supplier qualification, requirements placed on the supplier to be ISO 9000 certified, source inspection, receiving inspection, etc.

Supply chain management: The flow of items from raw materials to accepted products at the customer location. It is a methodology used to reduce cost, lead times, and inventory, while increasing customer satisfaction.

Surveys: A systematic way to collect information about a specific subject by interviewing people. Often, the interview takes the form of a series of questions that are presented to a target audience in either written or verbal form.

Sustaining sponsor: Individual/group who has the political, logistical, and economic proximity to the individuals.

SWOT analysis: Strengths, Weaknesses, Opportunities, and Threats analysis. It is used to help match the organization's resources and capabilities to the competitive environment that exists in their market segment. It is often used as part of the strategic planning process.

System: The organizational structure, responsibilities, procedures, and resources needed to conduct a major function within an organization or to support a common business need.

Systematic design: A very structured step-by-step approach to designing that was developed in Germany. It defines four main phases of the design process:

1. Clarification of the tests: Collect information, formulate concepts, identify needs.
2. Conceptual design: Identify essential problems and sub-factors.
3. Embodiment: Develop concepts, layouts, and refinements.
4. Detailed design: Finalize drawings, concepts, and generate documentation.

***t*-Test:** The *t*-test employs the statistic t with $n - 1$ degrees of freedom to test a given statistical hypothesis about a population parameter. It is used when the population standard deviation is unknown and is effective in small sample sizes (less than 30 items).

Tactic: How the strategies will be implemented.

Taguchi methods: Design of experiment approaches by Dr. Taguchi for use where the output depends on many factors without having to collect data using all possible combinations of values for these variables. It provides a systematic way of selecting combinations of variables so that their individual effects can be evaluated.

Takt time: Takt time is the rate at which a completed item leaves the last step in the process. It should be equivalent to the rate at which customers, internal or external, require the output. It drives the pull system, as it eliminates the need for in-process stock. The process should be designed so that each step in the process is operating at the same Takt time as the sales process. This is the ideal situation that keeps the process in continuous flow without buildup within the process or between processes.

Team: A team is a small group of people who work together that realize their interdependencies and understand that both personal and team goals are best accomplished with mutual support.

Team charter: It is preferable that the team charter is defined by the Six Sigma leadership team. It is the major contribution they can make by providing clear direction and expectations. The team charter does not map the route for the project but does provide the boundaries and destination. It includes project objectives, project process boundaries, limitations, key deliverables, outside resources, and indicators/targets.

Team management: The coordination and facilitation of the activities that go on within the team to ensure the effectiveness and efficiency of the team are optimized, and the desired results are accomplished on schedule within cost.

Theory of constraints (TOC): There is one point in every system that limits the flow through the process. The theory of constraints is used to identify these bottlenecks and eliminate them. This is a set of tools that examines the entire system to define continuous improvement opportunities. It consists of a number of tools. For example:

- Transition tree
- Prerequisite tree
- Current reality tree
- Conflict resolution diagram
- Future reality tree

Three-factor, three-level experiments: This provides a three-dimensional look at a process or problem. Often a three-factor experiment is required after screening a large number of variables. These experiments may be full or fractional factorials. Usually the positive and negative levels in two-level designs are expressed as 0 and 1 in the design categories. Three-level designs often use 0, 1, and 2.

Throughput yield (TPY): The yield that comes out of the end of a process after any errors that are detected have been scrapped or reworked and re-entered into the process. Effective rework procedures can often increase first-time yield from 10% to a throughput yield of 100%.

Time-related process capability index (Cpk): This takes into account the drift that a product will have over time caused by common variation. Caused by things like different operators, different set-ups, allowable differences in material, etc. For all customer impact measurements, a Cpk of at least 1.33 is the normal accepted standard unless the product is screened to protect the customer.

Tollgate: Process checkpoints where deliverables are reviewed and measured, and readiness to move forward is addressed. Usually if a total project has not completed all of its commitments that are due at a tollgate, the project does not progress to the next level until these commitments are met. Typically this is a management review to determine if the project should continue.

Total cost management: A comprehensive management philosophy for proactively managing an organization's total resources (material, capital, and human resources) and the activities that consume those resources.

Total Productive Maintenance (TPM): A methodology used to keep the equipment within the organization at peak operating efficiency, thereby eliminating equipment downtime.

Total productivity management: A methodology designed to direct the organization's efforts at improving productivity without decreasing quality. It is designed to eliminate waste by involving employees, effective use of information technology, and automation.

Total Quality Management (TQM): A methodology designed to focus an organization's efforts on improving the quality of internal and external products and services. ISO 8402 defines it as "a management approach of an organization, centered on quality, based on the participation of all its members and aiming at long-term success through customer satisfaction and benefits to the members of the organization and to society." TQM is a conceptual, philosophical, and structured group of methodologies that require management and human resource commitment to the embodiment of a philosophy where all of the management, employees, processes, practices, and systems throughout the organization understand their customers, both internal and external, and provide them with organizational performance that fulfills or exceeds the customers' expectations. It is part of the evolution from quality control to statistical quality control to total quality control, and it embodies all the criteria now included in all of the international, national, and local quality award systems.

Treatment combination: The set of levels of all factors included in a trial in an experiment is called a treatment or treatment combination.

Tree diagrams: A systematic approach that helps the user think about each phase or aspect of solving a problem, reaching a target, or achieving a goal.

Trial: An observation made with all of the variables set at predesigned levels and held constant during the duration of the observation.

Tribal knowledge: Any unwritten information that is not commonly known by others throughout the organization. It is part of a total organization's knowledge assets and one that is frequently lost as individuals change jobs or leave the organization. Unlike other forms of intellectual assets, tribal knowledge cannot be converted into company property unless it is transformed into a hard knowledge base.

TRIZ: A methodology that was developed in Russia and stands for "theory of innovative problem solving" originated by Genrich Altshuller in 1946. It is effective at identifying low-cost improvement solutions during the Define or Identify phase. It is also helpful in defining the root cause of defects. This approach expands on systems engineering methodologies and provides a powerful system management method for problem definition and failure analysis. It is an effective approach to generating innovative ideas and solutions to problems. TRIZ is a Russian acronym.

Types of data: There are basically two major groupings of data. They are:

- Attributes data: The kind of data that are counted, not measured. They are collected when all you need to know is yes or no, go or no go, accept or reject.
- Variables data: Variables data are used to provide a much more accurate measurement than attributes data provide. It involves collecting numeric values that quantify a measurement, and therefore requires a smaller sample to make a decision.

Types of teams: There are many different types of teams that are identified by different properties related to the team organization and objectives. Typical teams are:

- Department improvement teams, focusing on individual area improvement opportunities
- Quality circles, voluntary teams that form themselves
- Process improvement teams, typically working across functions, focusing on optimizing a total process typified by process re-engineering and process redesign
- Task forces, typified by an emergency that occurs within an organization
- Natural work teams, made up of individuals who are brought together to perform ongoing activities

***u*-Charts:** Plot the number of defects per unit with a varying sample size.

Value-added analysis (VA): A procedure for analyzing every activity within a process, classifying its cost as value-added, business-value-added, and no-value-added, and then taking positive action to eliminate the no-value-added cost and minimize the business-value-added.

Value stream: All of the steps/activities (both value-added, business-value-added, and no-value-added) in a process that the customer is willing to pay for.

Value stream mapping: This tool is used to help you understand the flow of materials and information as an item makes its way through the value stream. A value stream map takes into account not only the item but also the management and information systems that support the basic item. This is helpful in working with cycle time reduction problems and is primarily used as part of the Lean toolkit.

Variable control charts: A plot of variables data of some parameter of a process's performance, usually determined by regular sampling of the product, service, or process as a function (usually) of time or unit numbers or other chronological variable. This is a frequency distribution plotted continuously over time, which gives immediate feedback about the behavior of a process. A control chart will have the following elements:

- Center line (CL)
- Upper control limit (UCL)
- Lower control limit (LCL)

Variables data: The kind of data that are always measured in units, such as inches, feet, volts, amps, ohms, centimeters, etc. Measured data give you detailed knowledge of the system and allow for small, frequent samples to be taken. These are data that are equivalent to quantitative data. There are two types of variables data: discrete (count-type data) and continuous data.

Variation: A measure of the changes in the output from the process over a period of time. It is typically measured as the average spread of the data around the mean and is sometimes called noise.

Vision: A description of the desired future state of an organization, process, team, or activity.

Vision statement: A group of words that paints a clear picture of the desired business environment 5 years in the future. A vision statement should be between two and four sentences.

Visual controls: Visual controls are tools that tell employees "what to do next," what actions are required. These often eliminate the need for complex standard operating procedures; promote continuous flow by eliminating conditions that would interrupt flow before it happens.

Visual factory/visual office: A system of signs, information displays, layouts, material storage, and equipment storage. It uses color coding and error proofing devices. The five S's are part of visual controls and the visual office. Typical tools used in the visual office or control center would be a continuously updated electronic sign indicating the number of clients that are waiting for their phone call to be answered, or the length of time it takes to respond to a phone inquiry.

Vital few: The 20% of the independent variables that contribute to 80% of the total variation.

Voice of the business (VOB): The stated and unstated needs and requirements of the organization and its stakeholders.

Voice of the customer (VOC): The customer's expression of his or her requirements, in his or her own terms. It describes the stated and unstated needs and requirements of the external customer.

Voice of the employee (VOE): The term used to describe the stated and unstated needs and requirements of the employees within your organization.

Voice of the process (VOP): The term used to describe what the process is telling you about what it is capable of achieving.

Waste: Anything in your processes that your customer is unwilling to pay for; extra space, time, materials quality issues, etc.

Waste elimination: The ability to apply Lean Six Sigma concepts and tools to eliminate identified wastes.

Waste identification: The ability to "see" waste in your organization. This includes organizing waste into one of its nine defined categories.

Weibull distribution: A continuous probability distribution with the probability density function. It is frequently used in analyzing the field life data rate due to its flexibility. It can mimic the behavior of other statistical distributions, such as the normal and the exponential. An understanding of the failure rates often provides insight into what is causing the failure.

Work breakdown structure (WBS): A Gantt chart used in project management to monitor and plan the activities related to doing the project as well as defining their interrelationships and their present status.

Work flow monitoring: An online computer program that is used to track individual transactions as they move through the process to minimize process variation.

Work standards: When work standards are practiced, everyone in the organization is committed to performing the work in the same best way. Work standards include documentation methods and developing engineering standards to set the expectation and measurement matrix. They provide job aids and training to the employees that effectively communicate the best ways to perform an activity and set the minimum performance standard for the trained employee.

World-class operations benchmarking: A form of external benchmarking that extends the benchmarking approach outside the organization's direct competition to involve dissimilar industries.

***X*:** In Six Sigma *X*'s are all the inputs that are required to produce the output *Y*. It includes the materials, procedures, process, and suppliers.

***X* bar:** Simply the mean of a population.

***X* bar–*S* charts (*X* bar Sigma charts):** These control charts are often used for increased sensitivity to variation.

***X-MR* charts:** Individual readings are plotted on these charts, and a moving range may be used for short runs and in the case of destructive testing.

$\bar{X}$-*R* control chart: An important statistical tool that can be used to signal problems very early and thus enable action to be taken before large volumes of defective output have occurred.

***X-Y* axes graph:** A pictorial presentation of data on sets of horizontal and vertical lines called a grid. The data are plotted on the horizontal and vertical lines, which have been assigned specific numerical values corresponding to the data.

***Y*:** In Six Sigma *Y*'s are the outputs from a process.

Yellow Belt (YB): A Yellow Belt typically has a basic understanding of Six Sigma, but does not have the experience, training, or capability to lead projects on his/her own. They work on special assignments to assist Green Belts and Black Belts in developing and implementing Six Sigma projects. As Yellow Belts gain experience, they become good candidates for Green Belt training.

Z: Data points positioned between the mean and another location as measured in standard deviations. *Z* is a measure of process capability and corresponds to the process sigma value.

Zero defects: This was a complete system directed at eliminating all defects from a product. It was originated by Phil Crosby on a military contract and spread throughout the world. It sets a higher standard for performance than Six Sigma by 3.4 defects per million opportunities. It focused on perfection, which is impossible to reach but should be our objective.

Zmin: The distance between the process mean and the nearest spec. limit (upper or lower) measured in standard deviation (sigma) units.

Appendix B: The Six Sigma Body of Knowledge

The following is a list of the Six Sigma body of knowledge. Under the columns marked "Green," "Black," and "Master" LSS Belts, the following symbols are used:

- "A" means they are almost *always used.* At least 90% of the projects will use these tools. (The related belt must be trained on how to use these tools or already have been trained in the use of these tools.)
- "O" means *often used.* It is used in more than 50% of the projects. (The related belt should be trained on how to use these tools or already have been trained on these tools.)
- "S" means *sometimes used.* It is used in 25 to 49% of the projects. (The related belt should know what they are used for and know where to go to get more information on how to use them.)
- "I" means *infrequently used to never used.* It is used in less than 24% of the projects. (These tools are nice to know but not required and not part of the belt's training or certification test.)

	LSS Belts		
Body of Knowledge	**Green**	**Black**	**Master**
5S's	O	O	O
5M's (materials, machines, manpower, methods, measurements)	A	A	A
Acceptance decisions	I	S	S
Activity network diagrams	S	S	S
Affinity diagrams	S	O	O
Area activity analysis (AAA)	S	I	I
Automation	I	S	S
Axiomatic design	I	S	S
Bar charts/graphs	A	A	A
Benchmarking	S	O	O
Bessel function	I	S	O

(Continued)

Body of Knowledge		LSS Belts		
		Green	Black	Master
Binomial distribution		O	O	O
Bivariate distribution		I	S	O
Box plots		S	O	O
Brainstorming		A	A	A
Bureaucracy elimination		S	O	O
Business case development		A	O	O
Business process improvement		S	O	O
Business-value-added		S	A	A
Calibration		O	O	O
Cause-and-effect (fishbone) diagrams		O	O	O
Cause-and-effect matrix		O	O	O
Cellular manufacturing		S	O	O
Central limit theorem		O	O	O
Chi-square distribution		O	O	O
Coefficient of contingency (*c*)		I	S	S
Collecting data		A	A	A
Communication techniques		O	O	A
Confidence interval for the mean/proportion/variance		O	O	O
Conflict resolution		O	O	O
Continuous flow manufacturing (CFM)		S	O	O
Control charts				
	X bar–*R* charts	O	O	O
	Run charts	O	O	O
	MX bar-*MR* charts	S	O	O
	X-*MR* charts	S	O	O
	X bar–*S* charts	S	O	O
	Median charts	I	S	O
	Short-run charts	S	S	O
	p-Charts	O	O	O
	r-Charts	S	O	O
	u-Charts	S	O	O
	CuSum control charts	I	S	S
Correlation coefficient		O	O	O
Cp		O	O	O
Cpk		O	O	O
CQFA (cost, quality, features, and availability)		S	S	O

Body of Knowledge		**LSS Belts**		
		Green	**Black**	**Master**
Critical to quality (CTQ)		A	A	A
Critical path method		O	O	O
Culture roadblocks		I	O	O
Cumulative distribution function		S	O	O
Current state mapping		O	O	O
Customer requirements		A	A	A
Customer surveys		S	O	O
Cycle time analysis		S	O	O
Design for maintainability and availability		I	S	S
Design for Six Sigma (DFSS)		I	S	O
Design for *X* (DFX)		I	S	O
Design of experiments				
	Three factor, three level experiment	I	O	O
	Randomized block plans	I	S	O
	Latin square designs	I	O	O
	Graeco-Latin designs	I	S	O
	Full factorial designs	I	O	O
	Plackett-Burman designs	I	S	O
	Taguchi designs	I	O	O
	Taguchi's robust concepts	I	S	O
	Mixture designs	I	S	O
	Central composite designs	I	S	S
	EVOP evolutionary operations	I	I	S
DMADV (Define, Measure, Analyze, Design, Verify)		S	O	O
DMAIC (Define, Measure, Analyze, Improve, Control)		O	O	O
Effort/impact analysis		I	O	O
Equipment certification		S	S	S
Error proofing		O	O	O
Exponential distribution		I	O	O
External and internal customers		O	O	O
F-distribution		I	S	S
Facilitation of teams		I	O	O

(Continued)

Body of Knowledge		LSS Belts		
		Green	Black	Master
Factorial experiments		I	S	O
Failure mode and effects analysis		O	O	O
Fast-action solution team (FAST)		O	O	S
First-time yield (FTY) or rolled-through yield (RTY)		O	O	O
Five whys (5Ws)		O	O	O
Flowcharts		O	O	O
Focus groups		S	O	O
Force field analysis		O	O	O
Frequency distribution		O	O	O
Future state mapping		O	O	O
Gantt charts		O	O	O
Gaussian curves		I	S	S
General surveys		O	O	O
Histograms		O	O	O
History of quality		S	S	S
Hypergeometric distribution		I	S	O
Hypothesis testing				
	Fundamental concepts	S	O	O
	Point and interval estimation	I	S	O
	Tests for means, variances, and proportions	I	O	O
	Paired comparison tests	I	O	O
	Analysis of variance	O	O	O
	Contingency tables	S	O	O
	Nonparametric tests	S	O	O
Interrelationship diagraphs (IDs)		S	O	O
Interviewing techniques		O	O	O
IT applications		S	O	O
Just-in-time		I	S	S
Kaizen and you		A	A	A
Kaizen and process troubleshooting		S	O	O
Kaizen and teams		S	O	O
Kaikaku (transformation of mind)		I	S	O

Body of Knowledge	LSS Belts		
	Green	Black	Master
Kanban	S	S	S
Kano model	S	O	O
Kendall coefficient of concordance	I	I	S
Knowledge management	I	S	O
KPIs	O	O	O
Kruskal-Wallis one-way analysis	I	S	S
Lean thinking	S	O	O
Levene test	I	S	S
Lognormal distribution	I	S	S
Loss function	S	O	O
Management theory history	I	S	O
Mann-Whitney *U*-test	I	S	S
Market segmentation	S	S	O
Matrix diagrams	O	O	O
Measure of dispersion	O	O	O
Measurement error	O	O	O
Measurement systems analysis (MSA)	S	O	O
Measurement tools	O	O	O
Method of least squares	O	O	O
Mistake proofing (poka-yoke)	S	S	S
Mood's median test	I	S	S
Motivating the workforce	S	O	O
Multivari analysis	S	O	O
Multiple linear regression	S	O	O
Negotiation techniques	O	O	O
Nominal group technique	O	O	O
Normal distribution	O	O	O
Normal probability plots	S	O	O
Null hypothesis	S	O	O
Opportunity cycle (protection, analysis, correction, measurement, prevent)	S	S	S
Project management	A	A	A
Organizational change management	O	O	O
Organizational culture diagnosis	S	O	O

(Continued)

Body of Knowledge	LSS Belts		
	Green	Black	Master
Overall equipment effectiveness (OEE)	S	S	S
Pareto diagrams	A	A	A
Pattern and trend analysis	I	S	S
Plan-Do-Check-Act (PDCA)	S	S	S
Plan, Understand, Streamline, Implement Continuous Improvement (PUSIC)	S	O	O
Point of use storage (POUS)	O	O	O
Poisson distribution	S	O	O
Poka-yoke (mistake proofing)	S	S	S
Poor-quality cost	S	S	O
Portfolio project management	I	S	O
Prioritization matrices	O	O	O
Probability concepts	O	O	O
Probability density function	I	S	O
Probability plots	O	O	O
Process capability studies	O	O	O
Process decision program charts (PDPC)	S	O	O
Process elements	O	O	O
Process fail points matrix	S	S	S
Process mapping	O	O	O
Process performance matrix	S	O	O
Process redesign	S	O	O
Program evaluation and review technique (PERT)	I	S	S
Poisson series	I	S	O
Project decision analysis	I	S	O
Project financial benefits analysis	A	A	A
Project selection matrix	I	S	A
Pugh concept selection	I	I	I
Pull systems	S	O	O
Quality @ source	O	O	O
QFD (quality function deployment)	S	O	O
Qualitative factor	O	O	O
Quantitative factor	O	O	O
Re-engineering	I	S	O

Body of Knowledge	LSS Belts		
	Green	**Black**	**Master**
Regression analysis	S	O	O
Reliability analysis	I	S	O
Response surface methodology (RSM)	I	S	S
Rewards and recognition	S	O	O
Risk analysis	A	A	A
Risk assessment	A	A	A
Robust design approach	S	O	O
Root cause analysis	A	A	A
Sampling	O	O	O
SCAMPER	S	S	S
Scatter diagrams	O	O	O
Seven basic tools	O	O	O
Sigma	O	O	O
Sigma conversion table	O	O	O
Signal-to-noise ratio	I	S	O
Simple language	S	O	O
Simple linear regression	S	O	O
Simplification approaches	O	O	O
Simulation modeling	I	S	O
Single-minute exchange of die (SMED)	I	S	O
Six Sigma metrics	I	S	O
Spearman rank correlation coefficient	I	S	O
Stakeholders	O	O	O
Standard work	A	S	S
Statistical process control	O	O	O
Statistical tolerance	S	S	O
Stem and leaf plots	I	S	O
Strengths, Weaknesses, Opportunities, and Threats (SWOT) analysis	S	O	O
Structural roadblocks	S	O	O
Student's *t*-distribution	I	S	O
Supplier controls	S	O	O
Supplier, inputs, process, outputs, customers (SIPOC) diagrams	O	O	O

(Continued)

Body of Knowledge		LSS Belts		
		Green	Black	Master
Systematic design		S	S	O
Takt time		S	S	O
Team building		O	O	O
Team charter		A	A	A
Team management		A	A	A
Theory of constraints		S	S	O
Throughput yield (TPY)		I	S	S
Tollgates		O	A	A
Total Productive Maintenance (TPM)		S	S	O
Tree diagrams		O	O	O
TRIZ		I	I	I
Types of data		O	O	O
Types of teams		I	S	O
Value/No-value-added activities		S	O	O
Value stream analysis (mapping)		S	O	O
Value stream management		S	O	A
Variance (σ^2, s^2)		I	O	O
Variation analysis				
	Rational subgroups	S	O	O
	Sources of variability	S	O	O
	Randomness testing	S	O	O
	Precontrol techniques	I	S	O
	Exponentially weighted moving average (EWMA)	I	S	S
	Moving average	S	O	O
Visual factory/visual office		S	O	O
Voice of the customer (VOC)		A	A	A
Voice of the supplier (VOS)		O	O	O
Waste identification		A	A	A
Waste elimination		A	A	A
Weibull distribution		I	O	O
Wilcoxon-Mann-Whitney rank sum test		I	S	S
Work breakdown structure		O	O	O
Work standard		I	S	O
Z-value		I	S	O

Appendix C: Six Sigma Green Belt Tools

The following is a list of basic nonstatistical tools that make up the Six Sigma Green Belt toolkit:

- Affinity diagrams
- Brainstorming
- Cause-and-effect diagrams
- Cause-and-effect matrix
- Check sheets
- Flowcharting
- Force field analysis
- Graphs
- Histograms
- Kano model
- Nominal group technique
- Pareto analysis
- Plan-Do-Check-Act
- Project management
- Root cause analysis
- Scatter diagrams
- 5S's
- 5W's
- 5W's and 2H's

As an SSGB, you should have been introduced to the following statistical tools:

- Analysis of variance (ANOVA)
- Attributes control charts
- Basic probability concepts
- Basic statistical concepts
- Binomial distribution
- Chi-square test

- Correlation coefficient
- Cpk—Operational (long-term) process capability
- Cpk—Using the *K*-factor method
- CPU and CPL—Upper and lower process capability indices
- Data accuracy
- Data, scale, and sources
- Histograms
- Hypothesis testing
- Mean
- Median
- Mode
- Mutually exclusive events
- Normal distribution
- Poisson distribution
- Probability theory
- Process capability analysis
- Process capability study
- Process elements, variables, and observations concepts
- Range
- Sampling
- Six Sigma measures
- Standard deviation
- Statistical process control
- Statistical thinking
- Variables control charts
- Variance

Index

B

D

E

L

M

N

Q

T

X

Z

Printed in the United States
by Baker & Taylor Publisher Services